LABORATORY MANUAL TO ACCOMPANY

MODERN ELECTRONIC COMMUNICATION

by **JEFFREY S. BEASLEY** and **GARY M. MILLER**

NINTH EDITION

MARK E. OLIVER
JEFFREY S. BEASLEY
DAVID H. SHORES

PEARSON
Prentice
Hall

Upper Saddle River, New Jersey
Columbus, Ohio

Editor-in-Chief: Vernon Anthony
Editorial Assistant: Lara Dimmick
Production Editor: Rex Davidson
Production Manager: Matt Ottenweller
Design Coordinator: Diane Ernsberger
Cover Designer: Kellyn Donnelly
Cover Art: Superstock
Director of Marketing: David Gesell
Marketing Manager: Jimmy Stephens
Marketing Assistant: Les Roberts

This book was set in Times Roman by Carlisle Publishing Services and was printed and
bound by Bind-Rite Graphics. The cover was printed by Coral Graphic Services, Inc.

Pearson Education Ltd.
Pearson Education Singapore Pte. Ltd.
Pearson Education Canada, Ltd.
Pearson Education—Japan

Pearson Education Australia Pty. Limited
Pearson Education North Asia Ltd.
Pearson Educación de Mexico, S.A. de C.V.
Pearson Education Malaysia Pte. Ltd.

10 9 8 7 6 5 4 3 2 1

ISBN-13: 978-0-13-156855-6
ISBN-10:   0-13-156855-8

PREFACE

This laboratory manual is designed to accompany *Modern Electronic Communication,* Ninth Edition, by Jeffrey Beasley and Gary Miller.

The 37 laboratory experiments and Electronics Workbench (EWB) experiments herein support the most important and applicable areas in the field of electronic communication. Each of these experiments has been thoroughly tested in the laboratory, and more than half have been used successfully for many years with electrical engineering technology students at Monroe Community College. These experiments are of varied length. Depending on the laboratory schedules used at your institution, some of these experiments may require several sessions for completion. We strongly recommend that instructors determine which steps of the procedure might be eliminated if lab time is at a premium. Also, a few of the steps might better be completed as a demonstration rather than being done by each laboratory group. This is true especially if your institution is lacking some of the required test equipment.

We have tried to keep the equipment costs as low as possible. However, we found it impossible to do an adequate job without the following test equipment at each workstation:

1. Dual-trace oscilloscope

   Frequency response: 100 MHz

   Vertical sensitivity: 59 mV/div.

   Y vs. X capability for sweeping

2. Low-voltage power supply

   Two required unless a dual-tracking supply is available.

   Voltage/current: 0 to 30 V dc at 1 A

3. "Deluxe" function generator

   Functions: sine, square, triangle

   Frequency range: 1 Hz to 2 MHz

   VCG capability for sweeping

   DC offset capability

4. "Regular" function generator

   Functions: sine, square, triangle

   Frequency range: 1 Hz to 600 kHz

5. Frequency counter: 10 Hz to 2 MHz

6. Prototype test board

Some of the laboratory experiments have optional steps requiring students to use a distortion analyzer or spectrum analyzer. If either one is available, we recommend using it for demonstration or for use by all groups on a rotating basis. Experiments 27, 28, and 30 require a pulse generator in order to produce a positive-going pulse with a narrow pulse width. Perhaps this could be borrowed from the digital electronics laboratory at your institution.

Many of the laboratory experiments will work out better if the circuits are fabricated ahead of time by the students or by the lab technician serving your department. Radio communication circuits are prone to problems such as unplanned oscillations, stray capacitance, lead inductance, circuit loading by measuring instruments, and so on. With wiring errors added to this problem list, the lab experience can be very disconcerting to students, especially if the circuit is quite extensive and time is limited in the laboratory. We recommend that "prebuilts" be used for Experiments 12–31. Students will then be able to spend their lab time taking data using test equipment and, most important, learning how it works!

On the other hand, students must still get a dose of troubleshooting and hands-on fabrication work to become proficient in these areas. Experiments 1, 3, 5–8, 11, and 13 are good candidates for these objectives to be met. The circuitry is less complicated and easier to build and troubleshoot. Experiments 2, 4, 9, and 10 contain very little circuitry and are designed to stress theoretical and instrumentation concepts rather than troubleshooting or building skills. Thus, depending on your course objectives, time factors, and test equipment availability at your institution, we recommend an appropriate balance of these three types of experiments to meet your needs.

We have left areas in the procedural steps of each experiment for the recording of data and the sketching of pertinent waveforms. However, many of the steps will require the students to furnish their own paper to answer the questions, draw more detailed pictures, and plot graphs. We recommend that laboratory instructors discuss with their students which experiments should be written up as formal reports and which should be completed without extensive writing. It is important to establish a balance between good report writing techniques and an understanding of the objectives of each experiment. It may not be possible to accomplish both of these objectives in the time available for each experiment.

The two fiber optic experiments require some rather expensive hand tools and supplies. Most of these tools and supplies are available from Fiber Instrument Sales, Inc. (**www.fiberinstrumentsales.com**). We have included part numbers for most of the equipment and supplies listed in Experiments 30 and 31.

Making lab experiments relevant to theory is a challenging job for an instructor. New devices and applications become increasingly complex. Nevertheless, there are several opportunities to make the instructor's job more manageable. Here are a few:

1. Excellent test equipment is becoming less expensive. One example is the Instek GSP-810 spectrum analyzer. The cost of this product is about $3,200 (including the tracking generator option). The minimum resolution is 3 kHz, which is adequate for a wide variety of relevant lab experiments.

2. Many of the older lab experiments operate at frequencies below 10 kHz, which is too low for an inexpensive spectrum analyzer. Several companies, including Mini-Circuits®, offer complete RF modules that make the demonstration of many electronic communication concepts very easy. The disadvantage of these modules is that the student cannot see what is going on inside.

3. An alternative to the modules is the use of various types of specialized integrated circuits. There are several manufacturers for these ICs, including MAXIM. In order for these ICs to work at sufficiently high frequencies to be analyzed by low-cost spectrum analyzers, the student will need to use customized circuit boards. The use of prototype boards is no longer acceptable because they introduce large parasitic circuit elements. One

exception, for which prototype boards are adequate, is certain basic digital circuits, which work even when there are substantial parasitic circuit elements.

4. Electronic simulation of lab experiments is another effective way for students to develop insight into electronic communication concepts.

Experiments 32–37 use all four of the "opportunities" described above to illustrate concepts presented in each chapter. Experiments 1–26 represent a vast array of ideas that can be used by instructors and students to deepen their understanding of electronic communication concepts. Experiments 27–31 help the student develop a better understanding of RF and optical transmission lines.

The files for the EWB experiments are on the CD packaged with the *Modern Electronic Communication* textbook. They are also available to download from the text's Companion Website at *www.prenhall.com/beasley*. Click on Companion Website, then select the lab you want.

ACKNOWLEDGMENTS

We would like to thank the following companies for permission to include their drawings and data sheets in the appendix of this manual: Amidon Associates, Inc.; AMP Incorporated; Analog Devices, Inc.; EXAR Corporation; Fiber Instrument Sales, Inc.; Harris Semiconductor Corporation; Hewlett-Packard Company; Maxim Integrated Products, Inc.; Mini-Circuits; Motorola Semiconductor Products; Murata-Erie North America, Inc.; National Semiconductor Corporation; Raytheon Company; Signetics Corporation.

Thanks to Gary Miller for helping develop the original Laboratory Manual; William J. Mooney for allowing us to include many of his original designs; and Donald Russ at Microwave Data Systems for his help with two of the data communications experiments. Last, but not least, we thank our families for their patience and support.

Mark E. Oliver
Jeffrey S. Beasley
David H. Shores

CONTENTS

PART II: ELECTRONICS WORKBENCH (EWB) MULTISIM EXPERIMENTS

APPENDIX: MANUFACTURER DATA SHEETS 279

PART I

LABORATORY EXPERIMENTS

ACTIVE FILTER NETWORKS

OBJECTIVES:

1. To become acquainted with the frequency response characteristics of the four basic types of active filter networks:
  a. Low-pass
  b. High-pass
  c. Bandpass
  d. Notch

2. Using a calculator, to become capable of predicting the theoretical frequency response of popular active filter designs using "cookbook" equations.

3. To build and measure the frequency response of popular active filter designs.

TEST EQUIPMENT:

Dual-trace oscilloscope

Function generator

Low-voltage power supply (2)

Frequency counter

COMPONENTS:

Resistors ($\frac{1}{2}$ watt): 1.5 kΩ, 4.7 kΩ (3), 9.4 kΩ, 12 kΩ (2), 22 kΩ, 56 kΩ, 220 kΩ

Capacitors: 0.005 μF (2), 0.01 μF (3), 0.05 μF, 0.1 μF

741 Operational amplifiers (2)

Prototype board

THEORY:

The active filter today is one of the more frequently used applications of op amps in electronic communication systems. In these circuits, the op amp's large open-loop gain is sacrificed in order to achieve each of the following closed-loop frequency response characteristics:

1. Very steep high-frequency and/or low-frequency roll-offs.

2. Predictable 3 dB cutoff frequencies.

3. Voltage amplification at desired frequencies.

4. Stability (no tendency to oscillate).

3

PRELABORATORY:

1. Review electronic circuit theory on active filter design in your linear electronics text.

PROCEDURE:

In this experiment you will be assigned a number of the following active filter circuits to build and test. You should follow each of the steps of the test procedure for the circuit assigned.

Part I—Low-Pass Filter
Part II—High-Pass Filter
Part III—Bandpass Filter
Part IV—Notch Filter

Part I: The Butterworth Second Order Low-Pass Active Filter

1. The Butterworth Second Order Low-Pass Active Filter is designed to pass all frequencies below its cutoff frequency with a constant maximum voltage gain. Above this frequency, the input signal is attenuated at a rate of -12 dB/octave or -40 dB/decade. This means that as the frequency is doubled, the output voltage drops by 12 dB. For the particular Butterworth design given in Fig. 1-1, the maximum voltage gain at low frequencies is fixed at unity or 0 dB, with a phase shift of zero degrees. As the frequency is increased to approach the 3 dB cutoff frequency, the phase shift should increase negatively or as a lagging angle, and the voltage gain should drop appropriately. The design equations are given in Fig. 1-1.

2. Using your calculator, predict the theoretical 3 dB cutoff frequency and the voltage gain in decibels for this active low-pass filter at each of the following frequencies in Hz: 100, 200, 300, 400, 500, 600, 700, 800, 900, 1 k, 2 k, 3 k, 4 k, 5 k, 6 k, 7 k, 8 k, 9 k, and 10 k.

3. Build the active low-pass filter on the prototype board. Experimentally determine the 3 dB cutoff frequency and the decibel voltage gain at each

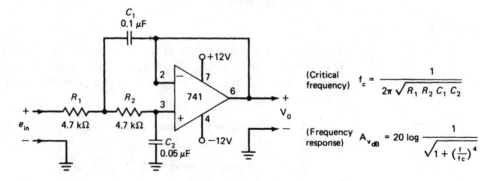

(Critical frequency) $f_c = \dfrac{1}{2\pi \sqrt{R_1 R_2 C_1 C_2}}$

(Frequency response) $A_{v_{dB}} = 20 \log \dfrac{1}{\sqrt{1 + \left(\frac{f}{f_c}\right)^4}}$

FIGURE 1-1

of the test frequencies given in step 2. Use a constant input voltage level of 0.5 V_{p-p}.

4. Prepare a table that clearly compares all theoretical and measured data in steps 2 and 3 in a neat orderly fashion.

5. Using two-cycle semilog graph paper, plot the theoretical and measured frequency response data. Connect the data points with a smooth curve. Identify the 3 dB cutoff frequency on the curves.

6. From the resulting curve of the measured data points, verify how close the roll-off rate approaches the −40 dB/decade or −12 dB/octave at high frequencies.

Part II: The Butterworth Second Order High-Pass Active Filter

1. The Butterworth Second Order High-Pass Active Filter is designed to pass all frequencies above its cutoff frequency with a constant maximum voltage gain. Below this frequency, the input signal is attenuated at a rate of +12 dB/octave or +40 dB/decade. This means that as the frequency is halved, the output voltage drops by 12 dB. For the particular Butterworth design given in Fig. 1-2, the maximum voltage gain at low frequencies is fixed at unity or 0 dB, with a phase shift of zero degrees. As the frequency is decreased to approach the 3 dB cutoff frequency, the phase shift should increase positively or as a leading angle, and the voltage gain should drop appropriately. The design equations are given in Fig. 1-2.

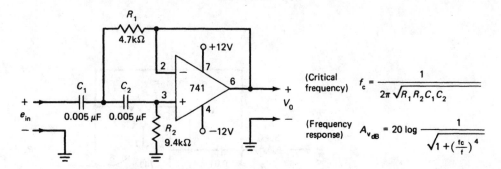

$$f_c = \frac{1}{2\pi \sqrt{R_1 R_2 C_1 C_2}}$$

$$A_{v_{dB}} = 20 \log \frac{1}{\sqrt{1 + \left(\frac{f_c}{f}\right)^4}}$$

FIGURE 1-2

2. Using your calculator, predict the theoretical 3 dB cutoff frequency and the voltage gain in decibels for this active high-pass filter at each of the following frequencies in Hz: 100, 200, 300, 400, 500, 600, 700, 800, 900, 1 k, 2 k, 3 k, 4 k, 5 k, 6 k, 7 k, 8 k, 9 k, and 10 k.

3. Build the active high-pass filter on the prototype board. Experimentally determine the 3 dB cutoff frequency and the decibel voltage gain at each of the test frequencies given in step 2. Use a constant input voltage level of 0.5 V_{p-p}.

4. Prepare a table that clearly compares all theoretical and measured data in steps 2 and 3 in a neat orderly fashion.

5. Using two-cycle semilog graph paper, plot the theoretical and measured frequency response data. Connect the data points with a smooth curve. Identify the 3 dB cutoff frequency on the curves.

6. From the resulting curve of the measured data points, verify how close the roll-off rate approaches the +40 dB/decade or +12 dB/octave at low frequencies.

Part III: The Active Bandpass Filter

1. The Active Bandpass Filter is designed to pass all input signals within a given range, called the bandwidth, while rejecting those frequencies outside of this range. This makes it have a frequency response which resembles that of a resonant circuit. The center frequency, f_r, is the frequency within the bandwidth at which the voltage gain is a maximum. The center frequency gain, $A_{v(max)}$, can be larger than unity but cannot exceed $2Q^2$. The Q of the filter is defined exactly the same way as it is defined for a resonant circuit, that is:

$$Q = f_r/\text{BW}$$
where

$$\text{BW} = f_H - f_L$$
and

f_H and f_L are the frequencies at which the gain has dropped by 3 decibels from its $A_{v(max)}$ value.

For this particular design, the equations are valid only if the value of Q is kept below 10.

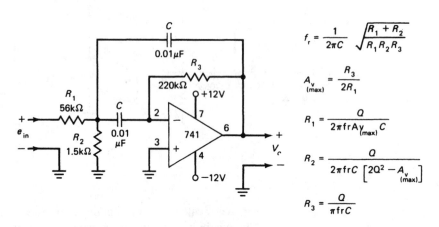

$$f_r = \frac{1}{2\pi C} \sqrt{\frac{R_1 + R_2}{R_1 R_2 R_3}}$$

$$A_{v(max)} = \frac{R_3}{2R_1}$$

$$R_1 = \frac{Q}{2\pi f r A_{v(max)} C}$$

$$R_2 = \frac{Q}{2\pi f r C \left[2Q^2 - A_{v(max)} \right]}$$

$$R_3 = \frac{Q}{\pi f r C}$$

FIGURE 1-3

2. Using the given circuit and design equations, determine each of the following theoretical values:
a. f_r
b. $A_{v(max)}$
c. Q
d. BW
e. f_H
f. f_L

3. Build the bandpass filter on the prototype board and measure the decibel voltage gain at each of the following test frequencies in Hz: 100, 200, 300, 400, 500, 600, 700, 800, 900, 1 k, 2 k, 3 k, 4 k, 5 k, 6 k, 7 k, 8 k, 9 k, and 10 k. Use an input voltage level of 0.5 V_{p-p}. Also, measure the actual values for f_r, f_H, and f_L.

4. Using two-cycle semilog graph paper, plot the measured frequency response data. Connect the data points with a smooth curve. Identify the 3 dB cutoff frequencies on the curves.

Part IV: The Active Notch Filter

1. The Active Notch Filter is designed to pass all frequencies outside of its bandwidth, while rejecting or sharply attenuating the frequencies within its bandwidth. Within its effective bandwidth is the null frequency, f_r, where the voltage gain is a minimum value. The cutoff frequencies, f_H and f_L, are the frequencies at which the voltage gain (outside the bandstop) drops 3 dB below $A_{v(max)}$. The Q of the notch filter is defined the same way as it is defined for a resonant circuit: as f_r/BW. The design equations are given in Fig. 1-4.

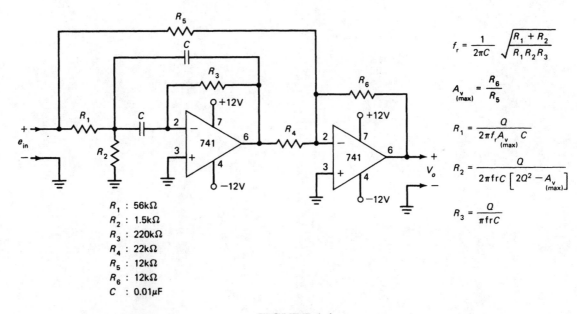

$$f_r = \frac{1}{2\pi C} \sqrt{\frac{R_1 + R_2}{R_1 R_2 R_3}}$$

$$A_{v(max)} = \frac{R_6}{R_5}$$

$$R_1 = \frac{Q}{2\pi f_r A_{v(max)} C}$$

$$R_2 = \frac{Q}{2\pi f r C \left[2Q^2 - A_{v(max)} \right]}$$

$$R_3 = \frac{Q}{\pi f r C}$$

R_1 : 56kΩ
R_2 : 1.5kΩ
R_3 : 220kΩ
R_4 : 22kΩ
R_5 : 12kΩ
R_6 : 12kΩ
C : 0.01μF

FIGURE 1-4

2. Using the given circuit and the design equations, determine each of the following theoretical values:
   a. f_r
   b. $A_{v(max)}$
   c. Q
   d. BW
   e. f_H
   f. f_L

3. Build the notch filter circuit and measure the decibel voltage gain at each of the test frequencies in Hz: 100, 200, 300, 400, 500, 600, 700, 800, 900,

1 k, 2 k, 3 k, 4 k, 5 k, 6 k, 7 k, 8 k, 9 k, and 10 k. Use an input voltage of 0.5 V_{p-p}. Also, measure the actual values for f_r, f_H, and f_L.

4. Using two-cycle semilog graph paper, plot the measured frequency response data. Connect the data points with a smooth curve. Identify the 3 dB cutoff frequencies on the curve.

REPORT / QUESTIONS:

Evaluate how well the theoretical and actual data compare in each of the active filters that you were assigned.

FREQUENCY SPECTRA OF POPULAR WAVEFORMS

OBJECTIVES:

1. To become acquainted with the Fourier series and its use in representing the frequency spectra of signals commonly used in communication systems.
2. To become familiar with frequency response of ceramic filters.
3. To become familiar with the use of spectrum analyzers.

REFERENCE:

Refer to Section 1-6 of the text.

TEST EQUIPMENT:

Dual-trace oscilloscope
Function generator
Frequency counter
Low-voltage power supply
Spectrum analyzer (if available)

COMPONENTS:

Ceramic filter: Murata-Erie CFM-455D
Resistors ($\frac{1}{2}$ watt): 1.5 kΩ (2)

THEORY:

The Fourier series is a mathematical tool used to represent any periodic function as an infinite series of sine or cosine functions. In electronics, waveforms of voltage or current are periodic functions which lend themselves to the use of the Fourier series. In the field of electronic communication, the Fourier series is often utilized to explain how signals are filtered and processed within the various blocks and stages that make up a communication system. In its general form,

9

$$f(t) = A_0 + \sum_{n=1}^{\infty} A_n \cos(n\omega_n t + \phi_n) + \sum_{n=1}^{\infty} B_n \sin(n\omega_n t + \phi_n)$$

Fortunately, most waveforms can be represented in much easier terms. The easiest waveform to represent mathematically is the sine wave itself.

$$e(t) = E_0 + E_{max} \cos(\omega t + \phi)$$

where E_0 = dc offset
E_{max} = peak value of the sine wave
ω = frequency, radians per second
ϕ = initial phase angle

Other popular waveforms that can be written as a Fourier series are the square wave and triangle wave shown, respectively, in Fig. 2-1(a) and (b).

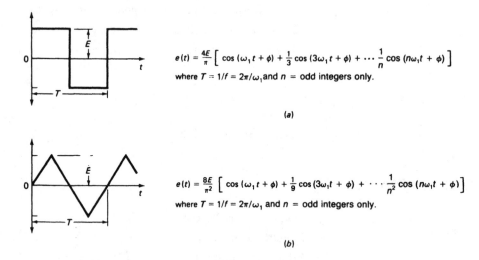

$$e(t) = \frac{4E}{\pi}\left[\cos(\omega_1 t + \phi) + \frac{1}{3}\cos(3\omega_1 t + \phi) + \cdots \frac{1}{n}\cos(n\omega_1 t + \phi) \right]$$

where $T = 1/f = 2\pi/\omega_1$ and n = odd integers only.

(a)

$$e(t) = \frac{8E}{\pi^2}\left[\cos(\omega_1 t + \phi) + \frac{1}{9}\cos(3\omega_1 t + \phi) + \cdots \frac{1}{n^2}\cos(n\omega_1 t + \phi) \right]$$

where $T = 1/f = 2\pi/\omega_1$ and n = odd integers only.

(b)

FIGURE 2-1 Popular waveforms and their Fourier series

If the Fourier series of a waveform is known, its spectral content can be represented on a spectrum diagram. A spectrum diagram is a sketch of voltage (or current) versus frequency in which its individual harmonic amplitudes are plotted as vertical lines (arrows) at each harmonic frequency. For example, for a 60-Hz sawtooth waveform:

$$e(t) = \frac{2E}{\pi}\left[\sin(377t) - \frac{\sin(2 \times 377t)}{2} + \frac{\sin(3 \times 377t)}{3} + \cdots \frac{(-1)^{n+1}\sin(n \times 377t)}{n} \right]$$

Figure 2-2 shows the spectrum of this signal if $E = 10$ V.

Notice that if the peak voltage is negative, it is still displayed as positive, since the negative sign merely denotes a 180° phase shift of that sinusoidal term. A spectrum diagram can be displayed for an unknown signal using a spectrum analyzer. Often, the spectrum analyzer's display is more useful in communication applications than is the more popular oscilloscope display of waveforms.

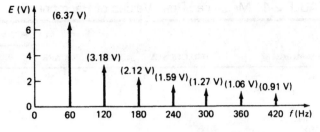

$$E_1 = \frac{2(10\ V)}{\pi} = +6.37\ V$$

$$E_2 = \frac{-2(10\ V)}{2\pi} = -3.18\ V$$

$$E_3 = \frac{+2(10\ V)}{3\pi} = 2.12\ V,\ \text{etc.}$$

FIGURE 2-2

PRELABORATORY:

Determine the theoretical peak values of the first nine harmonics of the two waveforms given in Fig. 2-1, using the given Fourier series and the procedures shown in the theory section. For both of these waveforms, let $E = 5$ V. Compile your data into a table similar to Table 2-1.

PROCEDURE:

1. Build the circuit shown in Fig. 2-3. Connect a function generator to TP_1 and apply a 10-$V_{p\text{-}p}$ 455-kHz sine wave. Connect an oscilloscope at TP_2 and measure V_o. Carefully fine-tune the frequency of the function generator until you produce a maximum output voltage. You should discover five peaks near 455 kHz, but the center peak should be slightly larger than the other two, as sketched in Fig. 2-4. Make sure that you are tuned to the larger center peak frequency. Measure the peak values of V_o and V_{in} and determine the insertion loss of the filter. You may discover your filter has a gain, rather than a loss due to impedance transformation properties.

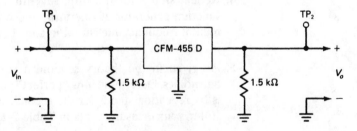

FIGURE 2-3

2. Now set the function generator to produce a ± 5 V amplitude 455-kHz square wave such as that shown in Fig. 2.1. Carefully fine-tune the function generator so as to peak the output voltage. Again, you will find five specific frequencies near 455 kHz, where V_o is a maximum. Use the center peak frequency that causes the largest of the five peak voltages to result at TP_2. Measure the largest peak value of V_o at TP_2. Also, measure the frequency of the function generator using a frequency counter. Record these values in Table 2-1.

3. To determine the peak value of the third harmonic, instead of retuning the ceramic filter to resonate at the third harmonic frequency, set the function

TABLE 2-1 Measured Peak Values of Harmonics

| | | OUTPUT VOLTAGE AT TP_2 | |
HARMONIC	FREQUENCY	SQUARE	TRIANGLE
First			
Second			
Third			
Fourth			
Fifth			
Sixth			
Seventh			
Eighth			
Ninth			

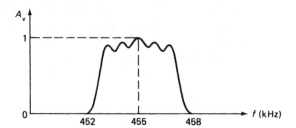

FIGURE 2-4 Typical frequency response of ceramic filter

generator to exactly one-third of the 455-kHz value measured in step 1. Again, you will find five specific frequencies near 455 kHz/3 where V_o is maximum. Again, use the center peak frequency that causes the largest of the five peaks to result in TP_2. Measure V_o at TP_2 and the frequency of the function generator. Record these values in Table 2-1.

4. Repeat step 3 for the fifth, seventh, and ninth harmonics by setting the function generator at one-fifth, one-seventh, and one-ninth of the fundamental frequency measured in step 1. Enter your resulting measurements in Table 2-1.

5. As given in the theory section, a square wave should exhibit no even harmonics. Determine how "perfect" your function generator's square wave is by repeating step 3 for the second, fourth, sixth, and eighth harmonics. Enter your measurements in Table 2-1.

6. Now set the function generator to produce a 10-V_{p-p} triangle wave such as that shown in Fig. 2-1(b). Repeat steps 2–5 to measure the peak amplitudes of the first nine harmonics of the triangle wave. Enter your measurements in Table 2-1.

7. If a spectrum analyzer is available, produce the spectral display of the signal at TP_1 for the square and triangle waveforms. Set the controls so as to clearly display the amplitudes of the first nine harmonics.

REPORT / QUESTIONS:

1. Compare the theoretical values of the prelaboratory section with the measured values obtained in steps 2–6. What factors may have led to discrepancies between theoretical and measured values?

2. Sketch a frequency spectrum diagram for each of the two waveforms shown in Fig. 2-1, using the theoretical values determined in the prelaboratory section.

3. Suppose that the waveform shown in Fig. 2-1(a) is applied to a low-pass filter that exhibits a break frequency of 455 kHz and -20 dB/decade roll-off. Calculate the amplitudes of the first nine harmonics which would be contained in the output signal of the low-pass filter. Sketch the spectrum diagram of the resulting output signal of the filter. Assume that this waveform has a fundamental frequency of 455 kHz and an amplitude, E, of 5 V.

NAME _____

TUNED AMPLIFIERS AND FREQUENCY MULTIPLICATION

OBJECTIVES:

1. To investigate the behavior of a negative clamper.
2. To study class C bias and amplification.
3. To understand the theory of frequency multiplication.

REFERENCE:

An application of frequency multiplication can be found in Section 5-5 of the text.

TEST EQUIPMENT:

Dual-trace oscilloscope
Sinusoidal function generator
Low-voltage power supply
Prototype board
Frequency counter

COMPONENTS:

Transistor: 2N2222 or equivalent
Signal diode: 1N914/1N4148 or equivalent
Resistors ($\frac{1}{2}$ watt): 1 kΩ, 120 kΩ
Capacitors: 3.3 nF, 1 μF (2)
Inductor: 10 mH

PRELABORATORY:

Determine the resonant frequency of the tank circuit shown in Fig. 3-4 using

$$f_r = \frac{1}{2\pi\sqrt{LC}} \quad f_r = \text{_____}$$

15

PROCEDURE:

1. Build the circuit of Fig. 3-1. Monitor V_{in} with channel A (1 V/div dc-coupled) of the dual-trace oscilloscope. Monitor V_o with channel B (1 V/div dc-coupled). Apply a 1-V_{p-p} 1-kHz sine-wave signal as V_{in}. You should notice that the waveforms of V_{in} and V_o are identical. Now slowly increase the amplitude of V_{in} and you should notice that V_{in} and V_o are no longer identical with respect to their dc offset. Record V_{in} and V_o waveforms when $V_{in} = 4$ V_{p-p}.

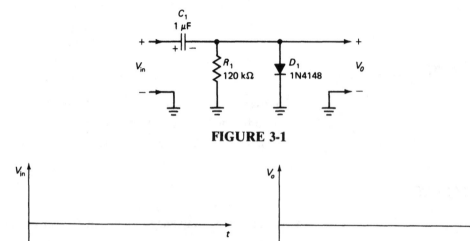

FIGURE 3-1

Also, determine at what critical voltage level of V_{in} the waveforms of V_{in} and V_o no longer possess the same dc offset. (This level may be a judgment call on your part.) This is the process of negative clamping action. The circuit of Fig. 3-1 can now be disassembled.

Critical level = _____

2. Negative clamping is often used to bias a transistor for class C amplification. Class C bias puts the Q-point at a point "beyond" cutoff. To create a class C amplifier the diode used in the circuit of Fig. 3-1 is replaced with the base-emitter junction of a transistor. Build the circuit of Fig. 3-2. Monitor V_{in} with channel A (1 V/div dc-coupled) and monitor V_b with channel B (1 V/div dc-coupled). Again, apply a 1-V_{p-p} 1-kHz sine wave as V_{in}. Slowly increase the amplitude of V_{in} and you should notice that V_{in} and V_b behave exactly as V_{in} and V_o did in step 1. Now move channel B (5 V/div dc-coupled) to monitor V_o. Sketch the V_b and V_o waveforms when $V_{in} = 1$ V_{p-p}. Repeat for V_{in} set at 2 V_{p-p} and 4 V_{p-p}. Measure the pulse-width of the negative pulse in V_o.

From these waveforms notice that it takes a certain initial V_{in} level to force the Q-point to move into the active region from cutoff. This is what is meant by "beyond cutoff" bias in class C amplifiers. The circuit of Fig. 3-2 can now be disassembled.

3. Notice that this amplifier has a voltage gain greater than unity and also has a fairly high efficiency due to the small average collector current level.

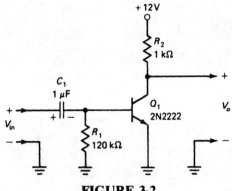

FIGURE 3-2

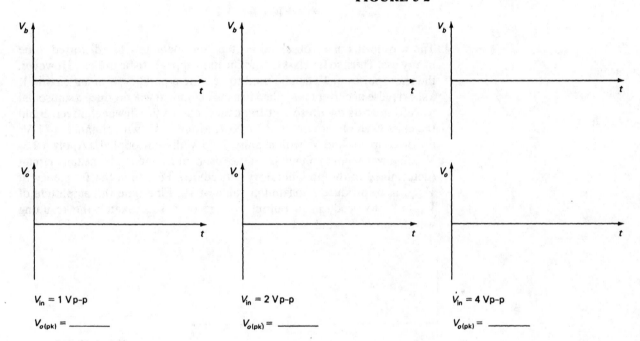

$V_{in} = 1\ \text{Vp-p}$

$V_{o(pk)} =$ _____

$V_{in} = 2\ \text{Vp-p}$

$V_{o(pk)} =$ _____

$V_{in} = 4\ \text{Vp-p}$

$V_{o(pk)} =$ _____

The small average collector current results from the small duty cycle resulting from negative clamping action as seen in V_o waveforms of step 2. Calculate the duty cycle for each of the waveforms of step 2 using Fig. 3-3 and the equation:

$$\% D = \text{duty cycle} = \frac{\Delta t}{T} \times 100\%$$

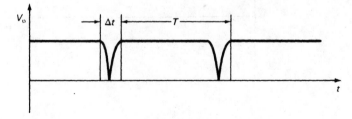

FIGURE 3-3

$V_{in} = 1\ \text{V}_{p\text{-}p};\ \% D =$ _____

$V_{in} = 2\ \text{V}_{p\text{-}p};\ \% D =$ _____

$V_{in} = 4\ \text{V}_{p\text{-}p};\ \% D =$ _____

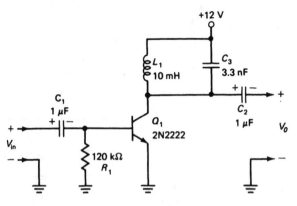

FIGURE 3-4

4. The waveforms of V_o observed in step 2 are definitely too distorted to be of any use. Thus, so far class C amplification appears to be useless. However, the short pulses of V_o can be used to activate a parallel resonant circuit. If a short pulse of current is applied to a tank circuit, it will produce a sinusoidal waveform at its own resonant frequency, due to the flywheel effect. Build the class C amplifier circuit of Fig. 3-4. Monitor V_{in} with channel A (1 V/div dc-coupled) and V_o with channel B (5 V/div dc-coupled). Apply a 1.5-V_{p-p} sine-wave input voltage set at the resonant frequency of the tank circuit (determined in the prelaboratory procedure). Fine-tune the frequency of V_{in} so as to produce a maximum value of V_o. Fine-tune the amplitude of V_{in} so as to produce an output voltage of 8 V_{p-p}. Sketch the resulting waveforms of V_{in} and V_o.

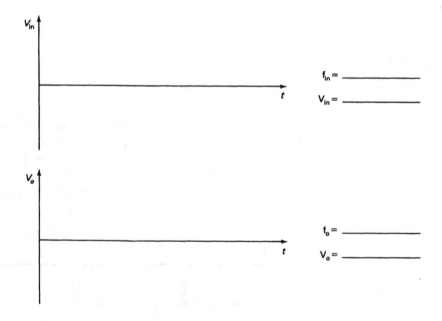

5. The main limitation of a class C amplifier is its narrow bandwidth of frequencies which can be amplified. This is due to the Q of the tank circuit. Fortunately, in most communication applications it is desired for an RF amplifier to have a narrow bandwidth so that it can filter out undesired frequencies. This amplifier is sometimes referred to as a "tuned amplifier" because of this characteristic. Determine the bandwidth of this amplifier by adjusting the frequency of V_{in} above and below the resonant frequency

so as to force V_o to drop 3 dB below its maximum value set in step 4. The bandwidth is simply the difference between the upper and lower 3-dB frequencies. Record these values.

$$f_+ = \text{_____} \qquad f_- = \text{_____} \qquad BW = \text{_____}$$

6. The Q of the tank circuit can be determined using the equation below. Determine this and record your results.

$$Q = \frac{f_r}{BW} \qquad Q = \text{_____}$$

7. Another use of a tuned class C amplifier is in frequency multiplication. In step 4, notice that for each cycle of the sine wave produced in the tank circuit by the flywheel effect, the tank circuit is recharged by another pulse of collector current due to negative clamping action and the fact that the input voltage is exactly matched to the resonant frequency of the tank circuit. However, it is possible to recharge the tank circuit on every other cycle of the sine wave by setting the frequency of V_{in} to exactly one-half of the resonant frequency. Do this with the circuit of Fig. 3-4. Again, fine-tune the frequency of V_{in} so as to produce a maximum value of V_o. Also, adjust the amplitude of V_{in} so as to produce $V_o = 8\,V_{p-p}$. Record the resulting waveforms of V_{in} and V_o. This is a frequency-doubler circuit.

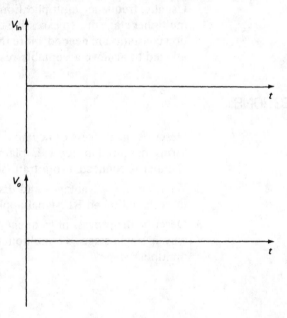

8. Try adjusting the frequency of V_{in} so as to produce ×3, ×4, and ×5 frequency multiplication. Record your resulting input and output voltage amplitudes and frequencies. Measure the input and output frequencies with a counter, if one is available. Also, record your waveform sketches for the ×5 multiplier.

TYPE	V_{in}	f_{in}	V_o	f_o
×3				
×4				
×5				

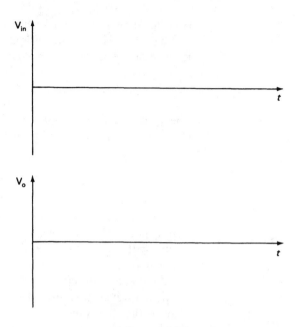

Note that the voltage amplitude decreases over the later cycles of V_o until the next current pulse restores the original amplitude. This distortion can be minimized by the use of higher Q components within the tank circuit. Usually, frequency multiplication greater than 5 is not possible in a single multiplier stage due to excessive distortion of this type. If larger multiplication constants are needed, more than one stage of multiplication is generally utilized to achieve acceptable results.

REPORT/QUESTIONS:

1. Describe the process of negative clamping action as observed in the waveforms measured in step 1. Explain how class C bias results when a negative clamper is connected to a transistor as observed in step 2.
2. What are the advantages and disadvantages of a class C amplifier design in both audio and RF signal applications?
3. Describe the process of frequency tripling in a ×3 multiplier stage. Explain why a single-stage ×7 multiplier stage does not work as well as a ×3 multiplier stage.

LOW-PASS IMPEDANCE TRANSFORMATION NETWORKS

OBJECTIVES:

1. To understand the concept of impedance matching for maximum power transfer.

2. To investigate the use of resonant LC networks to produce impedance transformation.

3. To design, build, and test an L-network.

REFERENCES:

Refer to Sections 1-7 and 2-6 of the text.

TEST EQUIPMENT:

Dual-trace oscilloscope

Function generator

Frequency counter

Prototype board

COMPONENTS:

Inductors: two selected design values from Table 4-1

Capacitors: two selected design values from Table 4-1

Resistors ($\frac{1}{2}$ watt): 50 Ω, 376 Ω, 426 Ω (use 470 Ω in parallel with 1800 Ω), and selected values needed to simulate a specific generator impedance (explained in procedure section)

THEORY:

Most RF circuits are required to deliver power to loads which have impedances quite different than the circuit load value that would optimize circuit output power. For example, a transistor amplifier might require an ac collector load resistance of 3000 Ω to deliver the desired power gain, but the actual load resistance available could well be a 50-Ω antenna. The problem then arises of finding a way to make the

50-Ω load resistance appear as a 3000-Ω load resistance to the transistor. If a resistive matching network is used, the resistors themselves produce a significant power loss in the system. This results even though they are providing the proper match for maximum power transfer. One solution frequently employed is to use a transformer. A second solution that works as long as the frequency of the signal remains relatively constant is to use an *LC* impedance transformation network. Ideal networks, being purely reactive, exhibit zero power loss. Realistic *LC* networks exhibit quite small values of power loss due to the real power dissipated in the pure resistance of the inductor. *LC* networks can be designed for operation at any desired frequency using fairly standard component values. They can be designed to exhibit either low-pass or high-pass filter characteristics. Low-pass filter characteristics are advantageous since they help eliminate unwanted harmonics from the system output.

In this experiment you will design, build, and test the simplest form of these networks, the two-element *LC* network, which is more commonly called the L-network. The name "L" is used because of its resemblance to an upside-down capital "L" when drawn out as a schematic, as shown in Fig. 4-1.

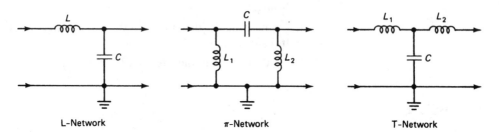

FIGURE 4-1 Typical impedance-matching networks

Other forms of these networks containing three or more elements are quite commonly found in communication systems. These include the T- and π-networks, which also resemble their names, as shown. In summary, these circuits will exhibit the following properties:

1. They will have either low- or high-pass frequency response.
2. They will exhibit close-to-zero power loss.
3. At a desired frequency, their input impedance will be purely resistive and of a predetermined value when loaded with a specified resistive load.

PRELABORATORY:

1. Review electrical circuit theory on series and parallel resonance and read the attached theory material on designing impedance-matching networks with low-pass frequency response characteristics.
2. Calculate the required values of inductance and capacitance in order to build the two specified L-networks in Table 4-2, given the values for Z_{in}, R_L, and f_r. Show all work. Use equations 4-5, 4-8, 4-9, and 4-10 to design your L-networks.
3. Check your L-network designs for accuracy by applying ac circuit theory to your resulting designs. For each of the two circuits designed in step 2, you should show all work as you do the following:

(a) Determine the reactances of each of the *L*'s and *C*'s in your circuit at the given match frequency of 218 kHz.

(b) Combine these reactances with the other resistances (series, parallel, etc.) to determine the input impedance of the circuit at the match frequency of 218 kHz. A few polar-to-rectangular (or vice versa) conversions may be required, so you may need to brush up on your complex-number arithmetic before completing this step.

(c) Your results should be very close to being a purely resistive impedance (small reactive component in rectangular form or a small phase angle in polar form). If this does not happen, you made a mistake in either your original design or your check.

4. The component values available in the laboratory are given in Table 4-1. Select values for *L* and *C* which are closest to your calculated values from this table.

TABLE 4-1 Standard Design Values

Inductors (µH)	1.0, 1.2, 1.5, 1.8, 2.2, 2.7, 3.3, 4.7, 6.8, 8.2, 10, 12, 15, 18, 22, 27, 33, 47, 68, 82, 100, 150, 220, 330, 470, 1000
Capacitors (nF)	0.1, 0.22, 0.33, 0.47, 0.68, 1.0, 1.5, 2.2, 3.3, 4.7, 6.8, 10, 15, 22, 33, 47, 68, 100, 220, 330, 470, 680, 1000

PROCEDURE:

1. Show your prepared design and test procedures to the lab instructor for verification that they are accurate and complete. If they are not correct, make corrections and modifications as necessary.

2. Determine the frequency response for both of your L-networks using the function generator to produce the sinusoidal input signal. It is important to simulate the proper internal impedance by adding the proper resistor in series. For example, if your signal generator has an internal impedance of 50 Ω and you want it to act like it has an internal impedance of 426 Ω, place a 376-Ω resistor (470 Ω, in parallel with 1.8 kΩ is close enough) in series with the generator as shown in Fig. 4-2.

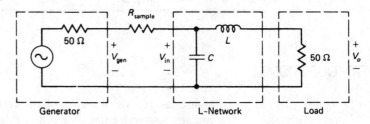

FIGURE 4-2

Consider V_{in} to be on the L-network end of the 376-Ω resistor. Take measurements of V_{in} and V_o at approximately 10 frequencies between 100 kHz and 1.0 MHz. Do not change the amplitude of the generator so as to produce a constant V_{in}, or else you will negate the effect of simulating the proper internal impedance and will not produce the valid frequency response. Calculate the relative decibel output voltage by comparing V_o to its maximum value and convert the ratio to decibels using

$$V_{o(dB)} = 20 \log \left| \frac{V_o}{V_{o_{(max)}}} \right|$$

3. Complete the following test procedures to determine experimentally the frequency at which the input impedance of the L-network is purely resistive. Determine the value of the purely resistive input impedance, R_{in}, at that frequency. Do this for both designs.

   (a) At the frequency at which Z_{in} is purely resistive, there cannot be any phase shift between V_{gen} and V_{in} as seen in Fig. 4-2. Thus, experimentally determine the frequency near 218 kHz at which the waveforms of V_{gen} and V_{in} exhibit zero phase difference. Use a frequency counter to measure this frequency and compare to the theoretical value of 218 kHz. For the step-up L-network, use a sample resistor equal to 50 Ω. For the step-down L-network, use a sample resistor equal to 426 Ω (rather than 376 Ω).

   (b) At this frequency, if Z_{in} is actually the design value given in Table 4-1, you should find that exactly half of V_{gen} is dropped across the sample resistor. The other half of V_{gen} is dropped across the input of the L-network. Verify that this is true. If you find that this relationship does not hold, measure the amplitudes of V_{gen} and V_{in} and solve for R_{in} using the voltage-divider equation:

$$V_{in} = V_{gen} \frac{R_{in}}{R_{gen} + R_{in}}$$

Compare your results with the theoretical value for Z_{in} from Table 4-2.

TABLE 4-2   Design Specifications: Low-Pass L-Section

FIGURE	NETWORK TYPE	Z_{in} (Ω)	R_L (Ω)	OPERATING FREQUENCY (kHz)
4-3	Step-down	426	50	218
4-5	Step-up	50	426	218

REPORT/QUESTIONS:

1. Using the results of step 2 of the test procedure, plot your resulting frequency response curves for both of the designed networks on graph paper by plotting $V_{o(dB)}$ versus frequency. It is not necessary to use semilog graph paper.

2. Write a design report describing what you did and how it worked out. Include all materials from the design procedure and test procedure. If your design procedure had to be modified, be sure to include both the original data and modified data. Explain what was wrong with the original design or procedure.

ADDITIONAL THEORY:

Impedance Transformation Using the L-Section

The L-section shown in Fig. 4-3 is an example of a low-pass tuned circuit, since at high frequencies the capacitor looks like a short and the inductor looks like an open, thus keeping the applied signal from ever reaching the load resistance, R_L.

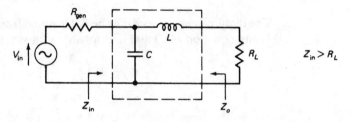

FIGURE 4-3

It is a tuned circuit because there is a particular frequency at which the inductive and capacitive reactances are equal and the circuit resonates, thus producing a maximum signal passed on to the load. If the values of L_p and C (shown in Fig. 4-4) are properly chosen to produce resonance at the frequency of the input signal, the generator will see a parallel resonant circuit in which the load resistance appears to be larger than it really is. Similarly, the load resistance will see a low-pass series resonant circuit made up of L and C. These principles are made obvious by redrawing the circuit of Fig. 4-3 into its parallel equivalent circuit, shown in Fig. 4-4.

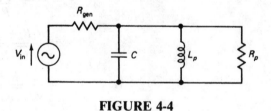

FIGURE 4-4

The series combination of L and R_L can be converted to an equivalent parallel combination by using the following series-parallel conversion equations:

$$R_p = R_L(1 + Q^2) \quad \text{where } Q = \frac{X_L}{R} \tag{4-1}$$

$$L_p = L\left(1 + \frac{1}{Q^2}\right) \tag{4-2}$$

At F_{in}, the frequency of the input signal, we want Z_{in} to be equal to some desired resistance value, R_d. If the values of L and C are chosen properly, X_c and X_{Lp} will cancel at $f = f_{in}$. The impedance of a parallel resonant circuit is very high. Thus at $f = f_{in}$,

$$R_d = Z_{in} = R_p = R_L\left(1 + \frac{X_L^2}{R_L^2}\right)$$

$$= \frac{X_L^2 + R_L^2}{R_L} \tag{4-3}$$

Solving equation (4-3) for X_L, we get

$$X_L^2 = R_L R_d - R_L^2$$
$$X_L = \sqrt{R_L(R_d - R_L)} \tag{4-4}$$

Finally, solving for the required inductance yields

$$L = \frac{\sqrt{R_L(R_d - R_L)}}{2\pi f_{in}} \tag{4-5}$$

The required capacitance may be determined from the fact that $X_c = X_{L,p}$ at resonance. Using equation (4-2) and multiplying both sides by $2\pi f_{in}$, we get

$$X_{L,p} = X_L\left(1 + \frac{1}{Q^2}\right) = X_L\left(1 + \frac{1}{X_L^2/R_L^2}\right)$$

$$= X_L\left(1 + \frac{R_L^2}{R_L^2}\right) = X_L\left(\frac{X_L^2 + R_L^2}{X_L^2}\right)$$

$$= \frac{X_L^2 + R_L^2}{X_L}$$

At $f = f_{in}$,

$$X_c = X_{L,p} = \frac{X_L^2 + R_L^2}{X_L} \tag{4-6}$$

Substituting equation (4-4) into (4-6) yields

$$X_o = \frac{\left[\sqrt{R_L(R_d - R_L)}\right]^2 + R_L^2}{\sqrt{R_L(R_d - R_L)}}$$

$$= \frac{R_L R_d - R_L^2 + R_L^2}{\sqrt{R_L(R_d - R_L)}}$$

$$= \frac{R_L R_d}{\sqrt{R_L(R_d - R_L)}} \tag{4-7}$$

Finally, solving for C gives us

$$C = \frac{1}{2\pi f_{in} X_o}$$

$$= \frac{1}{2\pi f_{in}\left|\dfrac{R_L R_d}{\sqrt{R_L(R_d - R_L)}}\right|}$$

$$= \frac{\sqrt{R_L(R_d - R_L)}}{2\pi f_{in} R_d R_L} \tag{4-8}$$

The L-section shown in Fig. 4-5 is another example of a low-pass tuned circuit for the same reasons as those stated for the circuit in Fig. 4-3. The main difference is that in this circuit if values of L' and C'_s (shown in Fig. 4-6) are properly chosen to produce resonance at the frequency of the input signal, the generator will "see" a series resonant circuit in which the load resistance will appear to be smaller than it is.

Again, a tool that can be used to demonstrate this phenomenon and derive the design equations is the use of series-parallel conversion equations and the series equivalent circuit shown in Fig. 4-6. However, before we get involved in another fairly long derivation as we did with the network of Fig. 4-3, compare the network

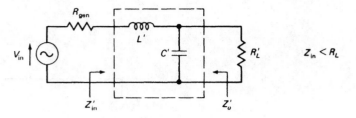

FIGURE 4-5

of Fig. 4-5 with the network of Fig. 4-3. A close inspection reveals that the network of Fig. 4-5 is nothing more than the network of Fig. 4-3 drawn in the reverse direction. Thus, if we think of the load resistance of Fig. 4-3 as the desired input impedance of Fig. 4-5, and if we think of the desired input impedance of Fig. 4-3 as the load resistance of Fig. 4-5, we can immediately devise the design equations. This can be done merely by substituting the new variable names for the old variable names in the design equations for Fig. 4-3.

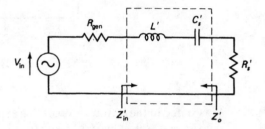

FIGURE 4-6

Since for Fig. 4-3,

$$L = \frac{\sqrt{R_L(R_d - R_L)}}{2\pi f_{in}} \tag{4-5}$$

then for Fig. 4-5,

$$L = \frac{\sqrt{R_d'(R_L - R_d')}}{2\pi f_{in}} \tag{4-9}$$

Similarly, since for Fig. 4-3,

$$C = \frac{\sqrt{R_L(R_d - R_L)}}{2\pi f_{in} R_d R_L} \tag{4-8}$$

then for Fig. 4-5,

$$C = \frac{\sqrt{R_d'(R_L - R_d')}}{2\pi f_{in} R_L' R_d'} \tag{4-10}$$

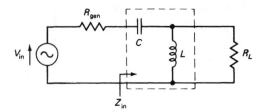

FIGURE 4-7

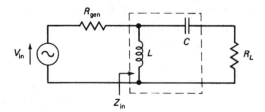

FIGURE 4-8

As stated in the introduction of this experiment, low-pass frequency response characteristics are usually important to eliminate the generation of harmonics from a nonpure sinusoidal input signal. However, this is not to say that high-pass filter L-networks are never found in practice. Examples of a high-pass filter L-network are shown in Figs. 4-7 and 4-8. At low frequencies, the capacitor looks like an open and the inductor looks like a short, thus keeping the input signal from ever reaching the load. The circuit of Fig. 4-7 functions to make the load resistance appear to be smaller than it really is, just as the low-pass filter network of Fig. 4-5 does. The circuit of Fig. 4-8 functions to make the load resistance appear to be larger than it really is, just as the low-pass filter of Fig. 4-3 does.

NAME _____

PHASE-SHIFT OSCILLATOR

OBJECTIVES:

1. To become familiar with the Barkhausen criteria for oscillation.

2. To analyze typical oscillator designs using operational amplifiers.

3. To predict if an amplifier will oscillate by first testing the circuit in the open-loop mode.

REFERENCE:

Refer to Section 1-8 of the text.

TEST EQUIPMENT:

Dual-trace oscilloscope
Low-voltage power supply (2)
Function generator
Volt-ohmmeter
Frequency counter
Prototype board

COMPONENTS:

741 Operational amplifier (2)
Resistors ($\frac{1}{2}$ watt): 3.3 kΩ (7), 6.8 kΩ (2)
Capacitors: 0.1 μF (2)

PROCEDURE:

1. Build the circuit of Fig. 5-1. Connect the two power supplies to produce $+V_{CC}$ and $-V_{CC}$, as shown in Fig. 5-2. Also, connect the function generator as V_1.

2. Set the power supplies to produce $+12$ V dc at their output terminals so as to provide ±12 V dc to the op amp. Connect R_9 to the output of the phase-shift network (C_2 and R_8) by connecting J_2 to TP$_4$. This sets up the

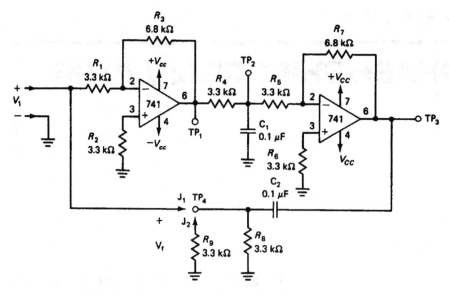

FIGURE 5-1

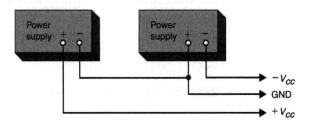

FIGURE 5-2

oscillator for open-loop measurements to be made. R_9 simulates the load that the amplifier would place on the lag network if the oscillator were running closed-loop. It is important that this load be included in the open-loop circuit so that realistic values of A_v and B are obtained.

3. Using the oscilloscope or VOM, perform a dc voltage check at TP_1 and TP_3 to verify that the op amp output voltage levels are at 0 V. If you find +12 V or −12 V at either test point, you probably have a defective op amp, which obviously means that a replacement is necessary.

4. Turn on the function generator and set the generator controls to produce a 100-mV$_{p-p}$ sine wave at a frequency of 1 kHz. Measure V_1 and V_f at approximately 10 equally spaced test frequencies ranging from 500 Hz to 2 kHz. Also determine the phase difference between V_1 and V_f at each of these 10 test frequencies. Do not forget to record whether the phase difference is positive (if V_f leads V_1) or negative (if V_f lags V_1).

FREQUENCY (Hz)	V_1 (V)	V_f (V)	ϕ (deg)	LEAD/LAG (+/−)
500				
2000				

STEP 4 DATA

5. The oscillator will now be tested for proper closed-loop operation. Disconnect the simulated load resistor, R_9, from TP$_4$. Close the loop by connecting J_1 to TP$_4$. Obviously, there is no longer any need for a generator hookup, so remove it from the circuit. This will remove the generator's internal impedance from acting as a load on the closed-loop system. Measure the frequency of oscillation by connecting the frequency counter at TP$_3$.

6. Sketch the resulting waveforms of the voltages at TP$_1$, TP$_2$, TP$_3$, and TP$_4$. Record with each sketch the voltage amplitude and phase. Let the voltage at TP$_4$ be the phase reference.

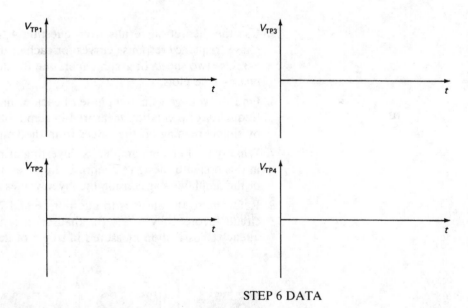

STEP 6 DATA

7. Turn off both of the power supplies. Check and make sure that all required measurements have been recorded. The circuit can now be disassembled.

REPORT / QUESTIONS:

1. Using the data gathered in step 4, calculate the oscillator's loop gain $(A_v \times B)$ by dividing the open-loop output voltage, V_f, by the applied input voltage, V_1. Do this at each of the 10 test frequencies used in step 4. Record your resulting voltage gain and phase data in tabular form.

2. At what frequencies does the oscillator's measured loop gain exceed unity? At what frequency does the open-loop phase difference between V_f and V_1 equal zero degrees? How do these frequencies compare to the frequency of oscillation measured in step 5?

3. What was the purpose of R_9 in the open-circuit test circuit? Why was 3.3 kΩ chosen as its resistance value?

4. Figure 5-3 shows the two phase-shift networks used in this oscillator design. What kind of filters are they (high-pass or low-pass)? What kind of phase-shift networks are they (lead or lag)? Using circuit theory or using SPICE on your PC, determine the theoretical voltage gain and phase difference between V_o and V_{in} for each of these two networks. Use the same ten test frequencies that were used in step 4 of the test procedure. Note that the circuit shown to the left in Fig. 5-3 cannot have a gain which exceeds 0.5 due to simple voltage-divider action between the two equal-value resistors.

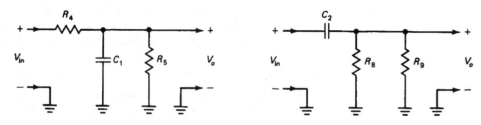

FIGURE 5-3

5. Use the theoretical results from question 4 to plot the voltage gain and phase frequency response curves for each of the two circuits shown in Fig. 5-3. Use two sheets of graph paper, one for the gain plots and one for the phase-angle plots.

6. Find the voltage gain and phase of each of the networks in Fig. 5-3 at the frequency of oscillation measured in step 5 of the test procedure. Do this by simply reading off the values from the graphs generated in question 5.

7. What type of op amp amplifiers (inverting or noninverting) are being used in this oscillator design? Theoretically determine the voltage gain of each of the amplifier stages using the given values for R_1, R_3, R_5, and R_7.

8. Referring to the answers to questions 6 and 7, determine if the oscillator circuit theoretically meets Barkhausen criteria for oscillation at the frequency of oscillation measured in step 5 of the test procedure.

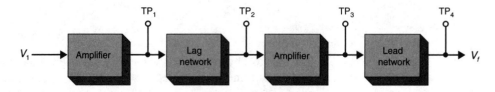

Loop gain $= A_{V1} \times A_{V2} \times A_{V3} \times A_{V4} \geq 1$

Total phase $= \phi_1 + \phi_2 + \phi_3 + \phi_4 = 0$ degrees

LC FEEDBACK OSCILLATOR

OBJECTIVES:

1. To reinforce the concepts of Barkhausen criteria for oscillation.
2. To analyze typical oscillator designs using operational amplifiers.
3. To become familiar with limiting and one application for this process.

REFERENCES:

Refer to Section 1-8 of the text.

TEST EQUIPMENT:

Dual-trace oscilloscope
Low-voltage power supply (2)
Function generator
Volt-ohmmeter
Frequency counter
Prototype board
Distortion analyzer (optional)

COMPONENTS:

741 Operational amplifier
Germanium diodes: 1N270, IN34A or equivalent (2)
Inductor: 1 mH
Capacitors: 2.2 nF (1)
Resistors ($\frac{1}{2}$ watt): 330 Ω, 470 Ω, 1 kΩ (2)

PRELABORATORY:

Determine the resonant frequency of the series resonant circuit of Fig. 6-1. Also, determine the theoretical voltage gain of the op amp stage using the given values of R_1 and R_2.

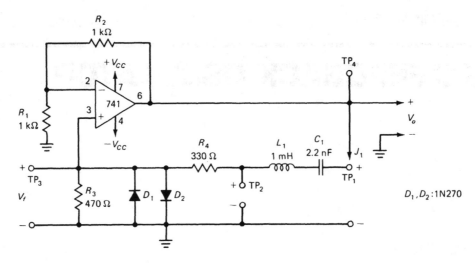

FIGURE 6-1

PROCEDURE:

1. Build the oscillator of Fig. 6-1. Use a pair of needle-nosed pliers to protect the glass body of each germanium diode from being cracked as the leads are bent. Connect the power supplies to the op amp as shown in Fig. 6-1. Adjust each power supply voltage level to 5 V dc to apply ± 5 V dc to the op amp.

2. To analyze this oscillator circuit design in the open-loop mode, connect the function generator to TP_1 and make sure that jumper J_1 is not connected to TP_1. Check the dc output voltage of the op amp. If the level is not approximately zero volts, you may have a wiring error or a defective op amp. Repair any faults that are found.

3. Apply a 250-mV$_{p-p}$ sine wave at TP_1 and tune the frequency to approximately the resonant frequency determined in prelab. Monitor V_{TP1} with channel A and monitor V_o with channel B of the oscilloscope. Fine-tune the frequency of the generator so as to produce a maximum signal amplitude at TP_4. Record the generator frequency and the amplitude of V_{TP_1}, V_f, and V_o. Also, record the phase angle of V_f and V_o, letting V_{TP_1} be phase reference.

$$V_{TP_1} = \underline{\hspace{1cm}} \quad V_f = \underline{\hspace{1cm}} \quad V_o = \underline{\hspace{1cm}}$$
$$f_{TP_1} = \underline{\hspace{1cm}} \quad \phi_f = \underline{\hspace{1cm}} \quad \phi_o = \underline{\hspace{1cm}}$$

4. Reconnect the generator at TP_2 and measure the same amplitudes and phase angles that were measured in step 3. From these measurements, determine the actual voltage gain of the op amp stage. Compare this value to the theoretical value determined in prelab.

$$V_{gen} = \underline{\hspace{1cm}} \quad V_f = \underline{\hspace{1cm}} \quad V_o = \underline{\hspace{1cm}}$$
$$A_v = \underline{\hspace{1cm}} \quad \phi_f = \underline{\hspace{1cm}} \quad \phi_o = \underline{\hspace{1cm}}$$

5. Now increase the amplitude of the generator voltage until limiting just begins to occur in the waveforms of V_f and V_o. This is the point at which the V_f and V_o waveforms start to look distorted due to the diodes beginning to turn on. Measure the amplitudes and sketch the waveforms of V_{TP_1}, V_f, and V_o at this critical setting.

6. Now the oscillator will be tested for closed-loop operation. Do this by removing the generator connection from TP_1 to avoid loading by the genera-

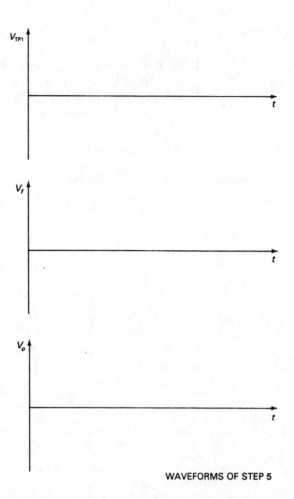

WAVEFORMS OF STEP 5

tor's internal impedance. Also, the jumper J_1 must be connected to TP_1. The circuit should now self-oscillate at the resonant frequency of the series resonant circuit. Measure the amplitude and phase angle of the voltages at TP_2, TP_3, and TP_4. Let the voltage at TP_4 serve as phase reference. Also, measure the frequency of oscillation and note how badly distorted the waveforms appear to be. Note that it may be necessary to use 10:1 probes on the oscilloscope to avoid having the scope's internal impedance from loading down the circuit and possibly keeping it from oscillating.

TP$_2$: Amplitude = _____   phase = _____

TP$_3$: Amplitude = _____   phase = _____

TP$_4$: Amplitude = _____   phase = _____

frequency = _____

disortion comments:

7. Short out the resistor R_4 by connecting a jumper between TP_2 and TP_3. With R_4 out of the circuit, the feedback factor, B, should be much larger.

Again, observe the waveforms at TP_2, TP_3, and TP_4. Note any changes that occur in the waveforms. Remove the R_4 short before proceeding.

8. If a distortion analyzer is available, measure the total harmonic distortion (% THD) of V_o at TP_4.

% THD = _____

REPORT/QUESTIONS:

1. What is the impedance of an ideal series resonant circuit at its resonant frequency? Using this answer and your observations for reference, explain why the criteria for oscillation for the *LC* feedback oscillator is met only at the resonant frequency.

2. Explain the results of step 7. What is the purpose of the germanium diodes in this oscillator design?

COLPITTS RF OSCILLATOR DESIGN

OBJECTIVES:

1. To investigate the theory of operation of a Colpitts oscillator.
2. To follow a "cookbook" design procedure in the fabrication of a working circuit.
3. To reinforce the concepts of Barkhausen criteria for oscillation.

REFERENCE:

Refer to Section 1-8 of the text.

TEST EQUIPMENT:

Dual-trace oscilloscope
Low-voltage power supply
Function generator
Volt-ohmmeter
Frequency counter
Prototype board

COMPONENTS:

2N2222 Transistor
Inductors: 8.2 μH, 27 mH
Capacitors: 0.1 μF (3), selected design values for C_1 and C_2 from Table 7-1
Resistors ($\frac{1}{2}$ watt): 680 Ω, selected design values for R_1, R_2, and R_3 from Table 7-1

THEORY:

In this experiment you will design, build, and test a Colpitts oscillator. This oscillator belongs to a class of oscillators called resonant oscillators. This name arises from the fact that they use LC resonant circuits as the frequency determining elements. Again, as in the oscillators of Experiments 5 and 6, Barkhausen criteria must be met for oscillations to occur. Specifically, the product of the loaded voltage gain of the active device's stage and the attenuation of the feedback network must be equal to or slightly larger than unity to sustain oscillations resulting from an undistorted sinusoidal output signal.

$$A_v \times B \geq 1$$

Also, the total phase shift that occurs around the closed loop must be close to 0° to ensure that positive feedback exists.

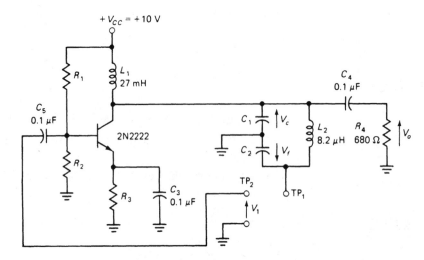

FIGURE 7-1

Figure 7-1 shows the basic Colpitts oscillator layout. It is easily recognized by the capacitor voltage division, which makes up the feedback network. Applying Barkhausen criteria to this circuit, we find

$$\frac{V_c}{B_1} \times \frac{V_f}{V_c} \geq 1$$

where

$$A_v = \frac{V_c}{B_1} \quad \text{and} \quad B = \frac{V_f}{V_c} = \frac{C_1}{C_2}$$

In other words, when the circuit is in the open-loop configuration, V_f must be equal to or slightly greater in amplitude to V_1. Also, V_f must be in phase with V_1. If these two conditions are met, then when the loop is closed, the circuit will oscillate. The amplitude criteria are fairly easily met. The voltage-divider action of C_1 and C_2 must yield an attenuation, B, that is the reciprocal of the voltage gain of the transistor amplifier stage, A_v.

$$A_v \geq \frac{1}{B}$$

Meeting the phase criteria is more difficult. The collector-to-emitter voltage of a common-emitter amplifier is 180° out of phase with the base-to-emitter voltage if the collector load is a pure resistance. The feedback scheme of obtaining V_c and V_f in opposite directions with respect to ground causes the additional 180° phase shift required to bring V_f to the desired 0° phase relationship. The load on the transistor's collector looks purely resistive when the circuit is resonant. Thus there is no additional phase shift created within the closed-loop system. At resonance, the tank circuit ideally looks like an infinitely large resistance in parallel with R_4. In practice, the equivalent parallel resistance is small enough to decrease the load slightly below 680 Ω, due to the Q of the inductor in the tank circuit, so circuit response may be slightly altered from what is ideally expected.

PRELABORATORY DESIGN:

Design the Colpitts oscillator circuit of Fig. 7-1 using the following design procedures. Draw a schematic of the final design in your lab report along with each of your design calculations.

1. *General design rules*: For the final design, use only a single component in each component location. Pick the value of the component from the list of standard value components available in the lab given in Table 7-1. Do this by selecting the standard value that comes closest to each of your design values.

TABLE 7-1 Standard Values of Resistors and Capacitors Available for Use as Design Values

Resistors (Ω):	10, 15, 22, 27, 33, 47, 68, 82,
	100, 150, 220, 270, 330, 470, 680, 820
	1 k, 1.5 k, 2.2 k, 2.7 k, 3.3 k, 4.7 k, 6.8 k, 8.2 k
	10 k, 15 k, 22 k, 27 k, 33 k, 47 k, 68 k, 82 k,
	100 k, 150 k, 220 k, 270 k, 330 k, 470 k, 680 k, 820 k
Capacitors (nF):	0.1, 0.22, 0.47, 1.0, 2.2, 4.7, 10, 22, 47, 100, 220, 470

2. *Dc design*: Determine the values of R_1, R_2, and R_3 to meet the following conditions. Assume that $V_{BE} = 0.5$ V.
   (a) The emitter current, I_E, is approximately 3.75 mA.
   (b) The current through the voltage-divider resistors, R_1 and R_2, is approximately one-tenth the value of I_E.
   (c) $V_{CE} = 5.5$ V. Assume that the 27-mH inductor is ideal (negligible winding resistance).

3. *Ac design:* Assume that the collector load is just R_c. Assume that the tank circuit impedance is too large to produce any noticeable loading on the amplifier stage. Also, assume that the amplifier's input impedance, reflected back through the tank circuit, does not produce any appreciable loading on the output of the amplifier stage.
   (a) Determine A_v of the amplifier stage using the values of R_1, R_2, and R_3 determined in the dc design. Use the equations given below to approximate the transistor's base-emitter ac resistance and the voltage gain of the amplifier stage:

$$A_v = -\frac{r_c}{r_e'} = -\frac{R_4}{r_e'} \quad \text{where } r_e' \cong \frac{0.025}{I_E}$$

   (b) Select single standard values of C_1 and C_2 to achieve a value of B such that $A_v \times B = 10$, and to cause the frequency of oscillation to be 1.8 MHz ± 200 kHz.

PROCEDURE:

1. Show the completed design and drawing to the lab instructor for verification that it is complete and accurate. The design and drawing should be completed before coming to the lab and is due for inspection at the beginning of the laboratory session.

2. Assemble the circuit on the prototype board. Leave the circuit in the open-loop configuration by leaving the jumper between TP_1 and TP_2 disconnected. Connect the power supply, generator, and oscilloscope to the circuit to make open-circuit measurements.

3. Using the volt-ohmmeter, measure V_{BE}, V_{CE}, and V_E. Determine if the transistor is biased close to the theoretical values used in the design procedures. Document the original data, changes to the circuit design, and final test data.

4. The generator should initially be tuned close to the theoretical resonant frequency of 1.8 MHz. Apply a 20 mV_{p-p} signal at TP_2 and determine the frequency that causes V_c to be of maximum amplitude. Record this frequency. You should use 10:1 probes on the scope to avoid having the scope load down the circuit.

5. At this test frequency, measure the voltage amplitude and frequency of the voltages V_1, V_c, and V_f. Also, record the phase angles of these voltages with respect to V_1.

VOLTAGE	AMPLITUDE	FREQUENCY	PHASE
V_1			
V_c			
V_f			

6. Using the data gathered in step 5, determine if $A_v \times B$ exceeds unity. Show your calculations.

7. If $A_v \times B$ is less than 1, make necessary changes to the circuit in order to cause the product to exceed unity. If the product does exceed unity, close the loop by disconnecting the generator and hooking up a jumper between TP_1 and TP_2. Measure the amplitude, frequency, and phase of the three voltages V_1, V_c, and V_f. Again, make sure that these waveforms are being measured with 10:1 scope probes.

8. Carefully measure and record the waveshape of the oscillator's output voltage. Does it look at all distorted? If it does, consult the troubleshooting chart in Fig. 7-2 and make changes to A_v and B to produce the most undistorted output voltage waveform. An easy way to decrease A_v without changing the bias conditions is to place an unbypassed emitter resistance between the emitter resistor and transistor's emitter. Try values between 10 and 100 Ω.

   (a) If this resistor is too large, the oscillator will not oscillate.
   (b) If this resistor is too small, the output voltage waveform will remain distorted.

9. Determine a final set of component values that will yield an apparently undistorted sinusoidal signal of desired frequency and record the final schematic in your report.

Symptom #1:   With loop closed, the oscilloscope shows no output signal.

Possible causes:

   (a)   No d.c. power supplied.
   (b)   1:1 scope probe loading down the circuit.
   (c)   Insufficient voltage gain in amplifier stage.
   (d)   Feedback voltage division too small.

Symptom #2:   With loop closed, the oscilloscope display as shown below:

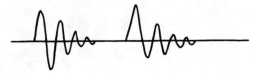

Possible causes:
   (a)   Way too much voltage gain.
   (b)   Feedback voltage division way too large.

Symptom #3:   With loop closed, the oscilloscope display as shown below:

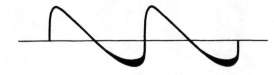

Possible causes:
   (a)   Slightly too much voltage gain.
   (b)   Feedback voltage division slightly too large.

If the output signal of the oscillator is an undistorted sinewave, then $A_r \times B$ is probably somewhere between 1.0 and 2.0.

FIGURE 7-2   Troubleshooting chart

10. Demonstrate to the lab instructor that your Colpitts oscillator is working properly in producing an undistorted output waveform in the closed-loop configuration.

REPORT/QUESTIONS:

1. Draw the ac and dc equivalent circuits for your final working Colpitts oscillator.

2. Explain how your Colpitts design operates, using the results of your measurements and conclusions that you have devised based on positive-feedback concepts and Barkhausen criteria.

HARTLEY RF OSCILLATOR DESIGN

OBJECTIVES:

1. To investigate the theory of operation of a Hartley oscillator.
2. To follow a "cookbook" design procedure in the fabrication of a working circuit.
3. To become familiar with the characteristics of toroid coils and the procedures associated with hand-winding coils.
4. To reinforce the concepts of Barkhausen criteria for oscillation.

REFERENCE:

Refer to Section 1-8 of the text.

TEST EQUIPMENT:

Dual-trace oscilloscope
Low-voltage power supply
Function generator
Volt-ohmmeter
Frequency counter
Prototype board

COMPONENTS:

2N2222 Transistor
Capacitors: 0.001 μF, 0.1 μF (3)
Resistors ($\frac{1}{2}$ watt): selected design values for R_1–R_5 from Table 7-1
Toroid core: iron powder type T-106, mix 2 (Polamar or Amidon)
Magnet wire: AWG No. 20 varnish-coated, 4-ft length

THEORY:

The Hartley oscillator design, shown in Fig. 8-1, operates in the same manner as the Colpitts, except that the voltage division is accomplished at the inductive half of the

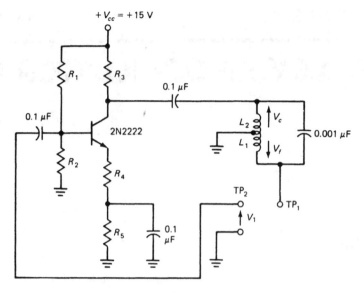

FIGURE 8-1

tank circuit instead of at the capacitive half. To make the results agree more closely with theory, the RF choke has been replaced with a fixed collector resistor, R_3. The tank circuit inductance, shown in Fig. 8-1 as L_1 and L_2, is fabricated by winding the apropriate number of turns on an iron-powder toroid core. The voltage division is accomplished by connecting a tap at the correct location between the ends of the coil. Iron-powder toroids are circular, doughnut-shaped devices fabricated by bonding very fine iron particles together. They make it possible to build a certain value inductor with fewer turns than would be required without an iron core. They also produce very predictable values of resulting inductance. The toroid shape confines the magnetic field within its circular border and therefore reduces the magnetic coupling from accidentally resulting between coils in neighboring circuits. Winding a single coil on one core simplifies construction and makes the desired voltage division in the feedback network much easier to produce. To find the number of turns needed for type T-106 mix 2 cores, the following relation holds:

$$N = 100 \sqrt{\frac{L \times 10^6}{135}}$$

where L is the design value of total inductance in henrys.

The location of the tap can be obtained by applying transformer principles. Since the upper and lower parts of the coil are wound on the same core, the flux cutting each turn is the same. Therefore, the voltage per turn is constant, allowing the following relationship to hold for the feedback attenuation factor:

$$B = \frac{V_f}{V_c} = \frac{N_1}{N_2}$$

where

N_1 = number of windings for L_1

N_2 = number of windings for L_2

$N = (N_1 + N_2)$ = total number of windings in the toroid coil

The major drawback to toroids is that the powdered iron exhibits loss, which appears as a resistor in parallel with the tuned tank circuit. For the T-106 mix 2 variety, the equivalent parallel resistance, R_P, is given by

$$R_P = \frac{K}{N^2}$$

where $K = 619.2 \times 10^3$ Ω-turns^2

N = total number of turns

This resistance, R_P, is in parallel with R_c in the ac equivalent circuit. The required tap location can be calculated once the voltage-divider ratio, B, is determined. Three turns is about the smallest practical number for obtaining predictable voltage division. When designing the transistor amplifier portion of the oscillator, the best results are usually achieved with gains in the range 5 to 20. Low values of voltage gain create a problem of obtaining the desired voltage division, B, which is very difficult to adjust. High values of voltage gain are hard to obtain repeatedly or predictably from the transistor at high frequencies. They also require the voltage division, B, to be unreasonably small and difficult to achieve.

PRELABORATORY DESIGN:

Design the Hartley oscillator circuit of Fig. 8-1 using the following design procedures. Draw a schematic of your final design in your lab report along with each of your design calculations.

1. *General design rules:* For the final design, use only a single component in each component location. Pick the value of the component from the list of standard value components available in the lab given in Table 7-1 of the Colpitts oscillator experiment. Do this by selecting the standard value that comes closest to each of your design values.

2. *Ac design:*
   (a) Determine the value of inductance, L, required to obtain oscillation at 1.8 MHz $\pm$ 200 kHz.
   (b) Determine the total number of turns required to achieve this total inductance.
   (c) Select a tap location that will yield a value of B somewhere in the range 0.18 to 0.25.
   (d) Calculate the equivalent parallel resistance, R_P, of the coil at resonance.
   (e) Calculate the value of voltage gain, A_v, of the amplifier to create $A_v \times B$ to be between 1.0 and 1.1.
   (f) Select a trial value of R_4 so that R_4 is approximately 15 times greater than the transistor's base-emitter dynamic resistance, r_e'. Use a dc emitter current value of 3.75 mA. Approximate r_e' using

$$r_e' \cong \frac{0.025}{I_E}$$

   (g) Calculate the value of ac collector resistance, r_c, required to yield the desired A_v, using

$$A_v = \frac{r_c}{r_e' + r_e} = \frac{R_P \| R_3}{r_e' + R_4}$$

(h) Using the r_c found in step (g) and the equivalent parallel resistance, R_P, found in step (d), calculate a trial value of R_3.

$$r_c = \frac{R_3 R_P}{R_3 + R_P}$$

3. *Dc design:*
   (a) Select values of R_1, R_2, and R_5, which when used with R_3 and R_4 from the ac design yields the following dc conditions. Assume that $V_{BE} = 0.5$ V.
      (1) Dc emitter current: $I_E = 3.75$ mA.
      (2) $V_{CE} = 6.0$ V.
      (3) The current through the voltage-divider resistors, R_1 and R_2, is approximately one-tenth the value of I_E.
   (b) Make sure that R_5 is set at least 10 times as large as R_4.

PROCEDURE:

1. Show the completed design and drawing to the lab instructor for verification that they are complete and acceptable.

2. Assemble the oscillator circuit using the prototype board. Set up the circuit in the open-loop configuration by leaving TP_1 and TP_2 disconnected.

3. Using the volt-ohmmeter, measure V_{BE}, V_{CE}, and V_E to determine if the transistor is biased properly for amplification to occur. Document the original data, any changes made to the circuit design, and final dc data.

4. The generator should be initially tuned close to the theoretical resonant frequency of 1.8 MHz. Apply a 200-mV$_{p\text{-}p}$ signal at TP_2 and determine the frequency at which V_c is of maximum amplitude. Use 10:1 scope probes to make sure that the scope's internal impedance does not alter the waveforms being observed due to loading.

5. At this frequency, measure the voltage amplitude, frequency, and phase of V_c, V_f, and V_1, using V_1 as phase reference.

6. Using the data gathered in step 5, decide if $A_v \times B$ is greater than unity.

7. If $A_v \times B$ is less than 1, make necessary changes to the circuit design to cause the product to exceed unity and repeat steps 2–6. If $A_v \times B$ is greater than 1, close the loop by connecting a jumper between TP_1 and TP_2 and disconnect the generator from the circuit. Measure the amplitude, frequency, and phase angle of the three voltages as was done in step 5. Again, use 10:1 scope probes.

8. Carefully measure and record the waveshape of the oscillator's output voltage. If the output waveform is not a clean sine wave, but shows distortion, or if there is no output waveform present, consult the troubleshooting chart in Fig. 7-2. Make appropriate modifications. Record all changes made to your design. An easy way to change the voltage gain of the amplifier stage without changing the bias conditions or the resonant frequency is to alter the value of R_4.

9. Determine a final set of values for the circuit that will yield an apparently undistorted sinusoidal signal of the desired frequency.

10. Demonstrate to the lab instructor that your Hartley oscillator is working properly in producing an undistorted output waveform in the closed-loop configuration.

REPORT / QUESTIONS:

1. Draw the ac and dc equivalent circuits for the oscillator stage.

2. Explain how your circuit operates, using the results of your measurements and conclusions you have devised based on positive-feedback concepts and Barkhausen criteria.

SWEPT-FREQUENCY MEASUREMENTS

OBJECTIVES:

1. To use the VCG swept-frequency capability of a signal generator to obtain the frequency response of a circuit.

2. To use a detector in a swept-frequency measurement setup.

3. To use a swept-frequency measurement procedure to obtain the insertion loss of a ceramic filter in decibels.

REFERENCE:

Ceramic filters are discussed in Section 4-3 of the text.

TEST EQUIPMENT:

Dual-trace oscilloscope: must be able to produce a Y versus X display with variable sensitivity on both axes

Function generators (2); one must have a VCG input jack

50-Ω Selectable attenuator pad; refer to Fig. 9-7

Low-voltage power supply

Frequency counter

COMPONENTS:

Ceramic filter: Murata-Erie Type CFM-455D

Resistors ($\frac{1}{2}$ watt): 50 Ω, 1.5 kΩ (2), 10 kΩ (3)

Capacitor: 0.1 μF

Germanium diode: 1N270, 1N34A, or equivalent

THEORY:

Swept-frequency measurement techniques are used extensively in the design and testing of RF and microwave components. This technique allows the technician to observe the entire frequency response of a circuit in real time on oscilloscopes.

A swept-frequency measurement setup can be thought of as one where constant amplitude signals of different test frequencies are applied to a test circuit in successive time intervals. The results of each of these successive measurements are then observed on an oscilloscope.

Implicit in this technique is the assumption that the rate at which the test frequency is changed can be ignored. This criterion can be met in almost all cases where the sweep frequency is no greater than one-hundredth of the lowest test frequency. In some noncritical applications, the sweep frequency can be as high as one-tenth the lowest test frequency.

The simplest form of a swept-frequency setup is shown in Fig. 9-2. This form of measurement displays the complete test sine wave and achieves a waveform such as the bandpass filter in the left column of Fig. 9-4. The peak values of the sine waves form the frequency response curve with respect to the zero volt axis for the filter under test. The portion of the display below the zero volt line is simply the mirror image of the top-half.

The major drawback to this display method is that the display instrument used for the resulting display must have a frequency response significantly greater than that of the highest frequency to be displayed. This method is obviously useless with an X-Y recorder at RF frequencies; however, it may prove useful for a high-quality oscilloscope at low RF frequencies.

A less ambiguous display and one that is compatible with low-frequency oscilloscopes and X-Y recorders is shown in the right column of Fig. 9-4. This display is obtained by inserting a diode peak detector at the output of the circuit under test. This peak detector generates a voltage amplitude that is proportional to the peaks of the RF sine waves passed by the test circuit and creates the desired bandpass filter response display. In this case the oscilloscope used does not have to display the RF frequencies, only the low-frequency sweep signal.

Within the procedure of this experiment, you will use both methods to display the bandpass characteristics of a ceramic RF filter. Limitations and calibration of the display will be illustrated. The ceramic bandpass filter used as a sample test device in this experiment is designed for use in a standard AM broadcast band receiver's intermediate-frequency circuits. The device has a specified center frequency of 455 kHz and a total-5-dB bandwidth of 30 kHz. It exhibits a typical 2-dB ripple and 2-dB insertion loss in the bandpass region and a very large (greater than 50 dB) attenuation cutoff in the stop band. These filters are fairly expensive and fragile, so handle them with care. These filter characteristics are specified between a 1.5-kΩ source and a 1.5 kΩ load, so make sure that the circuit given in Fig. 9-2 is used for testing this device.

PROCEDURE:

1. Build the test circuit shown in Fig. 9-1. Most function generators that are designed for sweeping capability will provide the user with an input jack. The name of the jack varies with the manufacturer of the generator. Some of the more popular names used are VCG (voltage-controlled generator) input, VC input, or FM (frequency-modulated) input. As the names imply, the instantaneous voltage level that is applied to this input jack will determine a certain amount of deviation in the output frequency of the generator. The exact amount of frequency change that occurs will also vary from one generator to another.

2. Adjust the RF generator to produce a 400-mV$_{p-p}$ sine wave at a test frequency of 450 kHz. Now apply a series of eight different positive dc voltage

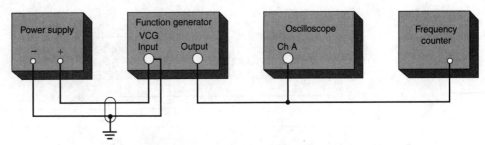

FIGURE 9-1 Determination of the deviation rate

levels to the VCG input jack and note the resulting output frequency of the RF generator. Consult the manual of the generator to make sure that you do not exceed the maximum allowable VCG input voltage before nonlinearity or damage results in the RF generator.

VCG INPUT VOLTAGE (V)	OUTPUT FREQUENCY (kHz)

3. Repeat step 2 using a negative dc voltage applied to the VCG input jack. Simply reverse the leads of the power supply to create a negative voltage at the VCG input jack of the generator.

VCG INPUT VOLTAGE (V)	OUTPUT FREQUENCY (kHz)

4. Plot the graph of the output frequency versus VCG input voltage for your generator. It should be a near-linear function. From your graph, determine your generator's resulting deviation rate, K_f, by reading off the slope of the straight line or using the following equation:

$$K_f = \frac{\Delta f}{\Delta V_{VCG}} = \frac{f_{max} - f_{min}}{V_{max} - V_{min}}$$

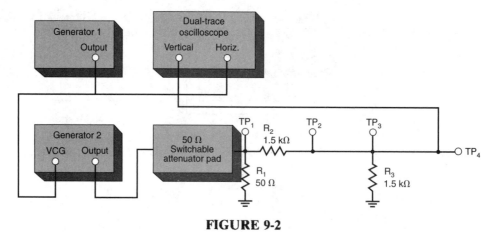

FIGURE 9-2

5. From the graph or the equation of step 4, determine the minimum and maximum dc voltages at the VCG input jack that would cause the output frequency to be 400 kHz and 500 kHz, respectively, assuming that the generator's frequency dial remains set at 450 kHz.

6. Calculate the peak-to-peak voltage swing necessary to sweep the output frequency of the generator between 400 and 500 kHz. Do this simply by finding the difference between the maximum and minimum VCG voltages in step 5.

7. Construct the test circuit of Fig. 9-2. In Fig. 9-2, we shall refer to generator 1 as the sweep generator since it controls the rate and amount of frequency deviation that occurs. We shall refer to generator 2 as the RF generator since its output frequency range must be kept significantly higher than that of the sweep generator.

   Adjust the 50-Ω attenuator pad switches so that no attenuation (0 dB) exists within the pad. Apply a 1-V_{p-p}, 450-kHz sine wave from the RF generator to the vertical input of the oscilloscope. Put the oscilloscope into its Y versus X mode of operation (time-base disabled). Make sure that the vertical and horizontal inputs are dc coupled. In this mode the scope is displaying the vertical input signal versus the horizontal input signal. Since the sweep generator of Fig. 9-2 is turned off, there should be no horizontal sweep, making the display a vertical line. Horizontally position the line in the middle of the screen. Also, adjust the vertical sensitivity so that a vertical line fills up most of the screen, but remains calibrated.

8. Turn on the sweep generator and adjust so as to produce a 5-Hz triangle output waveform. This should form a horizontally shifting vertical line display on your scope. If your sweep generator has a dc offset control, make sure that it is either turned off or set to exactly zero volt offset. Now adjust the amplitude of the sweep generator to the p-p value determined in step 5. Since this linearly changing triangle waveform is being applied to the VCG input of the RF generator, it must be allowing the output frequency of the RF generator to deviate in a linear fashion, from a minimum of 400 kHz to a maximum of 500 kHz. Since this triangle voltage is also being applied to the horizontal input of the oscilloscope, the horizontal axis must represent the frequency output signal of the RF generator. Thus the horizontal scale of the scope display can now be calibrated in terms of frequency. Increase the sweep frequency from 5 Hz to a value that causes the display to appear as a solid rectangle. Adjust the horizontal scale factor of your scope so that

the width of the resulting rectangular display just fills up the screen. You may need to uncalibrate the horizontal sensitivity control to do this.

9. Now insert the ceramic filter between TP$_2$ and TP$_3$ by removing the jumper and connecting the input of the filter to TP$_2$ and the output of the filter at TP$_3$. Now you should see a display similar to that given in Fig. 9-4(a). Sketch this display. Label the vertical axis with the appropriate voltage scale and the horizontal axis with the appropriate frequency scale.

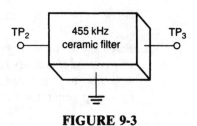

FIGURE 9-3

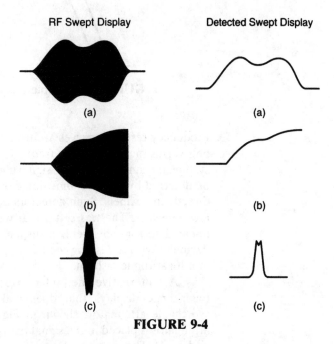

FIGURE 9-4

10. Repeat step 9 with the filter connected using each of the following ranges of frequencies to be swept. The full width of the horizontal scale should represent each of the following frequency ranges:
   (a) 430 kHz to 470 kHz
   (b) 200 kHz to 700 kHz

   You should produce displays similar to those given in the left column of Fig. 9-4(b) and (c). Before proceeding to step 11, restore the original display obtained in step 9.

11. Now remove the jumper between TP$_3$ and TP$_4$. Replace the jumper with the diode peak detector shown in Fig. 9-5. Make sure that the vertical input of the oscilloscope is dc coupled. You should now see a display similar to the right-hand column of Fig. 9-4(a). Sketch this display. Again, label the vertical axis with the appropriate voltage scale and the horizontal axis with the appropriate frequency scale.

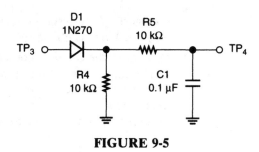

FIGURE 9-5

12. Increase the frequency of the sweep generator until the double image of the detected frequency response just becomes apparent. This display should be similar to that given in Fig. 9-6. Record this sweep frequency. Return the sweep generator back to 5 Hz before proceeding.

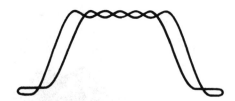

FIGURE 9-6 Double-image detected display

13. Frequency response curves are usually presented as graphs of gain in decibels versus the logarithm of frequency (Bode magnitude plots). In the ideal system, the amplitude of the signal coming out of the filter under test would be detected by a logarithmic detector and the resulting amplitude read off directly in decibels. Such detectors do exist and are referred to as square-law detectors. The test setup that we are using in this experiment uses a linear detector that yields a display in voltage amplitude rather than in decibels. We are also sweeping frequency in a linear fashion rather than in a logarithmic fashion.

An alternative method of vertical calibration for both linear and logarithmic displays that does make use of decibels uses the in-line switchable attenuator, shown in Fig. 9-7. In this method the switchable attenuator is placed in the signal path between the generator and the circuit under test. The circuit under test is then temporarily removed and replaced by a short circuit between TP$_2$ and TP$_3$. With the attenuator first set at no attenuation (0 dB), the detected display is set for a straight horizontal line corresponding to 0 dB at the top of the display of the scope. Also, the vertical deflection controls are adjusted so that total attenuation is then displayed as a horizontal line at the bottom of the scope display. Then successive steps of attenuation are switched in and the corresponding shift downward of the horizontal line on the scope display is then noted by showing a series of horizontal lines at each downward position as shown in Fig. 9-8. In this way the vertical display is then calibrated directly in decibels.

The temporary short between the input and output of the circuit under test represents a 0-dB loss filter. After resetting the attenuator to 0-dB attenuation, the filter to be tested is then placed back into the test circuit between TP$_2$ and TP$_3$ instead of the short and its vertical response is

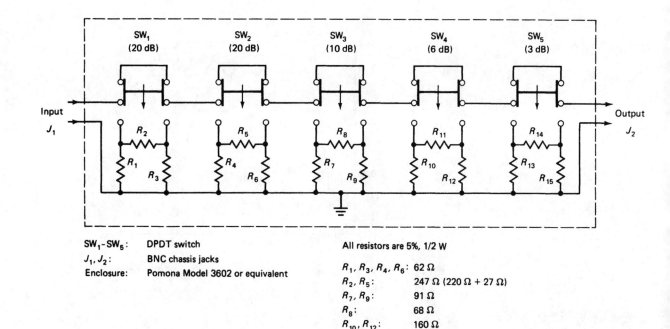

SW₁–SW₅: DPDT switch

J_1, J_2:   BNC chassis jacks

Enclosure: Pomona Model 3602 or equivalent

All resistors are 5%, 1/2 W

R_1, R_3, R_4, R_6:	62 Ω
R_2, R_5:	247 Ω (220 Ω + 27 Ω)
R_7, R_9:	91 Ω
R_8:	68 Ω
R_{10}, R_{12}:	160 Ω
R_{11}:	39 Ω
R_{13}, R_{15}:	300 Ω
R_{14}:	18 Ω

FIGURE 9-7   Schematic of a 50-Ω selectable attenuator pad

read out directly in decibels. This is done by superimposing the successive decibel horizontal lines on the detected filter response. Refer to Fig. 9-8. Not only is the response of the filter displayed, but also its insertion loss and other important characteristics.

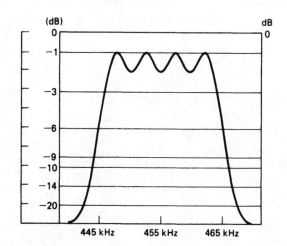

FIGURE 9-8   Sample detected frequency response

Complete the test procedure above for the 455-kHz ceramic filter. Calibrate the vertical axis in decibels, ranging from 0-dB maximum to a minimum of −20 dB of gain. Note that the decibel scale will not be linear but will resemble that shown in Fig. 9-8.

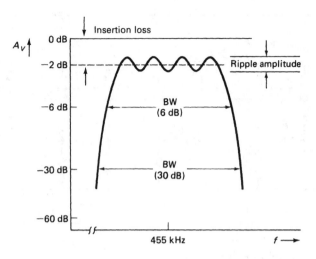

FIGURE 9-9 Example of a ceramic filter frequency response

REPORT / QUESTIONS:

1. Explain the purpose of the 50-Ω resistor at the output of the attenuator network in Fig. 9-2.

2. Refer to Fig. 9-9. Determine from your own measured swept-frequency display the following specifications for your 455-kHz ceramic filter.
   (a) Insertion loss in decibels
   (b) 6-dB bandwidth
   (c) Ripple amplitude

3. What is the hazard of using a swept-frequency display such as the one investigated in this experiment with respect to the validity of data?

4. Describe how the peak detector circuit operates. Why do the displays in steps 9 and 10 look filled in, whereas the one in Step 11 does not?

5. Write a short test procedure that a test technician could follow in order to measure the insertion loss of a 455-kHz filter using the swept-frequency display techniques investigated in this experiment.

NONLINEAR MIXING PRINCIPLES

OBJECTIVE:

To become acquainted with the process of creating new frequencies by mixing two signals of different frequencies together in a nonlinear device.

REFERENCE:

Refer to Section 2-2 of the text.

TEST EQUIPMENT:

Dual-trace oscilloscope
Function generator (2)
Frequency counter

COMPONENTS:

Resistors ($\frac{1}{2}$ watt): 1 kΩ, 1.5 kΩ (4)
Germanium diode: 1N270, 1N34A, or equivalent
Ceramic filter: CFM-455D or equivalent

THEORY:

When two sinusoidal signals of different frequencies, f_1 and f_2, are applied simultaneously to a nonlinear amplifier, a nonlinear mixing action occurs, resulting in the creation of several output frequencies which include:

1. The first and second harmonics of the original frequencies: $f_1, f_2, 2f_1$, and $2f_2$.
2. The sum and difference frequencies: $(f_1 + f_2)$ and either $(f_1 - f_2)$ or $(f_2 - f_1)$, whichever produces a positive result.
3. 0 Hz (dc offset).

One method of proving this principle is to feed the complex output signal of the mixer through a sharp bandpass filter tuned to each of the expected output frequencies, and then check for the presence of a sinusoidal signal of that particular frequency at the output of the bandpass filter. However, sharp filters, such as ceramic or crystal filters, are not available for all bandpass frequencies. Manufacturers of these devices have ample supplies available only for standard bandpass frequencies that are used in radio designs such as 455 kHz or 10.7 MHz. Thus, an alternate method that will be used in this experiment is to carefully select the two frequencies, f_1 and f_2, so that only one of the predicted output frequencies of nonlinear mixing ends up being near the standard bandpass frequency of 455 kHz. We shall use a 455 kHz ceramic filter to isolate this frequency from all the other frequencies produced by nonlinear mixing.

PRELABORATORY:

Using the theory of nonlinear mixing, list each of the frequencies contained in the output of a nonlinear amplifier, if the input signal frequencies being mixed together are set at each of the values listed in steps 2 and 3 of the test procedure for this experiment. For each output, circle the output frequency component that falls within the bandpass of the ceramic filter. Also, for each case, sketch a spectrum diagram showing what a spectrum analyzer would display as the output signal spectrum of the nonlinear mixer.

PROCEDURE:

1. Build the mixer stage shown in Fig. 10-1. The nonlinear I-V characteristics of the germanium diode make it function as a nonlinear mixer.

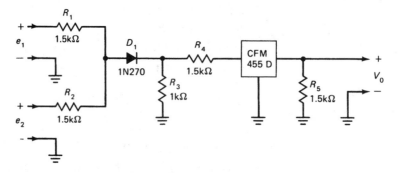

FIGURE 10-1

2. Set the amplitudes of e_1 and e_2 at 10 V_{p-p}. Set the frequency of e_1 to 455 kHz and the frequency of e_2 to 200 kHz. Monitor the output voltage using the oscilloscope. Carefully fine-tune the frequency of one of the two input generators so that the output voltage reaches a maximum amplitude. Once the amplitude is peaked, measure the output voltage amplitude and frequency.

3. Repeat step 2 using each of the following input frequencies:
   a. f_1 = 200 kHz and f_2 = 455 kHz
   b. f_1 = 227.5 kHz and f_2 = 300 kHz
   c. f_1 = 300 kHz and f_2 = 227.5 kHz
   d. f_1 = 355 kHz and f_2 = 100 kHz
   e. f_1 = 755 kHz and f_2 = 300 kHz
   f. f_1 = 295 kHz and f_2 = 80 kHz

REPORT/QUESTIONS:

1. Which outputs—$f_1, f_2, 2f_1, 2f_2, (f_1 + f_2)$, or $(f_1 - f_2)$—were of the largest amplitude?

2. If step 2 was completed using $f_1 = 682.5$ kHz and $f_2 = 227.5$ kHz, what problem would occur in looking for the output difference frequency component of $(f_1 - f_2) = 455$ kHz?

3. In step 3 (f), the only output frequency that would be within the bandpass of the ceramic filter would be $(f_1 + 2f_2)$, which is one of the smaller, usually ignored, output frequencies. How many decibels down is this output frequency compared to the simple sum and difference frequency outputs that are produced by nonlinear mixing action? How can these "extra" nonlinear modulation products be kept to a minimum amplitude level?

4. The frequency component referred to in question 3 is named a "third-order" frequency component since the sum of the two harmonics, $1f_1$ and $2f_2$, is 3. What three other third-order components could result from nonlinear mixing?

5. Research: What is meant by the third-order intercept spec of a nonlinear mixer?

AM MODULATION USING AN OPERATIONAL TRANSCONDUCTANCE AMPLIFIER

OBJECTIVES:

1. To become familiar with the 3080E operational transconductance amplifier (OTA) and how its variable gain feature makes it ideal for use as an AM modulator.

2. To observe typical AM waveforms that exhibit undermodulation, overmodulation, and 100% modulation.

3. To understand how a peak detector functions to extract the original intelligence from an AM waveform.

REFERENCE:

Refer to Chapter 2 of the text.

TEST EQUIPMENT:

Dual-trace oscilloscope: must have a Y vs X display capability

Function generator (2)

Low-voltage power supply (2)

Audio spectrum analyzer (if available)

COMPONENTS:

3080E Integrated circuit

Germanium diode: 1N270, 1N34A, or equivalent

Capacitors: 2.2 nF, 0.1 μF (3), 0.22 μF, 2.2 μF

Resistors ($\frac{1}{2}$ watt): 47 Ω (2), 4.7 kΩ (2), 10 kΩ (2)

Potentiometers: (10-turn trim) 10 kΩ (2)

PROCEDURE:

1. Build the circuit given in Fig. 11-1. This circuit is an AM modulator stage using a 3080E operational transconductance amplifier (OTA). Connect TP$_3$ to ground by placing switch S1 in position 1. Apply ± 6 V dc to the circuit.

61

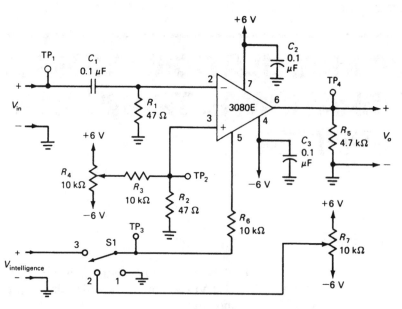

FIGURE 11-1 AM modulator stage using a 3080E OTA

Observe the dc voltage at TP_2 with an oscilloscope set in its most sensitive vertical scale factor. Adjust R_4 so as to force the voltage at TP_2 to be exactly 0 V dc.

2. Apply a 60-mV$_{p-p}$ 50-kHz sine wave at TP_1. This generator will simulate the RF carrier of an AM transmitter. Observe the waveform at TP_4. Adjust the function generator to make V_o exactly 2 V$_{p-p}$. Make small adjustments to R_4 to force V_o to have exactly 0 V dc offset.

3. Move S1 into position 2. This connects a 10-kΩ potentiometer between $+V_{cc}$ and $-V_{cc}$ with the wiper connected to TP_3. Determine the voltage gain of the OTA with the potentiometer set at seven settings between -6 and $+6$ V dc (use 2-V increments). _____ _____

4. Remove the 10-kΩ potentiometer used in step 3 from the circuit by placing S1 into position 3. Apply a 12-V$_{p-p}$, 200-Hz sine wave at TP_3. This signal simulates the intelligence signal of the AM transmitter. Make sure that you use a 10:1 scope probe to monitor the waveform at TP_3. Leave this signal on channel A of your scope. Observe the output waveform at TP_4 with channel B of the oscilloscope. The scope will need to be externally triggered with the intelligence signal in order to produce a stable display. You should see an AM waveform that exhibits approximately 100% modulation. Make minor adjustments to R_4 and to the intelligence signal amplitude in order to produce a symmetrically perfect AM waveform that has 100% modulation. Sketch the waveforms of the signals at TP_3 and TP_4.

5. An alternative method of checking the purity of the AM waveform is to use a swept-frequency display, sometimes called a trapezoidal display. Place the scope into Y vs X display mode. Let the AM waveform drive the vertical input of the oscilloscope. Let the intelligence signal at TP_3 drive the horizontal input of the oscilloscope. Make sure that the horizontal input is dc coupled. Adjust the position and sensitivity controls of the oscilloscope so as to center the display and use up as much of the screen as possible with the pattern. Sketch the resulting display (see Fig. 11-2). Determine E_{max} and E_{min}. Check the linearity of the top and bottom sides of the

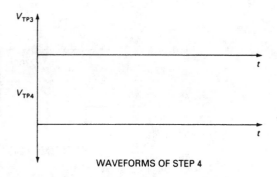

WAVEFORMS OF STEP 4

display. If the sides are not straight, the modulator is producing a distorted AM waveform.

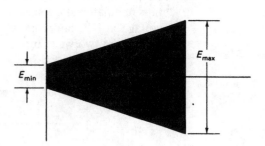

FIGURE 11-2 Typical trapezoidal display

6. If an audio spectrum analyzer is available, observe the spectral content of the AM waveform. Measure the amplitudes and frequencies of the carrier, upper side-frequency, and lower side-frequency components of the AM signal.

7. Return the controls of the oscilloscope to re-create the waveforms that were displayed in step 4. Now reduce the amplitude of the intelligence signal at TP_3 to produce exactly 50% modulation in the AM waveform at TP_4. This is done by adjusting so that E_{max} is three times as large as E_{min}. Sketch the resulting waveforms at TP_3 and TP_4.

8. Repeat steps 5 and 6 for 50% modulation. This is referred to as an undermodulated AM signal.

9. Again, adjust the oscilloscope to produce the waveforms displayed in step 4. Now increase the amplitude of V_{TP_3} to approximately 16 V_{p-p}. This should cause severe overmodulation to take place. Sketch the resulting waveforms of the signals at TP_3 and TP_4.

10. Repeat steps 5 and 6 for overmodulation. Take note of the additional side-frequency components that are produced in the spectral display. Do not disassemble the modulator before proceeding. Return the intelligence signal amplitude to approximately 12 V_{p-p}.

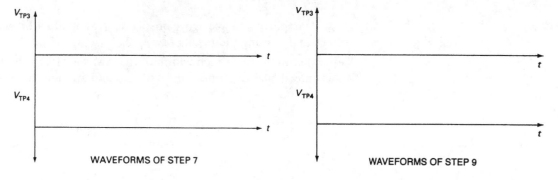

WAVEFORMS OF STEP 7   WAVEFORMS OF STEP 9

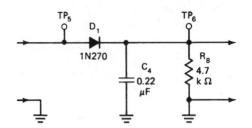

FIGURE 11-3 AM detector stage

11. Again, adjust the oscilloscope to produce the waveforms displayed in step 4. Now a simple peak detector will be used to detect the AM waveform. Build the detector circuit in Fig. 11-3. Connect a jumper between TP_4 of the modulator stage and TP_5 of the detector stage. Monitor the output voltage at TP_6 and compare it to the 12-V_{p-p} intelligence signal at TP_3. Sketch the two waveforms.

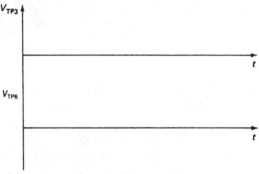

WAVEFORMS OF STEP 11

12. Increase the amplitude of the intelligence voltage at TP_3 to 16 V_{p-p} and notice the distortion that occurs with the re-created intelligence signal at TP_6. This is why overmodulation in an AM modulator stage must be avoided. Sketch the resulting waveform at TP_6. Return the amplitude at TP_3 to its original value before proceeding.

13. Try increasing and decreasing the value of the filter capacitor in the AM detector by substituting in a 2.2-nF and then a 2.2-μF capacitor for C_4. Note the distortion that occurs in the detected waveform at TP_6 if the filter capacitor is too large or too small.

REPORT/QUESTIONS:

1. Draw a graph of the voltage gain of the AM modulator versus the dc voltage at TP_3 using the results of step 3.

2. Verify that the percent modulation of the AM waveform observed in step 7 was indeed 50% using the values of E_{max} and E_{min} and the equation

$$\% \, m = \frac{E_{max} - E_{min}}{E_{max} + E_{min}} \times 100\%$$

3. For the 50% modulated AM waveform observed in step 7, assuming that the RF carrier frequency component was exactly 2 $V_{p\text{-}p}$ and the signal was measured across a 50-Ω antenna, calculate the upper and lower side-frequency component amplitudes. Also, determine:

(a) The total power of the RF carrier frequency component
(b) The total power of the upper side-frequency component
(c) The total power of the lower side-frequency component
(d) The total output power of the AM waveform

4. Describe briefly how the 3080E functions to produce an AM waveform and how the peak detector functions to extract the information from the RF carrier.

5. Explain why, when the oscilloscope is set up to display a swept frequency using the AM signal for vertical deflection and the intelligence signal for horizontal deflection, the scope displays a trapezoidal or triangular pattern.

6. Explain what caused the distorted waveforms in step 13 to result when C_4 was set too high or too low.

NAME _____

RF MIXERS AND SUPERHETERODYNE RECEIVERS

OBJECTIVES:

1. To analyze an RF mixer/IF amplifier system.

2. To determine experimentally the transconductance of a field-effect transistor using gathered frequency response data as reference.

3. To become familiar with the principles of operation of the superheterodyne receiver design.

REFERENCE:

Refer to Chapter 3 of the text.

TEST EQUIPMENT:

Dual-trace oscilloscope

Low-voltage power supply (2)

Function generators (3); one must be able to be amplitude modulated

Frequency counter

COMPONENTS:

Dual-gate FET: Type ECG222, 3N204, or equivalent (2)

Germanium diode: 1N270, 1N34A, or equivalent

Integrated circuit: LM386-3

Inductors: 33 μH (2)

Capacitors: Ceramic disk: 0.001 μF, 0.05 μF (2), 0.1 μF (3)
        silver mica: 0.0033 μF (2)
        electrolytic: 10 μF (3), 470 μF

Resistors: 4.7 Ω (2 watt), 10 Ω (1 watt), 22 kΩ (4), 100 kΩ ($\frac{1}{2}$ watt)

Potentiometer: (10-turn trim) 5 kΩ

Speaker: 8 Ω

THEORY:

This two-stage circuit is a typical RF mixer and first IF (intermediate-frequency) amplifier stage used in superheterodyne receiver designs. This circuit uses ECG222 dual-gate, N-channel field-effect transistors as the active elements. Q_1 serves as the mixer and Q_2 serves as a fairly selective amplifier.

PRELABORATORY:

Calculate the theoretical resonant frequencies for both tank circuits (the RF amplifier circuit and the IF amplifier circuit) in Fig. 12-1 using the given values of L and C. Also predict the frequency of V_{TP_5} if V_{LO} is at 1.7 MHz and V_{RF} is at 1.2 MHz.

PROCEDURE:

1. Build the circuit given in Fig. 12-1. Turn on the power supply and adjust for 3 V dc. Connect a function generator to TP_1. Apply a 50-mV$_{p\text{-}p}$ sine wave at TP_1. Use a test frequency that is approximately equal to the resonant frequency calculated in the prelab exercise. Temporarily short TP_2 to ground.

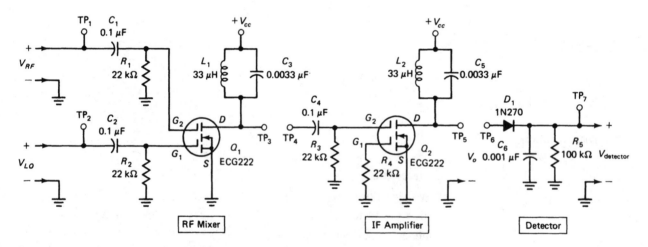

FIGURE 12-1 RF mixer/IF amplifier with detector

2. Test the RF mixer stage as an amplifier by measuring V_{RF} and V_0 at TP_1 and TP_3, respectively. Take a series of measurements at different frequencies to determine the frequency response of the mixer stage near the resonant frequency. Include enough measurements to be able to determine at what two frequencies the gain drops down 10 dB from its maximum value at the resonant frequency. Use approximately ten data points in addition to measuring the resonant frequency. Use a frequency counter to read each test frequency accurately. The output waveform at low frequencies may be distorted due to the mixer trying to resonate at harmonics of the input frequency. This effect is greatly magnified if the RF amplifier is overdriven.

3. Connect the function generator to TP_4. Now test the IF amplifier stage as an amplifier by measuring V_{in} and V_o at TP_4 and TP_5, respectively. Again, take a series of measurements at different test frequencies to determine the

frequency response of the IF amplifier stage near the resonant frequency. Again, include enough measurements to be able to determine at what two frequencies the gain drops down 10 dB from its maximum value at the resonant frequency. Determine the resonant frequency of the IF amplifier stage. If the resonant frequency is more than 10 kHz away from the resonant frequency of the RF mixer stage, fine-tune either tank circuit so as to produce exactly the same resonant frequency in both stages. Do this by adding a small parallel capacitance across the tank circuit in the amplifier that has the larger resonant frequency. Record the value of the added capacitance necessary to produce equal values of resonant frequency and which stage was modified.

4. Connect the generator back to TP_1 and connect a jumper between TP_3 and TP_4. Now test the RF-IF amplifier system as a cascaded amplifier by measuring V_{in} and V_o at TP_1 and TP_5, respectively. Use the same test procedure used in steps 2 and 3 when testing the stages individually. Also record the resonant frequency of the cascaded system. You should find that the cascaded amplifier offers a larger voltage gain at the resonant frequency and a smaller bandwidth.

5. Connect a second function generator to TP_2. Set V_{RF} (at TP_1) to an amplitude of 200 mV$_{p-p}$ at a frequency of 1.2 MHz. Set V_{LO} (at TP_2) to an amplitude of 2 V$_{p-p}$ at a frequency of 1.7 MHz. Monitor the output voltage at TP_5. Fine-tune the frequency of V_{LO} while keeping the frequency of V_{RF} constant, to produce a maximum output voltage at TP_5. Using the frequency counter, measure the frequencies of V_{RF}, V_{LO}, and the voltage at TP_5. Record these frequencies.

6. Increase and decrease the amplitude of V_{LO} to determine the value of V_{LO} that yields a saturated value of V_o. This means that further increases in V_{LO} do not significantly increase the amplitude of V_o. Do not, however, allow the amplitude of V_{LO} to exceed 10 V$_{p-p}$. Record this critical value of V_{LO}. Now, with V_{LO} set at this critical value, vary V_{RF} from 0 V to 200 mV$_{p-p}$ in 40-mV steps and record V_o at each setting. Record your measurements in tabular form.

7. Return V_{RF} back to 200 mV$_{p-p}$. Now vary V_{LO} from 0 V to the critical value measured in step 6, using approximately five increments. Record V_o at each setting.

8. In steps 9 to 12 we shall simulate the reception of an AM broadcast station by an AM receiver using the superheterodyne design. Hook up the test configuration as shown in Fig. 12-2. A jumper should be connected between TP_5 and TP_6. Make sure that the scope is set up to trigger on the intelligence signal (function generator 3). Also, make sure that the function generator that can be amplitude modulated is connected up as the RF generator at TP_1.

9. Again, set V_{RF} to 200 mV$_{p-p}$ at a frequency of 1.2 MHz and V_{LO} at 2 V$_{p-p}$ at a frequency of 1.7 MHz. Monitor the voltage at TP_5 and fine-tune the frequency of V_{LO} until the output amplitude is at a maximum. Now switch the RF generator over to the AM mode. Apply a 1-kHz sinusoidal intelligence signal to the AM input jack of the RF generator. Carefully increase the amplitude of the intelligence signal until the waveform of the RF generator output shows approximately 100% modulation. (The positive and negative envelopes should just barely touch.) You should see that the output of the IF amplifier at TP_5 is another AM waveform that has a carrier frequency of 500 kHz rather than the original 1.2 MHz. You should also see that the output of the diode detector is the original 1-kHz intelligence

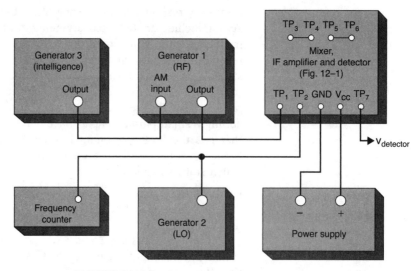

FIGURE 12-2 Superheterodyne receiver setup

signal mixed in with some noise. Sketch the waveforms observed at TP_5 and TP_7.

10. Build the audio amplifier in Fig. 12-3. Connect the input of the audio amplifier stage to TP_7. Connect the audio output to the speaker. Apply +12 V dc to this circuit using a second power supply. Adjust the volume control potentiometer to a comfortable listening level. You should be able to hear the original 1-kHz intelligence signal as the detected signal produced by the diode detector. Prove this by varying the frequency of the audio oscillator and note the shift in the frequency of the detected signal TP_7 and the audible output of the speaker.

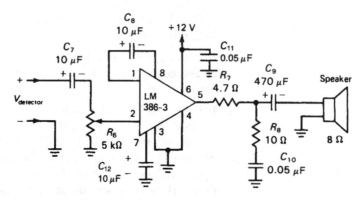

FIGURE 12-3 Audio amplifier stage

11. If you are close to a local AM broadcast transmitter, you may be able to receive the AM signal with this superheterodyne receiver system. Disconnect the AM signal generator from TP_1. In its place, connect an antenna. If there is no outside antenna connection available in your laboratory, try a 6- to 10-ft piece of unshielded wire instead. Carefully tune the frequency of the V_{LO} generator between 950 and 2150 kHz while observing the signal at TP_7. Try to peak the amplitude of the detected signal at TP_7 by fine-tuning the frequency of the V_{LO} generator. If a signal can be observed on

the scope display, you should be able to hear the audio signal with the speaker if the volume is increased high enough. Record the final frequency of V_{LO}. Using mixing principles, determine the carrier frequency of the received AM broadcast station.

12. Try to receive the same AM broadcast station received in step 11 by setting the local oscillator less than the RF carrier frequency by approximately 500 kHz.

REPORT / QUESTIONS:

1. The conversion gain of the mixer stage is defined as the output voltage in decibels (with respect to its input RF voltage level) with the local oscillator voltage held constant at some given value. Calculate the conversion gain using each of the values of V_o measured in step 6. Tabulate the results of step 6, including conversion gain in the tabulation.

2. Calculate the decibel voltage gain for each of the measured voltages in steps 2, 3, and 4. Tabulate your results.

3. Using a single sheet of linear graph paper, draw an overlay of each of the following plots. Calibrate the vertical axis in decibels.
   (a) The mixer-stage decibel voltage gain (down 10 dB on each side of the resonant frequency) vs. frequency.
   (b) The IF amplifier stage decibel voltage gain (down 10 dB on each side of the resonant frequency) vs. frequency.
   (c) The system decibel voltage gain vs. frequency. [Use the same frequency range as in parts (a) and (b).]

4. Using three sheets of linear graph paper, graph each of the following:
   (a) V_o versus V_{RF}, with V_{LO} held constant at its critical value. (step 6)
   (b) V_o versus V_{LO}, with V_{RF} held constant at 200 mV$_{p-p}$. (step 7)
   (c) Conversion gain in decibels versus V_{RF}, with V_{LO} held constant at its critical value.

5. Determine the transconductance, g_m, of the RF mixer FET using the frequency response data and the following procedure:
   (a) Using the frequency response data and assuming that the capacitor values are accurate, calculate the equivalent parallel inductance of the inductor.

$$f_r = \frac{1}{2\pi\sqrt{LC}}$$

   (b) Using the 3-dB bandwidth of the frequency response curve, calculate the Q of the tank circuit in the mixer stage.

$$f_r = \text{BW} \times Q$$

   (c) Using the results of step (b) and tank circuit theory, find the parallel resistance of the tank circuit in the mixer stage.

$$R_p = QX_L$$

   (d) Using the mixer stage's voltage gain at the resonant frequency and the results of step (c), find the transconductance of the FET.

$$A_v = g_m R_p$$

6. Summarize the principles of operation of the superheterodyne receiver investigated in steps 9–12.

7. Explain why it was possible to pick up the same radio station in steps 11 and 12 with two different local oscillator frequency settings. (Use example numbers to describe the mixing action that occurs in each of the two cases.)

CASCODE AMPLIFIERS

OBJECTIVES:

1. To build and test an RF cascode amplifier with simulated AGC (automatic gain control).
2. To obtain an understanding of cascode amplifier operation.
3. To compare and evaluate data of discrete and integrated circuit configurations which perform the same function.

REFERENCE:

An IF amplifier stage using the 3028 can be found in Section 3-5 of the text.

TEST EQUIPMENT:

Dual-trace oscilloscope
Low-voltage power supplies (2)
Function generator
Volt-ohmmeter
Prototype board

COMPONENTS:

Ceramic filter: CFM-455D or equivalent
3028A Cascode amplifier integrated circuit
CA3086 General purpose NPN transistor array
Transistors: 2N2222 (3)
Diode: 1N4001 (2)
Capacitors: 0.1 μF (10), 1 μF (2)
Resistors ($\frac{1}{2}$ watt): 470 Ω (2), 500 Ω, 680 Ω (2), 1 kΩ (5), 1.5 kΩ (5), 2.2 kΩ (3), 2.8 kΩ, 3.3 kΩ, 5 kΩ, 7.5 kΩ (4), 12 kΩ (4), 470 kΩ
Potentiometer: 1 kΩ, 10 kΩ (10-turn trim)

PRELABORATORY:

Using your textbook and other technical references, explain what the difference is between a cascaded amplifier configuration and a cascode amplifier. Explain what causes a cascode amplifier to offer a high input impedance and low input capacitance at high frequencies.

PROCEDURE:

1. Build the discrete amplifier configuration shown in Fig. 13-1. Turn on the dc supply and adjust V_{cc} for exactly +10 V dc. Take dc bias voltage measurements at each of the test points given. Record each of your measurements.

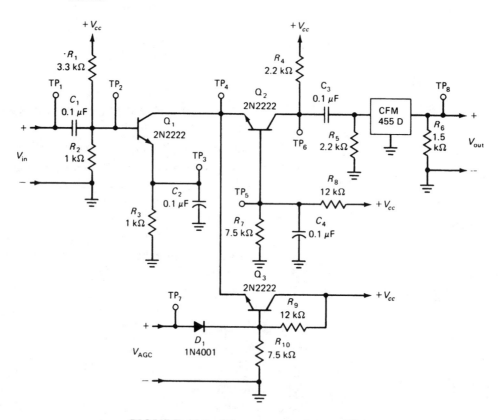

FIGURE 13-1 Discrete cascode amplifier

2. Apply a 455-kHz, 40-mV$_{p-p}$ sine wave at the RF input of the amplifier. Monitor the output voltage of the ceramic filter at TP$_8$. Fine-tune the generator so as to maximize the output voltage at TP$_8$. Sketch the waveforms observed at TP$_1$, TP$_6$, and TP$_8$.

3. Using approximately eight equal increments, increase V_{in} from 40 mV$_{p-p}$ to the value where any additional increase does not result in any appreciable increase in the filter's output voltage. For each increase, measure the amplitude of the voltages at TP$_1$, TP$_6$, and TP$_8$. Record your data in tabular form. Also, determine what level of V_{in} causes the voltage at TP$_6$ to become distorted due to clipping.

4. Repeat step 3, except this time decrease V_{in} from 40 mV$_{p-p}$ down to zero in four equal decrements. The voltage at TP$_6$ should remain unclipped at these settings.

5. Set V_{in} back to 40 mV$_{p-p}$. Turn on the AGC supply. Set the AGC voltage to approximately 4 V dc. Increase this voltage in 100-mV steps until you reduce the voltage gain of the stage to zero. At each new setting, measure

the voltage at TP_1, TP_6, and TP_8. You should end up with approximately five sets of data. If not, return the AGC voltage back to 4 V dc and take smaller increments until you end up with at least five AGC voltage settings before reducing the gain to zero. Note any AGC voltage settings that cause distortion or oscillations to occur in the voltage waveform at TP_6.

6. Build the amplifier configuration given in Fig. 13-2 or Fig. 13-3 depending if you have the 3028A (Fig. 13-2) or CA3086 (Fig. 13-3) integrated circuit. Turn on the two power supplies and adjust them for ±5 V dc. Adjust the 10-kΩ potentiometer so that the AGC voltage at pin 1 (3028A), pin 2 (3086) is close to zero. Take dc voltage readings at each of the eight pins of the 3028A integrated circuit. Record each of your measurements.

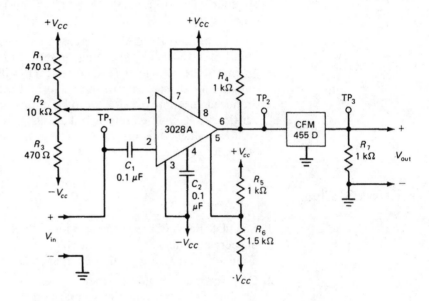

FIGURE 13-2 Integrated-circuit cascode amplifier

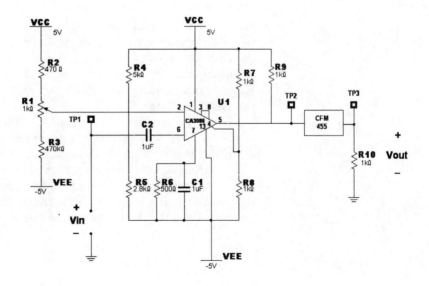

FIGURE 13-3 Integrated-circuit cascode amplifier using the CA3086

7. Apply a 455-kHz, 100-mV$_{p-p}$ sine wave at the RF input of this amplifier. Monitor the output voltage of the ceramic filter at TP$_3$. Fine-tune the generator to maximize V_o at TP$_3$. Sketch the waveforms observed at TP$_1$, TP$_2$, and TP$_3$.

8. Using approximately eight equal increments, increase V_{in} from 100 mV$_{p-p}$ to the value where any additional increase does not result in any appreciable increase in filter output voltage. For each increase, measure the amplitude of the voltage at TP$_1$, TP$_2$, and TP$_3$. Record your data in tabular form. Also, note what input voltage level causes the waveform at TP$_2$ to become distorted due to clipping.

9. Repeat step 8, except this time decrease V_{in} from 100 mV$_{p-p}$ down to zero in four equal decrements. The voltage at TP$_2$ should not be distorted at these settings.

10. Set V_{in} back to 100 mV$_{p-p}$. Increase the AGC voltage (positive polarity) at pin 1 of the 3028A by changing the setting of the 10 kΩ pot. Take measurements of the voltage at TP$_1$, TP$_2$, and TP$_3$ with the AGC voltage increasing in 200 mV dc steps up to the value that causes the voltage at TP$_3$ to be reduced to approximately zero. If this voltage drops drastically at one critical AGC voltage setting, take smaller increments of V_{AGC} around this critical value. Note any AGC voltages that cause distortion or oscillations to occur at TP$_3$.

REPORT/QUESTIONS:

1. Calculate the decibel voltage gain for both of the cascode amplifiers (including ceramic filter) at each input voltage setting used in steps 3, 4, 8, and 9. Tabulate your results so that comparisons can easily be made.

2. On three sheets of graph paper, plot each of the following for both of your amplifiers. Put both amplifier curves on the same graph:
   (a) A_v (in dB) versus V_{in} with AGC disabled
   (b) A_v (in dB) versus AGC voltage with V_{in} at initial setting
   (c) Ceramic filter input voltage versus V_{in} with AGC disabled

3. Describe and compare the voltage gain, power gain, saturation characteristics, AGC dynamic range, and maximum output power level for both of the cascode amplifier designs. Refer to your data and graphs. Conclude which amplifier appears to be the better design and defend your conclusion. Refer to the following definitions in formulating your evaluation.
   (a) Amplifier Saturation
      (1) A region of amplifier operation where an increase in the input signal voltage does not result in an appreciable increase in output voltage.
      (2) A region of amplifier operation where the amplifier's gain is decreasing; often specified as the range of V_{in} that causes the gain of the amplifier to be within 3 dB of its maximum value.
   (b) AGC Dynamic Range: the total change in amplifier decibel gain caused by varying the AGC voltage over its useful range.

SIDEBAND MODULATION AND DETECTION

OBJECTIVES:

1. To become familiar with the 1496 balanced mixer.

2. To build and evaluate a balanced modulator that produces a double-sideband suppressed carrier signal.

3. To build and evaluate a product detector that extracts the intelligence from a single-sideband suppressed carrier signal.

REFERENCE:

Refer to Chapter 4 of the text.

TEST EQUIPMENT:

Dual-trace oscilloscope: must have a Y vs X display capability

Function generator (2)

Low-voltage power supply (2)

Frequency counter

Spectrum analyzer (if available)

COMPONENTS:

1496P integrated circuit (2)

CFM-455D ceramic filter

Capacitors: 0.005 μF (3), 0.1 μF (4), 1.0 μF (2)

Resistors ($\frac{1}{2}$ watt): 47 Ω (5), 100 Ω (2), 1 kΩ (8), 3.3 kΩ (4), 6.8 kΩ (2), 10 kΩ (2)

Potentiometer: 10 kΩ (10-turn trim)

THEORY:

A balanced modulator (Fig. 14-1) is typically used to generate a double-sideband suppressed carrier signal in an SSB transmitter that uses the filter method of design. Nonlinear amplification causes the creation of first and second harmonics of the

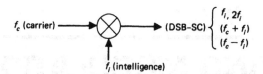

FIGURE 14-1 Balanced modulator function

intelligence input signal, the simple sum and difference frequency components, and a dc component. The first and second harmonics of the carrier input signal, which are normally produced in mixing action are suppressed in a balanced modulator. In the 1496, this is done by signal cancellation due to the symmetrical arrangement of the differential amplifier stage as shown in Fig. 14-2. A spectrum diagram of the output signal reveals the presence of the lower and upper sidebands but no RF carrier frequency component (Fig. 14-3).

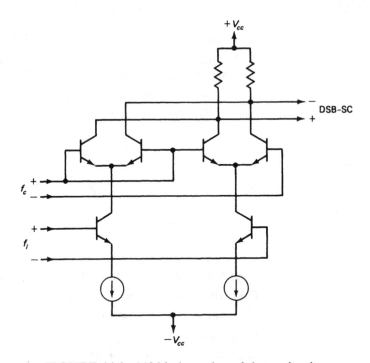

FIGURE 14-2 1496 balanced modulator circuitry

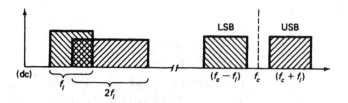

FIGURE 14-3 Balanced modulator typical output spectrum

To produce single sideband, a sharp filter that exhibits a fairly constant passband response and steep roll-off skirts is needed to attenuate all frequencies produced by a balanced modulator except for the desired sideband. A ceramic filter such as the one used in Experiments 2 and 9 may fit these design requirements.

A balanced modulator followed by a low-pass filter is typically used to re-create the original intelligence signal in an SSB receiver. Again, the balanced mixer's nonlinear amplification causes the creation of the first and second harmonics of the input SSB signal, the simple sum and difference frequencies of the two input signals, and a dc signal. Again, the first and second harmonics of the input carrier signal are suppressed by the balanced modulator. This time, however, since the frequencies of the SSB input and carrier input signals are fairly close to each other, the difference frequency components end up being much smaller than any of the other frequencies produced by mixing action. These difference frequencies are exactly equal to the original intelligence frequencies. For example, if upper sideband is being supplied to the SSB input of the balanced modulator, the frequency spectra shown in Fig. 14-4 will result in the output signal.

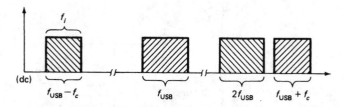

FIGURE 14-4 SSB detector output spectra before filtering

It is very easy to filter out the original intelligence from this complex signal because of the large frequency difference between the difference frequencies and all other output frequency components produced by mixing action. A similar result occurs if lower sideband is being supplied to the balanced modulator. All that is necessary is a simple *RC* low-pass filter that has a sufficiently large time constant at all RF frequencies. The balanced modulator, low-pass filter, and RF carrier oscillator make up what is known as a product detector in an SSB receiver.

PRELABORATORY:

Using the theory described above and in your text, draw the spectrum diagrams of the output signals that would result in each of the following designs.

1. The output double-sideband suppressed carrier signal if a 475-kHz sinusoidal RF carrier is mixed with a 20-kHz audio sine-wave signal in a balanced modulator.

2. The output double-sideband suppressed carrier signal if a 435-kHz sinusoidal RF carrier is mixed with a 20-kHz audio sine-wave signal in a balanced modulator.

3. The output signal if a USB signal having a frequency range of 455 kHz–460 kHz is mixed with a 460-kHz sinusoidal RF carrier in a balanced modulator.

4. The output signal if an LSB signal having a frequency range of 450 kHz–455 kHz is mixed with a 450 kHz sinusoidal RF carrier in a balanced modulator.

5. If the output signal of either design 3 or 4 above is passed through a low-pass filter that totally attenuates all signals above 100 kHz but passes all frequencies below 50 kHz, sketch the resulting output spectra of the low-pass filter.

PROCEDURE:

1. Build the balanced modulator circuit shown in Fig. 14-5. Apply ±10 V dc to the circuit. Apply a 2 V$_\text{p-p}$, 400-kHz sine wave at TP$_1$. This signal represents the input RF carrier. Monitor V_o at TP$_3$ with the oscilloscope. Adjust the carrier null potentiometer, R_2, for a minimum-amplitude 400-kHz sine wave. At the optimum precise setting it should null the carrier to zero amplitude and increase the carrier on either side of the null setting.

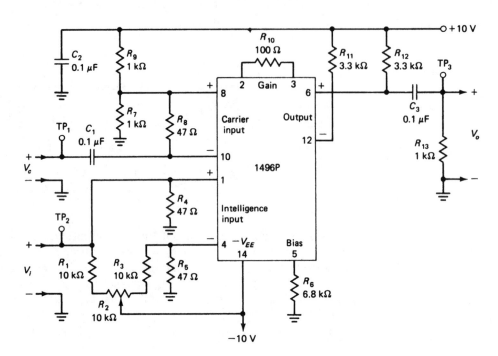

FIGURE 14-5 DSB-SC generator using a 1496P balanced modulator

2. Apply a 150-mV$_\text{p-p}$ 200-Hz sine wave at TP$_2$. This signal represents the input audio intelligence signal. Again, monitor TP$_3$ with the oscilloscope. Use 10:1 probes to avoid loading. The output signal at TP$_3$ should look like a "fuzzy" 400-kHz sine wave if the signal is being viewed with a small horizontal time scale, such as 0.5 μs/div, and is being triggered by the RF carrier signal. Sketch the observed signal at TP$_3$.

3. A better, more informative waveform can be displayed by placing the horizontal time scale at a much larger setting, such as 0.5 ms/div and by externally triggering the scope with the audio intelligence signal. Do this by observing the signal at TP$_3$ with channel A and the signal at TP$_2$ with channel B. The observed DSB-SC signal will probably be slightly distorted. Make small adjustments to the carrier null potentiometer, R_2, and to the intelligence and carrier amplitudes in order to produce a clean waveform such as that shown in Fig. 14-6. The two interwoven envelopes should be near-perfect sine waves that have a frequency of 200 Hz. The peak envelope voltages should all be equal to one another. Sketch the resulting waveform.

4. Determine the effect on the resulting DSB-SC waveform if changes are made in the frequencies of the carrier signal or the intelligence signal. Also, determine the effect of changing the gain resistance, R_{10}, to 47 Ω and to 1 kΩ. Return the display back to its original form as in step 3 before proceeding.

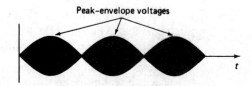

FIGURE 14-6 DSB-SC waveform showing no distortion

5. Observe and sketch the resulting swept-frequency display of the DSB-SC signal by applying the DSB-SC signal to the vertical input of the scope and applying the intelligence signal to the horizontal input of the scope. Place the scope in A versus B mode. Set the vertical and horizontal sensitivities to fill the screen with the display. You should note a double triangle or "bow-tie" shaped swept display. Notice the effect on the screen of changing the amplitude of the carrier and intelligence signals. Notice that when distortion occurs, the two triangles will no longer have straight sides. Try adjusting the value of the carrier null potentiometer. You should notice that the two triangles are symmetrical when the carrier has been nulled. Again, return the display to its original form as in step 3 before proceeding.

6. The double-sideband suppressed carrier signal will now be applied to a ceramic filter to produce true SSB. Recall from Experiment 9 that the CFM-455D has a 3-dB bandwidth of approximately 20 kHz and a center frequency of 455 kHz. Adjust the intelligence signal frequency to exactly 20 kHz and the RF carrier frequency to exactly 435 kHz. The DSB-SC signal should now have an upper sideband frequency of 455 kHz and a lower sideband frequency of 415 kHz. Build the circuit in Fig. 14-7 and connect it to the output of the balanced modulator circuit by connecting TP_3 and TP_4 together. Monitor the output voltage at TP_5. It should be a clean sine wave, since only the upper sideband frequency component makes it through the filter. Adjust the carrier frequency so that the voltage at TP_5 is at maximum amplitude. Measure its frequency with a counter.

$$f_{USB} = \text{_____} \qquad f_{carrier} = \text{_____}$$

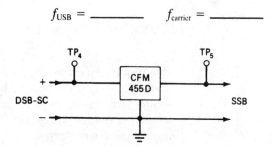

FIGURE 14-7 Ceramic filter to create SSB from DSB-SC

If a spectrum analyzer is available, observe the spectral content of the signals at TP_3 and TP_5. Sketch the spectral displays.

7. Repeat step 6, but this time allow only the lower sideband frequency component of the DSB-SC signal to make it through the ceramic filter. Since you cannot change the bandpass frequency range of the ceramic filter, you must change the frequency of the RF carrier signal entering the balanced modulator to align the lower sideband frequency with the bandpass.

$$f_{LSB} = \text{_____} \qquad f_{carrier} = \text{_____}$$

If a spectrum analyzer is available, observe the spectral content of the signals at TP$_3$ and TP$_5$. Sketch the spectral displays. Return to the scope display of step 6 before continuing.

8. The SSB signal produced at the output of the ceramic filter in steps 6 and 7 will now be applied to a balanced modulator. Build the circuit shown in Fig. 14-8. Connect the output of the ceramic filter to the detector's input by connecting TP$_5$ and TP$_6$ together with a jumper. The RF carrier driving the balanced modulator at TP$_1$ should also be connected to the RF carrier input of the detector at TP$_7$. Adjust the amplitude of V_c to approximately 3.5 V$_{p\text{-}p}$. Observe the output waveform of the product detector at TP$_8$ with channel A of the oscilloscope. Monitor the original intelligence signal with channel B of the oscilloscope. Use 10:1 probes.

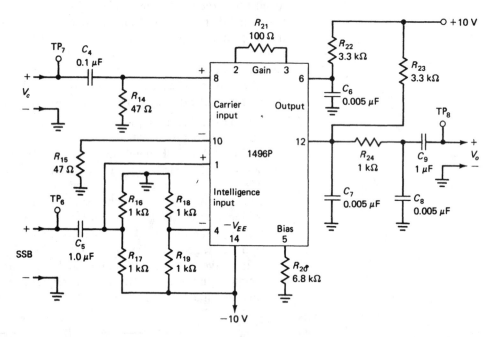

FIGURE 14-8 Product detector using a 1496P balanced modulator

9. Readjust the frequency of the intelligence for 200 Hz. Adjust the RF carrier frequency so that the detected output frequency signal is a maximum. You should find several frequencies within the bandpass of the ceramic filter where the output voltage at TP$_8$ peaks, but select the largest peak. Sketch the resulting waveforms.

10. Readjust the frequencies of the intelligence for 20 kHz. Again, adjust the RF carrier frequency so that the detected output signal is a maximum. You should find only two distinct frequencies within the bandpass of the filter where the output voltage at TP$_8$ is at a maximum. These frequencies are those that cause just the upper sideband or just the lower sideband to fall within the bandpass of the filter.

11. Adjust the audio function generator to produce a 20-kHz triangle wave instead of a sine wave. You should find that the output voltage at TP$_8$ remains as a sine wave. Also, you should find that the output voltage of the detector at TP$_8$ will peak at several settings of the RF carrier. This is because only part of the upper sideband or lower sideband frequencies produced at the balanced modulator output are being passed by the

20-kHz-bandwidth ceramic filter. This should also happen if a square-wave intelligence signal is being used.

12. Now adjust the intelligence signal to produce a 200-Hz triangle wave. Now the first 100 harmonics of the triangle wave's fundamental frequency can be within the 20-kHz bandwidth of either the upper sideband or lower sideband at the output of the balanced modulator. Thus, if we tune the RF carrier so that just the upper sideband or lower sideband frequency component is passed by the ceramic filter, the product detector should re-create the original triangle wave with minimal distortion. Tune the carrier to produce an output signal at TP_8 that is least distorted. Sketch the original and re-created intelligence signals.

13. Repeat step 11 using a square-wave intelligence signal.

14. Repeat step 12 using a square-wave intelligence signal.

REPORT/QUESTIONS:

1. Draw a sketch of the frequency spectra of the DSB-SC signals produced in steps 6 and 7. Label all important frequencies on the horizontal axis. Explain why a sine wave at 455 kHz was observed at the output of the ceramic filter in each case.

2. Draw a sketch of the frequency spectra of the DSB-SC signals produced in step 11 when a 20-kHz triangle wave or a 20-kHz square wave was used as the intelligence signal. Draw a sketch of the frequency spectra of the output signal of the product detector. Explain why the output signal of the product detector was not a close replica of the input triangle or square-wave intelligence signal in steps 11 and 13.

3. Draw a sketch of the frequency spectra of the DSB-SC signals produced in steps 12 and 14 when a 200-Hz triangle or square wave was used as the intelligence signal. Draw a sketch of the frequency spectra of the output signal of the product detector. Explain why the output signals of the product detector in steps 12 and 14 were close replicas of the input triangle or square-wave intelligence signals.

NAME _____

FREQUENCY MODULATION: SPECTRAL ANALYSIS

OBJECTIVES:

1. To become familiar with spectral displays of frequency modulated carriers at different values of modulation index.
2. To calculate the frequency deviation of an FM waveform by viewing its waveshape on an oscilloscope.
3. To determine the bandwidth of an FM signal using a spectrum analyzer and using Carson's rule.

REFERENCE:

Refer to Chapter 5 of the text.

TEST EQUIPMENT:

Dual-trace oscilloscope

Function generator (2); one must have a VCG input to produce FM.

Audio Spectrum Analyzer (HP3580A or equivalent)

50 ohm Attenuator Pad (such as given in Figure 9-7)

THEORY:

Frequency modulation is defined as a type of angle modulation in which the instaneous frequency of a carrier is caused to vary by an amount proportional to the modulating signal amplitude. The spectral display of an FM signal is much more complicated than that of an AM or SSB signal. This is because each modulating frequency creates an infinite number of side-frequency components on either side of the carrier frequency component. Fortunately, inspection of the infinite series reveals that the higher-ordered side-frequency components have very small, negligible amplitudes. This effect can be shown by use of a mathematical tool known as the Bessel function, which is given in tabular form in Table 15-1. This table also allows us to predict the effective bandwidth of the FM signal. Carson's rule also can be used to predict the effective bandwidth of the FM signal. Carson's rule states that:

$$BW = 2 (\delta + f_i) \tag{15-1}$$

where δ represents the deviation in carrier frequency.

TABLE 15-1   FM Side Frequencies from Bessel Functions

x (m_f)	CARRIER J_0	J_1	J_2	J_3	J_4	J_5	J_6	J_7	J_8	J_9	J_{10}	J_{11}	J_{12}	J_{13}	J_{14}	J_{15}	J_{16}
0.00	1.00	—	—	—	—	—	—	—	—	—	—	—	—	—	—	—	—
0.25	0.98	0.12	—	—	—	—	—	—	—	—	—	—	—	—	—	—	—
0.5	0.94	0.24	0.03	—	—	—	—	—	—	—	—	—	—	—	—	—	—
1.0	0.77	0.44	0.11	0.02	—	—	—	—	—	—	—	—	—	—	—	—	—
1.5	0.51	0.56	0.23	0.06	0.01	—	—	—	—	—	—	—	—	—	—	—	—
2.0	0.22	0.58	0.35	0.13	0.03	—	—	—	—	—	—	—	—	—	—	—	—
2.5	−0.05	0.50	0.45	0.22	0.07	0.02	—	—	—	—	—	—	—	—	—	—	—
3.0	−0.26	0.34	0.49	0.31	0.13	0.04	0.01	—	—	—	—	—	—	—	—	—	—
4.0	−0.40	−0.07	0.36	0.43	0.28	0.13	0.05	0.02	—	—	—	—	—	—	—	—	—
5.0	−0.18	−0.33	0.05	0.36	0.39	0.26	0.13	0.05	0.02	—	—	—	—	—	—	—	—
6.0	0.15	−0.28	−0.24	0.11	0.36	0.36	0.25	0.13	0.06	0.02	—	—	—	—	—	—	—
7.0	0.30	0.00	−0.30	−0.17	0.16	0.35	0.34	0.23	0.13	0.06	0.02	—	—	—	—	—	—
8.0	0.17	0.23	−0.11	−0.29	−0.10	0.19	0.34	0.32	0.22	0.13	0.06	0.03	—	—	—	—	—
9.0	−0.09	0.24	0.14	−0.18	−0.27	−0.06	0.20	0.33	0.30	0.21	0.12	0.06	0.03	0.01	—	—	—
10.0	−0.25	0.04	0.25	0.06	−0.22	−0.23	−0.01	0.22	0.31	0.29	0.20	0.12	0.06	0.03	0.01	−	—
12.0	0.05	−0.22	−0.08	0.20	0.18	−0.07	−0.24	−0.17	0.05	0.23	0.30	0.27	0.20	0.12	0.07	0.03	0.01
15.0	−0.01	0.21	0.04	−0.19	−0.12	0.13	0.21	0.03	−0.17	−0.22	−0.09	0.10	0.24	0.28	0.25	0.18	0.12

n OR ORDER

Source: E. Cambi, *Bessel Functions*, Dover Publications, Inc., New York, 1948. Courtesy of the publisher.

The index of FM modulation, m_f, determines how many non-negligible side frequency components appear in the FM signal spectra. It can be determined by:

$$m_f = \frac{\delta}{f_i} \qquad (15\text{-}2)$$

In this experiment, you will be measuring the waveform and spectral display of an FM signal when the modulation index is set at various levels.

PROCEDURE:

1. Adjust the RF generator to produce a 20 kHz sine wave with an amplitude of 3 Vp-p. Adjust the audio generator to produce a 500 Hz sine wave. Then reduce its amplitude down to zero.

2. Connect the audio generator up to the FM input jack (VCG) of the RF generator as shown in Figure 15-1. Also connect the output of the RF generator to the oscilloscope and spectrum analyzer as shown. Let the scope trigger on the carrier signal. Increase the amplitude of the intelligence signal at TP1 while watching the RF carrier at TP3 begin to blur due to FM action taking place. You may need to place 20 to 30 dB of attenuation between TP2 and TP3 to keep the modulation index down to the low levels as requested in the following steps.

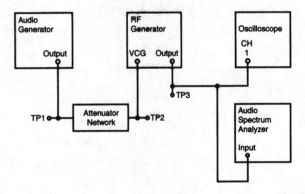

FIGURE 15-1

3. Tune in the RF carrier signal on the oscilloscope with the vertical sensitivity set at 0.5 V per division and the horizontal sensitivity set at 10 microseconds per division. This should allow you to see approximately two periods of the RF carrier waveform.

4. Carefully tune in the RF generator output signal with the spectrum analyzer so that you can see the carrier frequency component and the first six or seven side-frequency components on each side of the carrier. If you are using an HP-3580A spectrum analyzer, the following settings should suffice.

Center Frequency: 20.000 kHz	**Frequency Span/Div.:** 1 kHz/div.
Resolution BW: 100 Hz	**Sweep Mode:** repetitive
Display Smoothing: medium setting	**Amplitude Mode:** linear
Sweep Time/Div.: 2 sec /div.	**Amplitude Ref Level:** 0 dB
	Input Sensitivity: +10 dB

5. Fine-tune the spectral display so that the carrier frequency component is in the center of the display. Now disconnect the modulating signal from the VCG input of the RF generator. Measure the amplitude of the carrier frequency component in terms of vertical divisions. Sketch the resulting displays of your spectrum analyzer and oscilloscope. Make sure that you measure the period of the sine wave on the scope display.

6. Now reconnect the intelligence signal to the VCG input jack of the RF generator and slowly increase the amplitude from zero. You should see the amplitude of the carrier frequency component begin to diminish, while the first and second order side-frequency components begin to grow in amplitude. Continue to slowly increase the amplitude of the intelligence signal until the carrier frequency component reduces down to exactly zero amplitude while the side-frequency components continue to grow. You should now be able to see five upper and five lower side-frequency components on your spectral display. Sketch your resulting spectral and waveform displays. Carefully measure the amplitudes of each of the visible frequency components on your spectral display.

7. Measure the minimum and maximum periods of the blurry sine wave displayed on your oscilloscope, as shown in Figure 15-2. Invert these values to get the maximum and minimum values of carrier frequency.

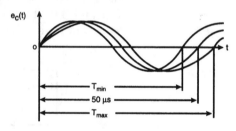

FIGURE 15-2

maximum period = _____   minimum frequency = _____

minimum period = _____   maximum frequency = _____

8. Calculate the index of modulation of this FM signal using:

$$m_f = \frac{\delta}{f_i}, \text{ where } \delta = \frac{f_{max} - f_{min}}{2}$$

9. Based upon the results of steps 7 and 8, determine what the maximum and minimum periods need to be in order to produce a modulation index of exactly 1.5.

minimum frequency = _____   maximum period = _____

maximum frequency = _____   minimum period = _____

10. Carefully adjust the amplitude of the signal at TP2 in order to produce these corresponding values of minimum and maximum periods. Sketch the resulting spectral and waveform displays. Accurately measure the amplitude and frequency of each frequency component in the spectral display.

11. This time, determine what the maximum and minimum periods need to be in order to produce a modulation index of exactly 5.0.

minimum frequency = _____   maximum period = _____

maximum frequency = _____   minimum period = _____

12. Carefully adjust the amplitude of the signal at TP2 in order to produce these corresponding values of minimum and maximum periods. Sketch the resulting spectral and waveform displays. Accurately measure the amplitude and frequency of each frequency component in the spectral display.

13. One more time, determine what the maximum and minimum periods need to be in order to produce a modulation index of exactly 8.0.

    minimum frequency = _____   maximum period = _____

    maximum frequency = _____   minimum period = _____

14. Carefully adjust the amplitude of the signal at TP2 in order to produce these corresponding values of minimum and maximum periods. Sketch the resulting spectral and waveform displays. Accurately measure the amplitude and frequency of each frequency component in the spectral display.

REPORT/QUESTIONS:

1. Make a table like that shown in Table 15-2. To enter each theoretical value, simply copy the value of J from the Bessel table given in Table 15-1. To calculate each value for J from your measured data, simply divide each spectral component's amplitude from steps 10, 12, and 14 by the unmodulated carrier's amplitude, which was measured in step 5. Enter each calculated result in your table.

TABLE 15-2

FREQUENCY COMPONENT	J-TERM	$m_f = 1.5$		$m_f = 5.0$		$m_f = 8.0$	
		THEORETICAL	CALCULATED	THEORETICAL	CALCULATED	THEORETICAL	CALCULATED
Carrier	J0						
First-Order	J1						
Second-Order	J2						
Third-Order	J3						
Fourth-Order	J4						
Fifth-Order	J5						
Sixth-Order	J6						
Seventh-Order	J7						
Eighth-Order	J8						
Ninth-Order	J9						
Tenth-Order	J10						
Eleventh-Order	J11						
Twelfth-Order	J12						

2. How close are your theoretical and measured values in question 1? Comment on any causes for discrepancies.

3. What are the theoretical values of m_f which cause the carrier amplitude to diminish to zero amplitude? In step 8, the smallest of these critical values of m_f should have been calculated. How close was it? What might have caused any discrepancies?

4. Consider the spectral display of step 12 to be the actual display of an FM radio station under test, where each vertical division represents 100 Vrms. What would be the resulting output power if this signal was measured across a 50 ohm load?

5. Repeat question 4 for the spectral display of step 5. The power level should be the same regardless of what m_f is set at. How close are they?

6. Calculate the resulting bandwidths for each of the spectral displays of steps 6, 10, 12, and 14 by use of Carson's rule. Compare your results to those revealed by spectral displays. Record your results in tabular form such as shown in Table 15-3.

TABLE 15-3

STEP	MODULATION INDEX	BANDWIDTH (SPECTRAL INSPECTION)	BANDWIDTH (CARSON'S RULE)
6	2.4		
10	1.5		
12	5.0		
14	8.0		

PHASE-LOCKED LOOPS: STATIC AND DYNAMIC BEHAVIOR

OBJECTIVES:

1. To become familiar with the phase-locked loop (PLL) and its major subsystem building blocks.

2. To study the static and dynamic behavior of the phase-locked loop.

REFERENCE:

Refer to Section 6-5 of the text.

TEST EQUIPMENT:

Dual-trace oscilloscope: must have a Y vs X display capability

Low-voltage power supply (2)

Function generator (2); one must have a VCG input for swept-frequency operation

Volt-ohmmeter

Frequency counter

Prototype board

COMPONENTS:

Integrated circuit: 565 phase-locked loop (2)

Capacitors: 0.001 μF (5), 2.2 nF, 0.1 μF, 10 μF

Resistors ($\frac{1}{2}$ watt): 680 Ω (5), 820 Ω, 2.2 kΩ, 3.3 kΩ (2), 4.7 kΩ (2), 10 kΩ, 33 kΩ (2)

Potentiometer: 10 kΩ

PRELABORATORY:

Calculate the free-running frequency of the PLL's voltage-controlled oscillator using the timing resistor and timing capacitor values given in Figs. 16-1 and 16-2. Use the equation given in the data sheet for the 565 PLL included in the Appendix.

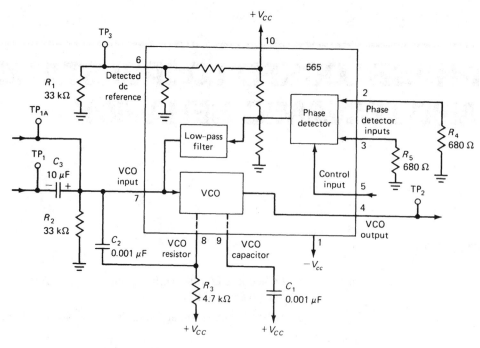

FIGURE 16-1 VCO static and dynamic test circuit

PROCEDURE:

VCO Static Tests

1. Build the circuit given in Fig. 16-1. Apply the power supply voltages and adjust for ±7 V dc. Monitor the VCO output voltage at TP_2 using the oscilloscope. Measure and record the waveform of the VCO output signal, noting the amplitude and frequency.

2. Note the effect on the free-running VCO output waveform by introducing each of the following changes, one at a time. Use a frequency counter to measure the frequency at TP_2 and observe the signal with an oscilloscope. Make sure that each alteration is restored back to its original form before proceeding to the next step.
   (a) Decrease the timing resistor, R_3, to 2.2 kΩ.
   (b) Increase the timing capacitor, C_1, to 2.2 nF.
   (c) Load the VCO output with a capacitative load by placing a series RC load between TP_2 and ground. Let $R = 680$ Ω and $C = 0.001$ μF.
   (d) Increase each of the phase detector input resistors, R_4 and R_5, one at a time, to 3.3 kΩ.
   (e) Decrease the load resistance at the detected dc reference output (TP_3) to 820 Ω.
   (f) Reduce each of the supply voltages by 2 V, one at a time.

3. Measure the reference dc voltage level at the phase detector output by connecting the scope through a 10:1 probe to TP_3. Apply a dc voltage level equal to the reference to TP_{1A} (no coupling capacitor). If a third low-voltage power supply is not available to do this, simply feed the V_{cc} power supply voltage through a 10-kΩ potentiometer acting as a voltage divider. Increase and decrease the applied voltage above and below the reference voltage level in 0.5-V steps. At each setting, measure the resulting VCO output frequency. Disconnect the dc voltage before proceeding.

VCO Dynamic Tests

4. Connect a function generator at TP_1. Apply a 2-Hz, 4-V_{p-p} sine wave as the VCO input voltage at TP_1. Sketch the resulting VCO output waveform displayed at TP_2. You will need to use your imagination in showing the motion of the resulting display. You are observing an FM waveform.
   (a) Carefully increase and decrease the amplitude of the function generator and note the effect upon the FM output signal at TP_2. Return to 4 V_{p-p} before proceeding.
   (b) Carefully increase and decrease the frequency of the function generator and note the effect on the FM output signal at TP_2. Return to 2 Hz before proceeding.

Phase-Detector Static Tests

5. Build the phase detector circuit shown in Fig. 16-2. Connect the test equipment to the phase detector test circuit as shown in Fig. 16-3. Use the scope to monitor the input voltage at TP_1 on channel A and to monitor the VCO output voltage at TP_4 on channel B. Use a VOM to monitor the difference voltage between TP_2 and TP_3 of the phase detector. Again, use power supply voltages of ± 7 V dc.

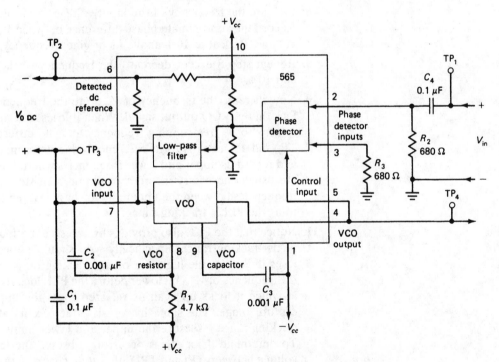

FIGURE 16-2 Phase-detector test circuit

6. With no voltage applied at TP_1, measure the difference voltage between TP_2 and TP_3 with the VOM. This value is the reference level of the phase detector output. Also, measure the free-running frequency of the VCO by measuring the frequency of the VCO output signal at TP_4 using the oscilloscope.

7. Apply a 0.5-V_{p-p} square wave at a frequency approximately equal to that of the VCO output at TP_4. Vary the frequency of this input signal and you

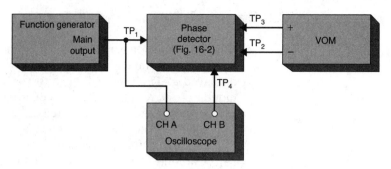

FIGURE 16-3 Phase-detector static test circuit configuration

should see that the VCO output signal will "track" the frequency of the input signal as long as it is fairly close to the free-running frequency. Also, the difference voltage measured with the VOM should vary as long as tracking is taking place. Adjust the frequency of the input voltage so that the difference voltage equals the reference voltage value measured in step 6. Record the waveforms of the input voltage and the VCO output signal under these conditions. Note the amplitudes, frequencies, and approximate phase difference between the two waveforms.

8. Increase the frequency of the input voltage to 5% above its original value. Record the approximate phase difference between V_{in} and the VCO output voltage. Repeat at 10% above its original frequency.

9. Repeat step 8 except decrease the frequency of the input voltage by 5% and 10%.

10. Now increase the frequency of V_{in} up to the frequency at which it loses its lock on the VCO output signal. When it loses lock, the VCO output signal returns to its free-running frequency and the difference voltage between TP_2 and TP_3 returns to its reference value determined in step 6. Measure and record its upper and lower frequencies where it loses lock. A few trials are usually necessary in order to obtain consistent results. This range of frequencies between the upper and lower measured frequencies is referred to as the PLL's tracking range.

11. Notice that the PLL also provides hysteresis at both upper and lower limits of its tracking range. Specifically, you should notice that when the PLL loses its lock at the high end of the tracking range, you need to decrease the frequency of V_{in} a bit lower before the PLL locks back up. The frequency at which it locks up again is referred to as the upper end of the PLL's capture range. The same hysteresis effect occurs at the low end of the tracking range. Measure the upper and lower limits of the capture range. To determine if lockup is occurring, observe the level of the difference voltage between TP_2 and TP_3 rather than checking the stability of the VCO output signal. Note that when measuring the low end of the capture range, there may be a few false triggers before true capture actually occurs. Again, several attempts will be necessary to yield consistent results. Record your results.

Phase Detector Dynamic Tests

12. A swept-frequency display can be used to measure simultaneously the capture and tracking frequency ranges of a PLL. To do this, the frequency of V_{in} is slowly swept linearly above and below the complete tracking range.

The PLL responds by repeatedly locking up and losing lock as the applied frequency of V_{in} is swept. Refer to the test configuration of Fig. 16-4. The vertical deflection of the display represents the value of the detected dc voltage from the phase detector, which should vary linearly about its reference value. The sweep signal, being triangular instead of ramp, permits us to see hysteresis at both ends of the tracking range. This permits us to measure the capture and tracking ranges. Refer to Experiment 9 for procedures for frequency calibration of the horizontal axis when producing a swept-frequency display using a generator that provides a VCG input jack. Set up the test configuration of Fig. 16-4 and measure the tracking and capture ranges of the PLL. Adjust the sweep generator for a 2.5-V_{p-p} triangle wave at a frequency of 20 Hz. Set the RF generator for a 500-mV_{p-p} sine wave at the VCO's free-running frequency. Make sure that the scope inputs are dc coupled to avoid distorting the display. Your resulting display should look similar to the one shown in Fig. 16-5. Make slight alterations to the amplitude and frequency settings to produce optimum results. Sketch your resulting display.

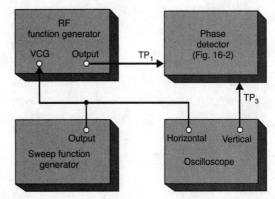

FIGURE 16-4 Phase-detector dynamic test configuration

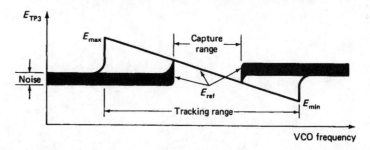

FIGURE 16-5 Typical swept-frequency display of phase detector

REPORT / QUESTIONS

1. Based on the results of each of the changes made in step 2, comment on the stability of a PLL's VCO when it is in the free-running state.

2. Plot the VCO output frequency versus dc input voltage for the measurements made in step 3. From your graph determine the VCO's deviation constant by determining the slope of the resulting linear graph.

$$K_f = \frac{\Delta f_{VCO}}{\Delta V_{TP1A}} = \frac{f_{max} - f_{min}}{V_{max} - V_{min}}$$

3. Explain the difference between the two frequency-modulated waveforms observed in step 4. What characteristic of the observed FM waveform was changing in each case? Refer to pages 271–273 of your text to help answer this question.

4. Plot the relative phase difference between V_{in} and the VCO output voltage versus frequency using the results of steps 8 and 9 of the phase detector static tests. From your graph determine the phase detector's deviation constant by determining the slope of the resulting linear graph.

$$K_\phi = \frac{\Delta_\phi}{\Delta f_{VCO}} = \frac{\phi_{max} - \phi_{min}}{f_{max} - f_{min}}$$

5. Explain what is meant by each of the following specifications of a PLL:
   (a) Free-running frequency
   (b) Tracking range
   (c) Capture range

NAME _____

FM DETECTION AND FREQUENCY SYNTHESIS USING PLLs

OBJECTIVES:

1. Further familiarization with phase-locked-loop operation.
2. To be acquainted with two popular applications of phase-locked loops: FM detection and frequency synthesis.

REFERENCE:

Refer to Sections 6-5 and 7-5 of the text.

TEST EQUIPMENT:

Dual-trace oscilloscope
Low-voltage power supply (2)
Function generators (2); one must have a VCG input to produce FM
Frequency counter

COMPONENTS:

Integrated circuits: 565 (2), 7493 (2), 7420
Transistor: 2N2222
Capacitors: 0.001 μF (4), 0.1 μF (3), 10 μF
Resistors ($\frac{1}{2}$ watt): 680 Ω (4), 4.7 kΩ (3), 10 kΩ, 33 kΩ (2)

PROCEDURE:

1. Build the FM detector circuit shown in Fig. 17-1. Apply ± 10 V dc to this circuit and measure the free-running frequency of the VCO part of the 565 PLL by measuring the frequency of the VCO output waveform at TP_4 with a frequency counter.

   free-running frequency = _____

2. Connect a function generator that has VCG capability to the phase detector input at TP_1. This generator will serve as the FM signal generator. Connect

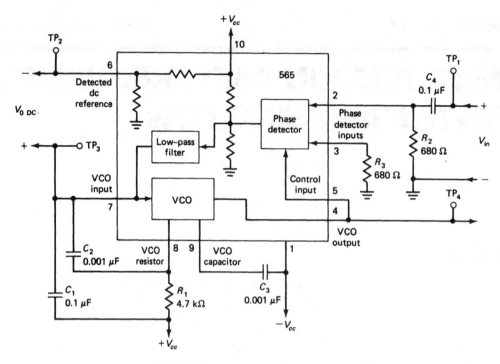

FIGURE 17-1 FM detector using a 565 PLL

channel A of the dual-trace oscilloscope to monitor the output of the FM generator. Connect channel B of the oscilloscope to monitor the VCO output signal at TP_4. Set the FM generator to produce a 500-mV$_{p-p}$ sine wave at the same frequency as the free-running frequency of the VCO part of the PLL. You should see the PLL lock up to the FM generator's frequency by observing both waveforms of the oscilloscope lock up at the same frequency.

3. Connect a second function generator as an intelligence signal by connecting it to the VCG input jack of the RF generator. Set the intelligence signal generator to produce a 100-mV$_{p-p}$ sine wave at a frequency of 20 Hz. You should see both waveforms on the oscilloscope frequency modulate. As long as the deviation of the FM signal does not cause the VCO output of the PLL to exceed its tracking range, the waveforms should remain frequency locked together.

4. Now connect channel A of the scope to monitor the original intelligence signal at the VCG jack of the RF generator. Connect channel B of the oscilloscope to monitor the detected dc reference signal at TP_3. Sketch the waveforms observed on the oscilloscope.

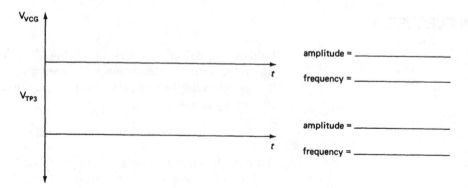

amplitude = _____

frequency = _____

amplitude = _____

frequency = _____

5. Increase and decrease the amplitude of the intelligence signal. Note the effect the amplitude changes have on the signals observed at TP_4 and TP_3. Return to the original amplitude setting before proceeding.

6. Increase and decrease the amplitude of the FM signal. Again, note the effect the amplitude changes have on the signals observed at TP_4 and TP_3. Return to the original amplitude setting before proceeding.

7. Increase the intelligence signal frequency and watch the waveform at TP_3. You will discover a critical frequency where the phase-locked loop can no longer follow the variations in frequency of the RF generator. Record this frequency and return the intelligence generator back to 20 Hz.

8. Set the intelligence generator to produce a 50-mV$_{p-p}$ 20 Hz square wave. Sketch the resulting detected signal at TP_3. It should exhibit overshoot and ringing. Measure the ringing frequency. It should be fairly close to the same frequency as the critical frequency measured in step 7.

9. Build the frequency synthesizer circuit shown in Fig. 17-2. Leave jumpers J_1, J_2, J_3, and J_4 disconnected from any test points. Apply ± 5 V dc to this circuit. Again, check the VCO output at TP_9 and measure the free-running frequency.

free-running frequency = _____

10. Connect a function generator to act as a master oscillator at TP_{10}. Connect channel A of the dual-trace oscilloscope at TP_{10} and channel B at TP_9. Connect jumper J_1 to ground. This causes the binary counter to be engaged to count up to its maximum value as a MOD-256 counter. This particular divide-by ratio, N, which is 256 in this case, determines the multiple that the output frequency is with respect to the input frequency. Even though the digital divider is dividing by 256, the value of $N = 256$ causes the input frequency to be multiplied by 256 to create the output frequency. Thus, by providing a constant input frequency through the design of a single, stable oscillator, we can produce a stable but variable output frequency simply by changing the value of N in the digital divider network. This can be done easily through the use of proper digital logic gates and appropriate switches which would be set properly either manually or perhaps by a microcontroller computer. This offers an alternative to the expensive use of multiple-crystal oscillators to create stable but multiple operating frequencies in receiver and transmitter designs.

   Apply a 0.5-V$_{p-p}$ sinusoidal signal to the synthesizer circuit at TP_{10}. Use an input frequency fairly close to the free-running frequency divided by 256. Observe the waveforms of V_{in} and V_o at TP_{10} and TP_9, respectively. Verify that f_o/f_{in} is equal to 256, using a frequency counter to measure f_o and f_{in}. You should notice that the loop locks up for output frequencies within a certain range about the free-running frequency. This range is again referred to as the tracking range. Measure the tracking and capture ranges of the closed-loop system, using the same techniques that were used in steps 10 and 11 of Experiment 16. The tracking and capture ranges should be measured with respect for f_o, not f_{in}. Also, measure the duty cycle of the output signal of the digital divider at TP_{11}.

duty cycle = _____

11. Connect jumpers J_1 to TP_1, J_2 to TP_2, J_3 to TP_7, and J_4 to TP_8. This makes the counter into a MOD-195 counter. If you are unfamiliar with MOD

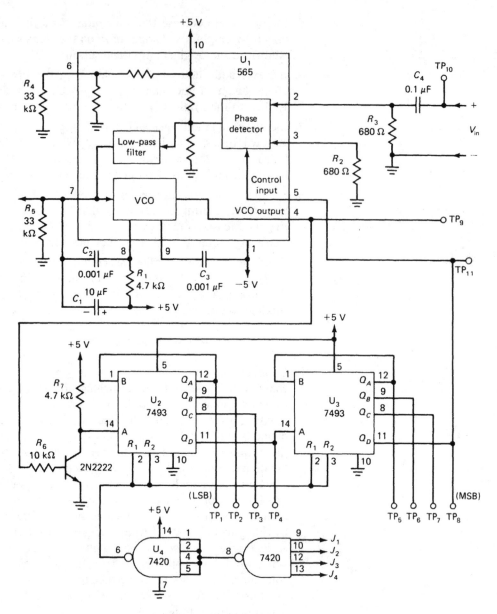

FIGURE 17-2 Frequency synthesizer using a 565 PLL

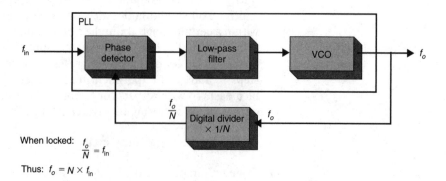

When locked: $\dfrac{f_o}{N} = f_{in}$

Thus: $f_o = N \times f_{in}$

FIGURE 17-3 Frequency synthesizer block diagram

counters and digital dividers, refer to a digital electronics text to review how the divide-by-195 function occurs with the 7420 NAND gate controlling the reset inputs of the dividers. Verify that $f_o = 195 \times f_{in}$ by measuring the frequencies, f_{in} and f_o, with a frequency counter when f_{in} is set at approximately 1/195th of the free-running frequency. Again, measure the tracking range of the loop and the duty cycle of the output signal of the digital counter at TP_{11}. Note that it may be possible for the PLL to lock up at input frequency ranges other than 1/195 of f_o. These "false" tracking ranges are usually quite small and unstable in nature.

12. Connect J_1 to TP_1, J_2 to TP_2, J_3 to TP_6, and J_4 to TP_8. Vary the frequency of V_{in} until the phase-locked loop locks up. Again, measure f_o and f_n with the counter and experimentally determine what the value of N now is for the digital divider. Determine theoretically what N is by following one of the procedures described below.
    (a) Review what signals cause the NAND gate to reset the counter, express this as a digital count in binary notation, and convert to decimal notation.
    (b) If you are patient and determined, you may be able to carefully count how many pulses of the input signal to the digital counter it takes to produce one complete pulse at the output. If your oscilloscope has delayed sweep capability, it should be helpful to you in your counting process!

    Again, measure the duty cycle of the output signal of the digital divider and the tracking range of the synthesizer. Be prepared to demonstrate to the lab instructor that you understand how the frequency synthesizer operates.

13. Try a small-value duty-cycle case. Connect the jumpers properly for a divide-by-139 counter. Calculate the free-running frequency divided by 139 and set f_{in} at this frequency value. Slowly vary the frequency, f_{in}, until you notice that the loop is back in lock. Again, measure f_o and f_{in} to verify that $N = 139$. Measure and record the duty cycle and tracking range as done in previous steps. You should find that at low duty cycles, the tracking range is very narrow, making the system more unstable.

REPORT/QUESTIONS:

1. Describe how the phase-locked loop can be used successfully to detect an FM signal. Explain how it works.

2. Explain why in step 6 there was very little if any effect of changing the amplitude of the FM signal on the amplitude of the detected output signal.

3. Describe how the phase-locked loop can be used successfully to synthesize a range of output frequencies by using a single stable oscillator and digital counter. Explain how the circuit works. Also, state its advantage over the use of multiple-crystal oscillators in a "channelized" radio.

4. Research: A limitation of this synthesizer design is its noncontinuous frequency ranges that can actually be synthesized. How can this limitation be removed through the addition of mixers? (Refer to Chapter 8 in your text.)

PULSE-AMPLITUDE MODULATION AND TIME-DIVISION MULTIPLEXING

OBJECTIVES:

1. To become familiar with pulse-amplitude modulation techniques.

2. To test and evaluate a simple PAM modulator and demodulator.

3. To test and evaluate a PAM communication system that utilizes time-division multiplexing.

REFERENCE:

Refer to Sections 9-5 and 9-6 of the text.

TEST EQUIPMENT:

Dual-trace oscilloscope
Low-voltage power supply
Function generator
Volt-ohmmeter

COMPONENTS:

Integrated circuits: 8038 (2), 4001 (2), 4029 (2), 4051 (2), 4066, LM311, LM2907

Diode: 1N4001

Capacitors: 0.001 μF (1), 0.0033 μF, .01 μF, 0.1 μF, 0.047 μF (1), 0.47 μF (3), 10 μF (2)

Resistors: 100 Ω, 470 Ω, 1 kΩ (4), 10 kΩ (15), 22 kΩ, 27 kΩ (2), 68 kΩ (2), 100 kΩ (2), 330 kΩ (2)

Potentiometers: (10-turn trim) 10 kΩ (2)

THEORY:

Pulse-amplitude modulation is one form of pulse modulation used in the transmission of digital signals in a message-processing format. In PAM, the RF carrier pulse's amplitude is directly proportional to the intelligence signal's amplitude (Fig. 18-1).

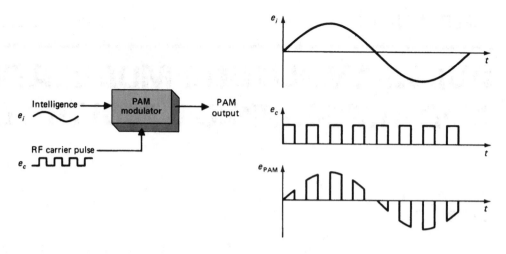

Figure 18-1   Generation of pulse-amplitude modulation

This is the least desirable of the various types of pulse modulation used in digital communication. This is due to the fact that amplitude variations are easily subject to noise interference, which is also an amplitude-varying phenomenon.

The simplest form of the PAM modulator is an analog switch that is turned on and off at the RF carrier pulse rate. As this switch changes state, the intelligence signal is connected and disconnected from the output. Thus the output PAM signal is a sampled version of the input intelligence signal. A spectral analysis of the complex PAM signal reveals that the intelligence signal frequency components are far removed in frequency from the other frequency components of the complex PAM waveform. This is true only if the sampling rate is considerably higher than the highest intelligence frequency being used. Thus if the sample rate is kept high enough in comparison to the intelligence signals being used, a simple low-pass filter can be used as a PAM demodulator to retrieve the original information.

The real use of PAM is in digital communication systems that employ time-division multiplexing. In these systems, more than one channel of communication can be sent through a single wire, fiber optic cable, or RF carrier frequency. A mechanical or electronic sampler is used in the transmitter to sample each one of a given number of intelligence signals. It covers each of the n-channels in one cycle and then proceeds to resample each channel continually in a periodic manner. Thus the output signal contains samples of each of the n-channels. A similar sampler is used in the receiver to separate the n-channels back into individual signals. Refer to Fig. 18-2.

To ensure successful demodulation of each channel of information, the transmitter and receiver samplers must be sampling at the same rate and in synchronization. The sync signal must also be part of the information sent from the transmitter to the receiver on the single cable or radio frequency. One possible synchronization technique is investigated in this experiment.

PRELABORATORY:

Using the data sheets provided for the 8038 function generator in the Appendix, determine the output frequency of the sine, triangle, and square-wave signals produced by the 8038 generators given in Figs. 18-3 and 18-4.

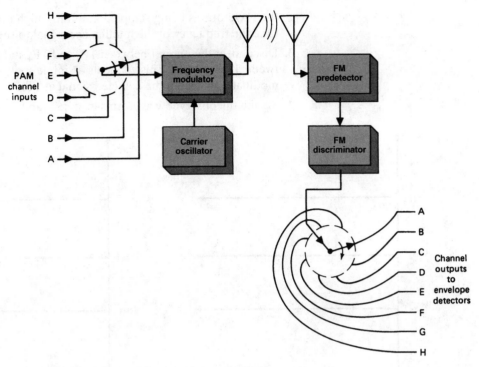

Figure 18-2 Time-division multiplexing

PROCEDURE:

1. Build the simple PAM modulator circuit shown in Fig. 18-3. Place a jumper between TP_3 and TP_4. Apply 10 V dc to the circuit. Observe the square wave at the output of the 8038 at TP_2. This signal will function as the RF carrier pulse. Measure its frequency and amplitude.

2. Apply a 6-$V_{p\text{-}p}$, 300-Hz sine wave at TP_1. This will function as the intelligence signal (data). Observe the signal at TP_1 with channel A and the resulting PAM signal at TP_4 with channel B. Let the intelligence signal serve as the trigger for the scope. Notice that the waveform at TP_4 appears to be a sine wave. If you spread out the horizontal time scale, you will notice that

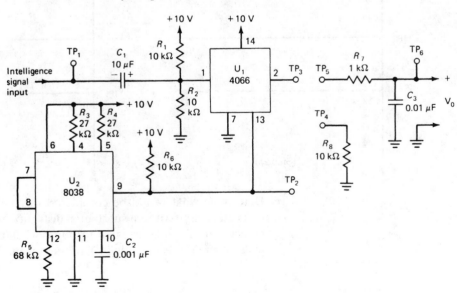

Figure 18-3 PAM modulator circuit

the waveform is being sampled at a fairly high rate. You may need to fine-tune the frequency of the intelligence signal to see the sampling action.

3. Disconnect the jumper between TP_3 and TP_4 and reconnect the jumper between TP_3 and TP_5. This connects a simple low-pass filter as the PAM demodulator. Observe the detected signal at TP_6.

4. Sketch the observed waveforms in steps 2 and 3.

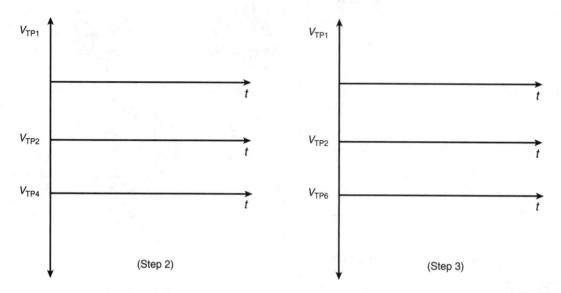

5. Repeat steps 2 and 3 using a 6-$V_{p\text{-}p}$, 3-kHz sine wave at TP_1. You should notice considerable distortion resulting from the intelligence signal frequency being too close to the sampling rate. This is known as aliasing distortion.

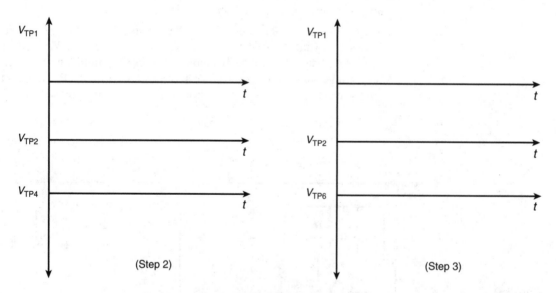

6. Determine what would be considered the highest intelligence frequency that can be used without substantial distortion resulting at the output of the detector at TP_6. Substantial distortion is when the step increment or decrement in amplitude exceeds 10% of the p-p amplitude of the waveform.

$$f_{\max} = \underline{\hspace{2cm}}$$

106   EXPERIMENT 18

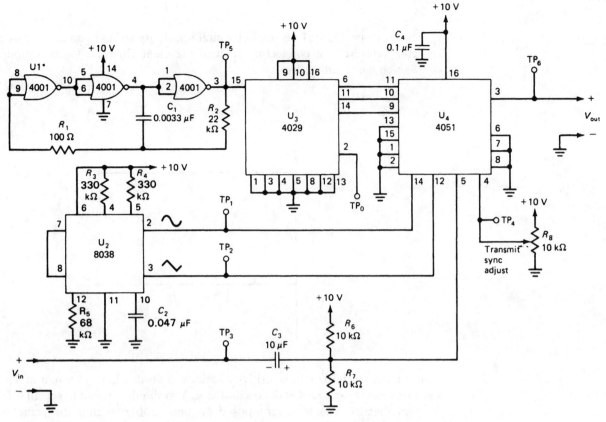

Figure 18-4 Time-division multiplexer

7. As stated in theory, the main use of PAM is in time-division multiplexing of multiple channels of information. Build the multiplexer circuit shown in Fig. 18-4. Here the 8038 serves as a function generator of two separate intelligence signals, a low-frequency sine wave and a low-frequency triangle wave. Apply 10 V dc to this circuit and observe these waveforms at TP_1 and TP_2. Measure the frequency of these two signals.

frequency = _____

8. Adjust the transmit sync potentiometer, R_8, to produce approximately 9 V dc at TP_4. Apply a 5-V, 30-Hz square wave at TP_3. The multiplexer should be sampling four channels of information: the sine wave and triangle wave at TP_1 and TP_2, the 30-Hz square wave at TP_3, and the 9-V dc sync level at TP_4. The sample rate is provided by the CMOS clock. Observe the clock output waveform at TP_5. Measure its frequency.

Clock frequency = _____

9. The clock drives a 4029 binary counter, which in turn supplies the 4051 multiplexer with a binary count from 1 to 8 in order for it to connect channels 1–8 to the output. To minimize circuit complexity, only channels 1, 3, 5, and 7 of the 4051 are used for data transmission. The channels 2, 4, 6, and 8 inputs of the multiplexer are tied to ground. Observe the output of the multiplexer at TP_6. Let the signal at TP_0 serve as the trigger signal for the

oscilloscope. Use 0.1-ms/div horizontal sensitivity so that you can see one complete cycle of connecting each of the eight channels to the output. Sketch the resulting display.

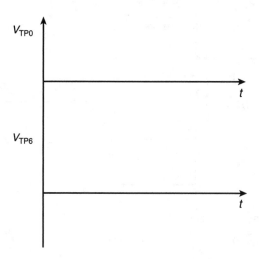

10. Adjust the horizontal sensitivity scale to 5 ms/div. Let the signal at TP_3 serve as the trigger for the oscilloscope. You should be able to see all four intelligence signals superimposed on one another within the complex waveform at TP_6. Sketch the resulting display.

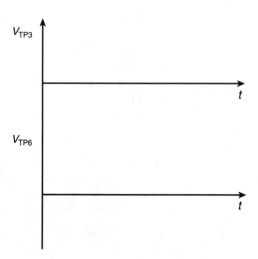

11. Do not disassemble the multiplexer. Now build the demultiplexer circuit given in Fig. 18-5. Apply 10 V dc to this circuit. Connect a jumper between TP_6 and TP_7. This connects the output of the multiplexer to the input of the demultiplexer. Also, connect jumpers between TP_{11} and TP_{12}, TP_{15} and TP_{16}, and TP_{19} and TP_{20}. The 9-V dc sync pulse must be detected by the comparator in order to provide the counter with the synchronizing reset pulse. Adjust the receive sync potentiometer, R_9, in order to produce approximately 8 V dc at TP_8. Only the 9-V dc sync pulse will ever exceed the

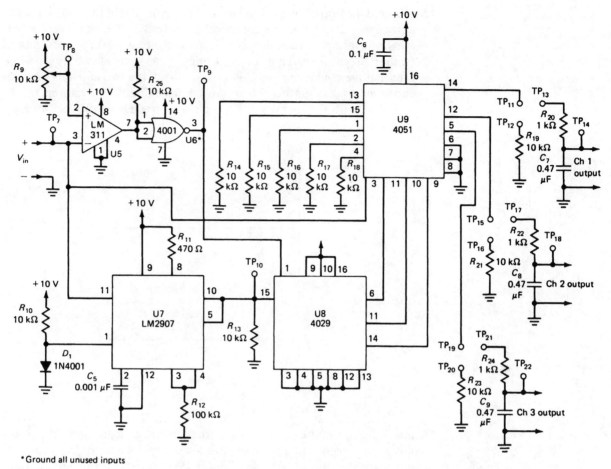

Figure 18-5 Time-division demultiplexer

*Ground all unused inputs

8 V dc comparator threshold voltage. This provides for the necessary synchronization to take place. Observe the signal at TP9. This is the resulting sync reset pulse. Also, observe the signal at TP6. Let the sync reset pulse serve as trigger for the oscilloscope. Sketch the resulting display.

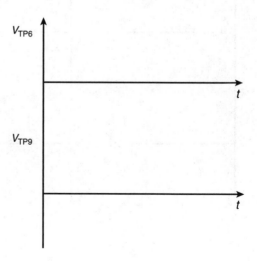

12. Observe the output of the LM2907 at TP_{10}. Notice that the 2907 serves as a frequency doubler. Its output trigger pulse fires each time that the multiplexer waveform crosses the reference voltage level of 0.7 V dc furnished by the 1N4001 diode. Thus the 4029 counter can cycle at exactly the same rate that the counter in Fig. 18-4 cycles. In fact, the two counters are synchronized to one another. Prove this by observing the waveforms at TP_7 and TP_{10}, with the signal at TP_{10} serving as trigger for the oscilloscope. Sketch the resulting waveforms.

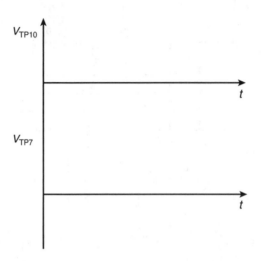

13. Since the transmitter and receiver counters are synchronized, the transmitter and receiver multiplexers must also be synchronized. The three output signals at TP_{12}, TP_{16}, and TP_{20} should be PAM versions of their respective intelligence signals. Prove this by observing the PAM waveforms at TP_{12}, TP_{16}, and TP_{20} along with their respective intelligence signals at TP_1, TP_2, and TP_3. You may need to reduce slightly the V_{cc} voltage to 9.5 V dc to guarantee that synchronization occurs. Sketch the resulting displays.

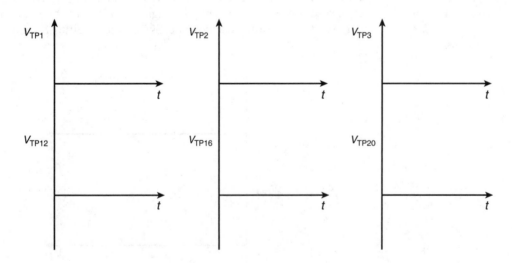

14. Connect the PAM demodulators by disconnecting the jumpers between TP_{11} and TP_{12}, TP_{15} and TP_{16}, and TP_{19} and TP_{20} and reconnecting the jumpers between TP_{11} and TP_{13}, TP_{15}, and TP_{17}, and TP_{19} and TP_{21}. Observe the detected intelligence signals at TP_{14}, TP_{18}, and TP_{22} along with their respective intelligence signals at TP_1, TP_2, and TP_3. Sketch the resulting displays.

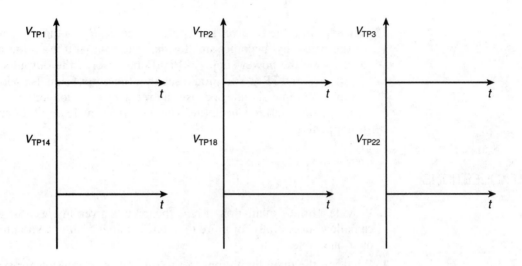

15. In step 14 you should see that the low-frequency sine wave is easily reproduced, but the low-frequency triangle wave and square wave are not reproduced without some distortion resulting. This is due to insufficient bandwidth within the time-division multiplexer. To remedy this, a higher sample rate using higher-speed components would be needed. Change the square-wave input signal at TP_3 to a 3-V_{p-p}, 8-Hz triangle wave and observe the detected display at TP_{22}. Sketch the resulting detected display. It should exhibit less distortion since the intelligence frequency has been reduced and a simpler waveform is being used.

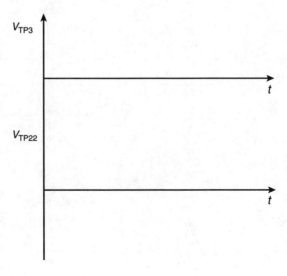

16. Now change the input waveform at TP_3 to a 3-V_{p-p} sine wave. Using the same criteria as in step 6, determine what would be considered the highest intelligence frequency that could be used in this communication system without substantial distortion resulting at the output of the detector at TP_{22}.

$$f_{max} = \underline{\hspace{2cm}}$$

17. Try adjusting the receiver sync potentiometer, R_9, so that the amplitude of the sync pulse becomes smaller than the comparator's reference voltage. Toggle the power supply off and on. Observe the output signals at TP_{14}, TP_{18}, and TP_{22}. You should see that the output signals no longer resemble the input signals because the receiver is no longer synchronized with the transmitter. Check the sync reset pulse at TP_9 and observe what is happening.

REPORT/QUESTIONS:

1. Provide a brief explanation of how the circuits given in Figs. 18-4 and 18-5 can allow successfully for more than one channel of information to be sent over one cable.

2. What are the main limitations of the time-division multiplexer system investigated in this experiment?

3. What allows the 4051 demultiplexer in the receiver stage of Fig. 18-5 to switch from channel to channel exactly in the same manner as does the 4051 multiplexer in the transmitter stage of Fig. 18-4?

4. Explain why in step 17, when the receiver sync potentiometer is adjusted so that the amplitude of the sync pulse is smaller than the comparator's reference voltage, the output waveforms no longer resemble their respective input waveforms.

PULSE-WIDTH MODULATION AND DETECTION

OBJECTIVES:

1. To observe two types of pulse-time modulation waveforms.

2. To test and evaluate a pulse-width modulator.

3. To test and evaluate a pulse-width demodulator.

REFERENCE:

Refer to Section 9-6 of the text.

TEST EQUIPMENT:

Dual-trace oscilloscope

Low-voltage power supply (2)

Function generator (2)

Volt-ohmmeter

Frequency counter

COMPONENTS:

Integrated circuits: 565, 1496P, 7486

Transistor: 2N2222

Capacitors: 4.7 nF (2), 0.01 μF (2), 0.1 μF (6), 0.47 μF (2), 10 μF, 470 μF

Resistors ($\frac{1}{2}$ watt): 47 Ω, 100 Ω, 390 Ω (2), 680 Ω, 1 kΩ (5), 3.3 kΩ, 4.7 kΩ (2), 5.6 kΩ, 10 kΩ (4), 33 kΩ, 47 kΩ

Potentiometers: (10-turn trim) 5 kΩ (2)

THEORY:

A type of modulation often used when transmitting low-frequency signals over a long distance via telephone lines or fiber optic scale is pulse-time modulation. In PTM, the amplitude of the intelligence signal is converted to variations in pulse length or variations in pulse position of the carrier signal. These two types of PTM are referred to as pulse-width modulation and pulse-position modulation, respectively. The wave shapes of PWM and PPM are given in Fig. 19-1.

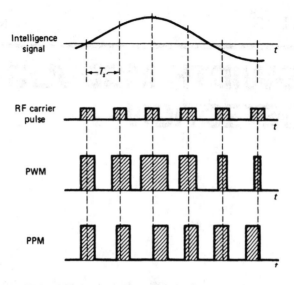

FIGURE 19-1 PWM and PPM waveforms

The amplitude of the carrier pulses remains constant in either PWM or PPM. Thus pulse-time-modulated signals exhibit the same advantages that are exhibited by frequency-modulated signals: noise immunity and low distortion. In addition, pulse signals are easily reproduced if they do get noisy and distorted. This is not possible if analog signals are used.

One possible method of generating PWM and PPM uses the phase-locked loop and digital Exclusive-Or gate as given in Fig. 19-2. The PLL is designed to lock up

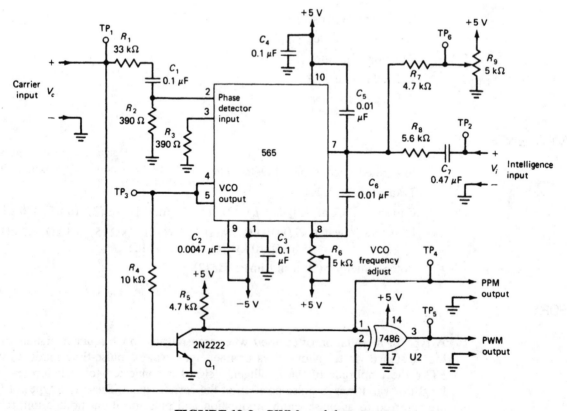

FIGURE 19-2 PWM modulator stage

near the frequency of the carrier. The carrier frequency is applied to the phase detector input, thus causing it to lock up at the carrier frequency. The intelligence signal is applied to the reference input of the VCO and causes the VCO's output signal to shift its phase with respect to the reference phase by an amount proportional to the amplitude of the intelligence signal. This fits the definition of pulse-position modulation. If the VCO output signal is fed into a switching transistor, the output of the switching transistor can be described as being a TTL-compatible PPM signal. To create PWM, the original RF carrier pulse and the PPM signal are fed into the two inputs of an Exclusive-Or gate. Its output will be "high" only during the time when the RF carrier pulse and PPM signals are at different logic states. Since the two signals are synchronized by the PLL, the Exclusive-Or output signal will be "high" only during the time of phase difference between the RF carrier pulse and the PPM pulse. This creates PWM. Refer to Fig. 19-1 to verify this.

A method that successfully demodulates the PWM signal uses a 1496 balanced modulator as given in Fig. 19-3. If the 1496 product detector that was investigated in Experiment 14 is slightly altered so that the differential amplifier transistors act as switching transistors, the pulse signal produced at its output will exhibit a dc offset that will vary as a function of the phase difference between its two inputs. Thus if the two input signals are the PWM signal and the original RF carrier pulse, the output voltage of the product detector will resemble the original intelligence signal. Fortunately, all of the other frequency components produced by the mixing action of the switching transistors in the 1496 are much higher frequencies than the original intelligence signal. Therefore, if the complex output signal of the 1496 is passed through a simple low-pass filter, the original intelligence signal will be re-created.

PRELABORATORY:

Calculate the free-running frequency of the PLL found in the PWM modulator of Fig. 19-2 if R_6 is adjusted to 3 kΩ using the information given in the Appendix for the 565 integrated circuit.

PROCEDURE:

1. Build the circuit given in Fig. 19-2. Apply ± 5 V dc and measure the VCO output frequency at TP$_3$. Adjust R_9 so as to produce approximately 3.5 V dc at TP$_6$. Adjust R_6 to produce exactly 15 kHz at TP$_3$.

2. Apply a 3 V, 15 kHz positive square wave at TP$_1$. This will function as the RF carrier pulse that is to be modulated by this circuit. Observe the waveform at TP$_3$. Use 10:1 scope probes to avoid loading of the signal by the scope. You should be able to see the PLL lock up as the RF carrier frequency is adjusted near 15 kHz. Measure the PLL's tracking and capture ranges as in steps 10 and 11 of Experiment 16.

	minimum	maximum
capture range:	_____	_____
tracking range:	_____	_____

3. Set the RF carrier back at 15 kHz. This should be approximately in the middle of the tracking range. If not, readjust R_6 and repeat step 2 to make

it so. Now apply a 2-V_{p-p}, 2-kHz sine wave at TP$_2$. This will function as the intelligence signal. Observe the signal at TP$_3$. Notice that as the intelligence signal amplitude is varied, the display will blur. This is due to the intelligence signal introducing phase shift into the PLL's VCO's output signal. This is pulse-position modulation.

4. Observe the waveform at TP$_4$ with channel A of the oscilloscope. This is a TTL-compatible PPM waveform. Now observe the intelligence signal at TP$_2$ with channel A and observe the PPM signal at TP$_4$ with channel B. Trigger the oscilloscope with the intelligence signal. Increase the amplitude of the intelligence signal to approximately 5 V_{p-p}. Carefully adjust the trigger level of the oscilloscope and the frequency of the intelligence signal in order to produce a stable display. Sketch the resulting display. It should look similar to the sketch given in Fig. 19-1.

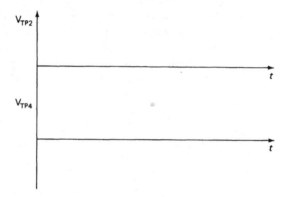

5. Temporarily disconnect the intelligence signal from TP$_2$. Observe the output of the Exclusive-Or gate at TP$_5$ with channel B and observe the original RF carrier pulse at TP$_1$ with channel A. Let the RF carrier pulse serve as the trigger signal for the scope. The waveform at TP$_5$ should be a square wave at exactly twice the frequency of the RF carrier pulse. Fine-tune the frequency of the RF carrier pulse. It should vary the duty cycle of the output waveform at TP$_5$. Adjust the frequency such that a duty cycle of approximately 50% is produced. Measure the frequency of the RF carrier pulse. It should be approximately in the center of the PLL's tracking range.

6. Reconnect the intelligence signal at TP$_2$. Adjust the amplitude between zero and 5 V_{p-p}. You should notice that the intelligence signal causes the pulse duration of the output signal of the Exclusive-Or gate to vary. This is seen as a blurring effect on the waveform at TP$_5$. This is pulse-width modulation.

7. Move the channel A probe to TP$_2$. Observe the waveform at TP$_5$ with channel B with the intelligence signal serving as the trigger signal for the oscilloscope. Adjust the amplitude of the intelligence signal to approximately 2 V_{p-p}. As in step 4, carefully adjust the trigger level of the scope and the frequency of the intelligence signal in order to produce a stable display. Sketch the resulting PWM display. It should look similar to the sketch given in Fig. 19-1. Do not disassemble this circuit before proceeding to step 8.

8. Build the PWM demodulator circuit given in Fig. 19-3. Connect the output signal of the PWM modulator to the input of the PWM demodulator by connecting a jumper between TP$_5$ and TP$_8$. Also, provide the balanced

modulator with the RF carrier pulse input by connecting a jumper between TP_1 and TP_7. Apply $\pm 5V$ dc to both circuits. The RF carrier should still be set at 3-V amplitude at a frequency near 15 kHz that causes approximately 50% duty cycle to exist at the modulator output at TP_5. The balanced modulator may capacitively load the RF carrier pulse, so do not expect the waveform at TP_1 to remain as a clean square wave as was observed in preceding steps.

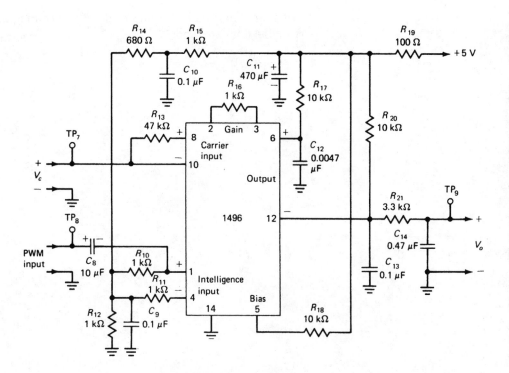

FIGURE 19-3 PWM demodulator

9. Adjust the amplitude of the intelligence signal to 1 V_{p-p} at a frequency of 30 Hz. Observe the waveform at TP_9. It should be a clean replica of the original intelligence signal. Fine-tune the amplitude and frequency of the carrier signal to produce optimum results. Verify that the system is working properly by temporarily disconnecting each of the two jumpers connected in step 8, one at a time. The output waveform of the demodulator at TP_9 should disappear if either of the two jumpers is disconnected. With the jumpers reconnected, adjust the frequency of the RF carrier pulse above and below the PLL's tracking range as was measured in step 2. You should see that when the PLL goes out of lock, the output signal of the demodulator at TP_9 disappears again. Readjust the frequency of the RF carrier pulse such as to produce 50% duty cycle at TP_5 before proceeding.

10. Determine the range of the intelligence frequencies that can be reproduced successfully by the PWM demodulator without any resulting distortion. Take data so as to be able to sketch the frequency response of the PWM digital communication system, i.e. V_{TP8} vs. frequency.

11. Determine the maximum and minimum amplitudes of both the intelligence signal and the RF carrier pulse that allow the output signal at TP_9 to remain undistorted.

REPORT/QUESTIONS:

1. Provide a brief explanation of how the intelligence signal applied at TP_2 of Fig. 19-2 is encoded as variations in pulse width in the output waveform at TP_5.

2. How does the 1496 balanced modulator stage shown in Fig. 19-3 function to detect the original information signal from the encoded PWM signal applied at TP_8?

3. What are the main limitations of the PWM communication system investigated in this experiment?

　　　　　　　　　　NAME _____

DIGITAL COMMUNICATION LINK USING PULSE-CODED MODULATION TECHNIQUES

OBJECTIVES:

1. To become familiar with the use of pulse-coded modulation (PCM) techniques to form error-free high-speed data communication links.

2. To build and test a working analog to digital converter (ADC) and digital to analog converter (DAC) to create a PCM data link.

3. To build and test a circuit that converts digital data from parallel to series format and then converts the serial data back to parallel format again.

4. To observe and compare serial data conforming to each of three standards: TTL, RS-232, and RS-422.

REFERENCE:

Refer to Section 8-3 of the text.

TEST EQUIPMENT:

Dual-trace oscilloscope

Function generator with offset capability

Square wave generator with offset capability

Low-voltage power supply (2)

COMPONENTS:

Integrated circuits: AD557, AD670, 74HC04, 74HC74, 74HC163, 74HC164, 74HC165, 74HC374, AM26LS31, AM26LS32, MAX232

Transistor: 2N3904

Resistors: 470 Ω (8), 560 Ω, 1 kΩ (2), 10 kΩ, 15 kΩ

Capacitors: 500 pF, 0.001 μF, 0.01 μF, 10 μF (4)

LEDs (8)

THEORY:

The coding of messages or signals in accordance with a specific set of rules is most always done in digital communication systems because it makes the data less sensitive to noise, provides less co-channel interference and distortion levels, and allows faded signals to be more easily re-created without any errors. Coded data communication links have greater transmission efficiencies. Coding techniques such as ASCII and EBCDIC are often used, and these systems are referred to as pulse coded modulation (PCM) systems.

In the first part of this experiment, a fairly elementary coding scheme using an 8-bit analog to digital converter (ADC) and an 8-bit digital to analog converter (DAC) is analyzed. The specific code being used is the simple process of letting the digital data's binary value represent the amplitude of the applied analog signal when the sample is taken by the ADC. The reconstructed analog output voltage level of the DAC is then simply based on the binary value of the applied digital data. With 8-bit ADCs and DACs, there are 2^8, or 256, possible binary values and specific analog levels.

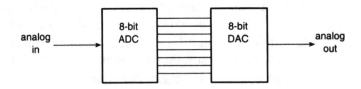

In the second part of this experiment, the 8-bit parallel data provided by the ADC is converted to a serial data format and then converted back to 8-bit parallel format before being subsequently processed by the DAC. This allows the data in the serial format to be sent more easily over long distances using two conductor cables.

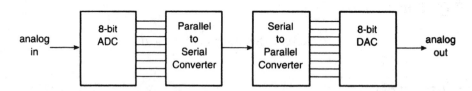

In the last part of the experiment, the TTL serial data is converted to either the RS-232 format or the RS-422 format before being sent over the two-conductor cable. Then at the other end of the cables, the signal is reconverted back to TTL before being subsequently converted back to parallel data. The RS-232 format allows for even longer cable lengths. The RS-422 format allows for the longest cable lengths before signal degradation and noise cause errors to occur in the communication link.

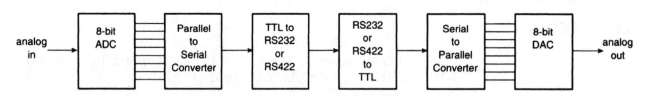

PROCEDURE:

1. Build the circuit given in Fig. 20-1. Connect a 5 V, 10 kHz positive square wave to the clock input at TP3. Observe the waveform at TP3 with channel 1 of the oscilloscope and the waveform at TP4 with channel 2. Trigger

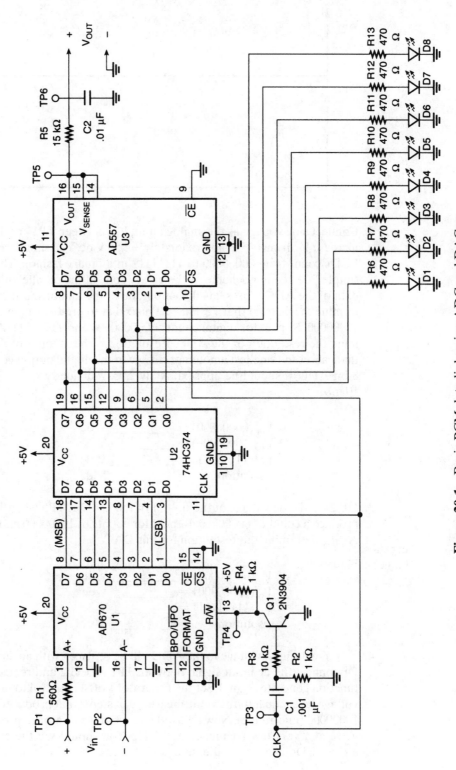

Figure 20-1 Basic PCM data link using ADC and DAC

on channel 2 with a negative slope. Make sure that the R/W input to the ADC is a negative-going pulse with a pulse-width of approximately 1.5 μs. Sketch the resulting waveforms.

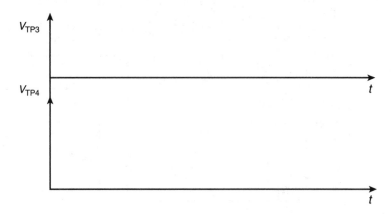

2. Connect a dc voltage as the analog input, V_{in}, between TP1 and TP2. Slowly increase V_{in} from 0 V to approximately +2.55 V dc. You should see the eight LEDs count from 00000000 to 11111111 in a binary fashion. This is a very simple PCM coding scheme, as the binary code of the parallel output digital data of the ADC represents the amplitude of the analog input voltage. Determine the dc level that V_{in} must be placed at to produce a digital count of 10000000. Repeat for digital counts of 00000000 and 11111111. Also, determine the resolution of the ADC. Resolution is the amount of dc variation that causes the smallest amount of change in the digital output code. For this 8-bit ADC, it should be approximately 2.55 V divided by $2^8 - 1$, which is 10 mV.

$$V_{in} \,(00000000) = \underline{\hspace{1cm}} \text{ V dc}$$
$$V_{in} \,(10000000) = \underline{\hspace{1cm}} \text{ V dc}$$
$$V_{in} \,(11111111) = \underline{\hspace{1cm}} \text{ V dc}$$
$$\text{resolution} \quad\;\; = \underline{\hspace{1cm}} \text{ V dc}$$

3. Observe the dc voltage at the output of the DAC at TP5 when V_{in} is set to produce a count of 00000000. Repeat for counts of 10000000 and 11111111. Again, determine the resolution for this DAC.

$$V_o \,(00000000) = \underline{\hspace{1cm}} \text{ V dc}$$
$$V_o \,(10000000) = \underline{\hspace{1cm}} \text{ V dc}$$
$$V_o \,(11111111) = \underline{\hspace{1cm}} \text{ V dc}$$
$$\text{resolution} \quad\;\; = \underline{\hspace{1cm}} \text{ V dc}$$

4. Now the dc input voltage to the ADC will be replaced with an analog signal. Disconnect the dc input signal between TP1 and TP2 and replace it with a function generator signal. Set the generator's offset control to produce a dc voltage level equal to the value measured in step 2 that produced a count of 10000000 (mid-range). Now adjust the function generator to produce a sine wave at a very low frequency of 0.1 Hz. You should see the count of the eight LEDs change as the sine wave's instantaneous amplitude varies.

Slowly increase the amplitude of the sine wave until it causes the LEDs to count from a minimum of 00000000 to a maximum of 11111111, to represent the negative and positive peaks of the sine wave, respectively. Measure the amplitude of V_{in} that causes this to happen.

$$V_{in}\,(\text{max}) = \underline{\hspace{2cm}} \text{ Vp-p}$$

5. Increase the frequency of V_{in} to 1 kHz. The LEDs should all blink very rapidly. Readjust the clock at TP3 to approximately 12.5 kHz. Connect channel 1 of the scope to TP1 and channel 2 to TP5. Trigger the scope on channel 1 using a positive scope. Slightly readjust the clock rate around 12.5 kHz until the digital steps of the waveform at TP5 appear to be almost stationary. Sketch the resulting waveforms at TP1 and TP5.

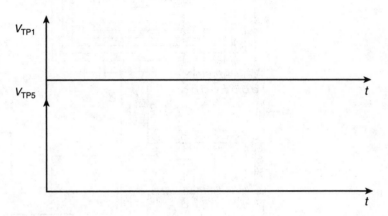

6. The digital steps of the waveform at TP5 can be filtered out with the low-pass filter made up of R_5 and C_2. Move channel 2 of the scope to TP6. Sketch the resulting waveforms.

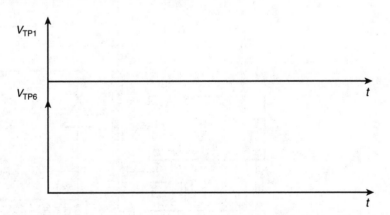

7. Notice that if the clock rate (frequency) at TP3 is reduced, the waveform at TP5 is no longer an accurate representation of the analog input signal. Also, the waveform at TP6 is quite distorted. Try resetting the clock rate to approximately 5 kHz so that the digital steps at TP5 appear to be almost stationary. Sketch the resulting waveforms at TP1, TP5, and TP6.

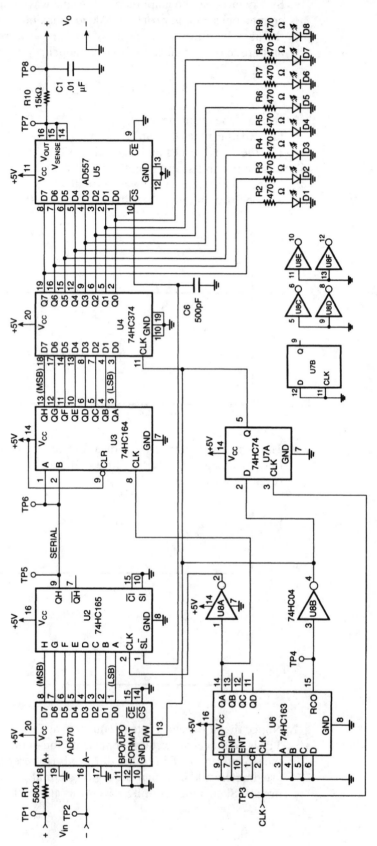

Figure 20-2 PCM data link with added parallel-serial-parallel conversion

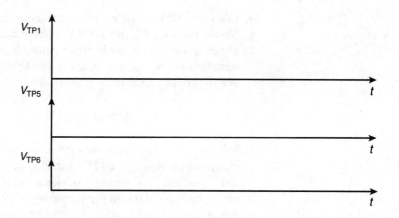

8. Now adjust the clock rate (frequency) at TP3 to approximately 50 kHz. Now the TP5 and TP6 waveforms should appear as close replicas of the analog input signal at TP1. Sketch the resulting waveforms. Observe what happens if V_{in} is changed to 1 kHz triangle wave and 1 kHz square wave of the same amplitude as before.

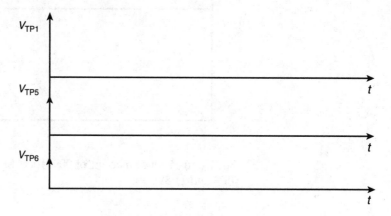

9. Build the circuit given in Fig. 20-2. Connect a 5 V, 10 kHz positive-going square wave at TP3. Observe the waveform at TP3 with channel 1 of your scope and the waveform at TP4 with channel 2. Trigger your scope with channel 2 in positive slope. The signal at TP4 is referred to as the "sample clock" signal and is what triggers the parallel digital data to be converted to serial data. Sketch the resulting waveforms.

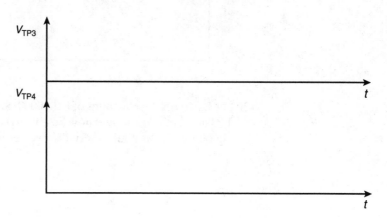

10. Connect a dc voltage as the analog input signal, V_{in}, between TP1 and TP2. Slowly increase V_{in} from 0 V to approximately 2.55 V dc as in step 2. You should again see the eight LEDs count from 00000000 to 11111111 in a binary fashion. Again, determine the dc level that V_{in} must be placed at in order to produce a count of 10000000.

$$V_{in} (10000000) = \text{_____} \text{ V dc}$$

11. With V_{in} set at this level, we shall now observe the serial data at TP5. Move channel 1 of the scope to TP5 and channel 2 to TP4. Again, trigger on channel 2. Reduce V_{in} down to produce a count of 00000000 while watching the changing serial data on the scope. Notice how the serial data on channel 1 resembles the binary code displayed by the eight LEDs. Set V_{in} to produce a count of 00000001. Sketch the resulting waveforms of TP4 and TP5.

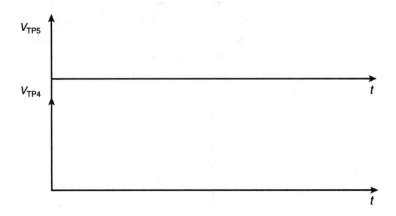

12. Set V_{in} to produce a count of 00001000. Sketch the resulting waveforms at TP4 and TP5.

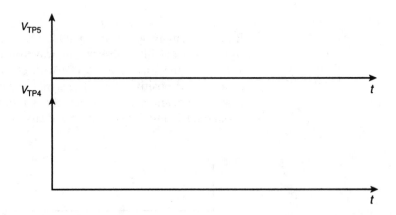

13. Set V_{in} to produce a count of 01000000. Sketch the resulting waveforms at TP4 and TP5. You should now be able to state whether the serial data coming out of the parallel to serial converter is MSB first or LSB first.

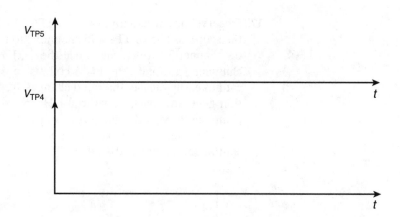

14. Now we shall check to see if the DAC is working as well as it did in the first circuit. Observe the dc voltage at the output of the DAC at TP7 when V_{in} is set to produce a count of 00000000. Repeat for counts of 10000000 and 11111111.

$$V_o \ (00000000) = \underline{\hspace{1.5cm}} \text{V dc}$$
$$V_o \ (10000000) = \underline{\hspace{1.5cm}} \text{V dc}$$
$$V_o \ (11111111) = \underline{\hspace{1.5cm}} \text{V dc}$$

15. Again, the dc input voltage to the ADC will be replaced with an analog signal. Connect a function generator between TP1 and TP2. Set the generator's offset control to produce a dc voltage equal to the value measured in step 10 that produced a count of 10000000 (mid-range). Now adjust the function generator to produce a very low frequency of 0.1 Hz. Slowly increase the amplitude of the sine wave unit it causes the LEDs to count from a minimum of 00000000 to a maximum of 11111111, to represent the negative and positive peaks of the sine wave, respectively. Measure the amplitude of V_{in} that causes this to happen.

$$V_{in} = \underline{\hspace{1.5cm}} \text{Vp-p}$$

16. Increase the frequency of V_{in} to 1 kHz. Again, the LEDs should all blink very rapidly. Readjust the clock at TP3 to 125 kHz. Connect channel 1 of the scope to TP1 and channel 2 to TP7. Trigger on channel 1. Slightly adjust the clock rate around 125 kHz until the digital steps of the waveform at TP7 appear to be almost stationary. Also, observe the output of the low-pass filter at TP8.

17. Observe the serial data at TP5 or TP6. Do this by connecting channel 1 of the scope to TP5 or TP6 and channel 2 to the sample clock at TP4. Trigger on channel 2. Notice that since the 1 kHz input sine wave amplitude is changing so rapidly, the PCM code is constantly changing and the scope cannot easily display the rapid changes. Reduce the frequency of the function generator back down to 0.1 Hz. Now you should be able to see the changing PCM code again as in steps 11–13. From the serial patterns observed in steps 11–13 sketch the waveform at TP5 that results for an instantaneous count of 01011010.

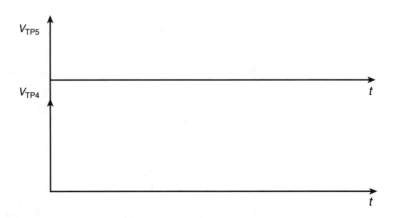

The serial data at TP5 and TP6 is a TTL signal that, when sent very long distances down data transmission lines, usually becomes susceptible to noise signals or loading and subsequent errors. Two popular data signal standards that greatly reduce these undesirable effects are known as RS-232 and RS-422.

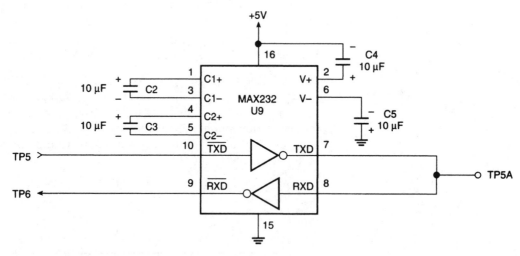

Figure 20-3 RS-232—TTL serial data conversion

18. We will now add an RS-232 driver receiver to the serial connection between the parallel to series converter and the series to parallel converter. Disconnect the jumper between TP5 and TP6 and replace it with the circuit given in Figure 20-3. Return the frequency of V_{in} to 1 kHz and observe the waveforms at TP1, TP7, and TP8. If they still appear as they did in step 16, the data link is still functional with the RS-232 additions. Now return the frequency of V_{in} to 0.1 Hz and observe the RS-232 data at TP5A. Do this by connecting channel 1 of the scope to TP5A and channel 2 to the sample clock at TP4. Trigger on channel 2. Sketch the waveform at TP5A for an instantaneous count of 01011010, as used in step 17.

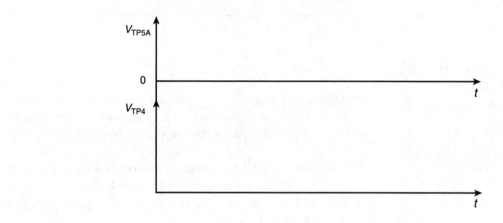

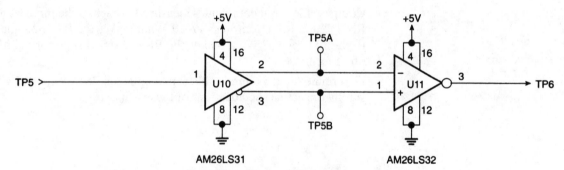

Figure 20-4 RS-422—TTL serial data conversion

19. We will now replace the RS-232 driver/receiver with an RS-422 driver/receiver. Disconnect the circuit of Fig. 20-3 from your data link and replace it with the circuit of Fig. 20-4. Again, return the frequency of V_{in} to 1 kHz and observe the waveforms at TP1, TP7, and TP8. If they still appear as they did in step 16, the data link is still functional with the RS-422 additions. Then return the frequency of V_{in} to 0.1 Hz and observe the RS-422 signals at TP5A and TP5B. Do this by connecting channel 1 of the scope to TP5A and channel 2 to TP5B, and external trigger the scope on the sample clock signal at TP4. Sketch the waveforms at TP5A and TP5B for an instantaneous count of 01011010, as used in step 17.

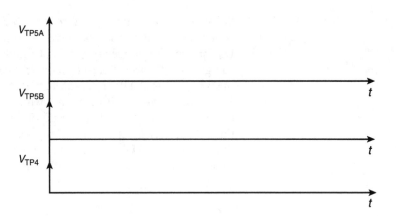

REPORT/QUESTIONS:

1. What would have been the resolution of the ADC and DAC in steps 2 and 3 if only 6 bits were used in the encoding and decoding schemes? Would the DAC output signal appear more or less distorted in steps 5 and 6? Explain why.

2. How important is it that the clock rate be set high in the encoding of the digital data? What are the limitations of the clock?

3. Was the serial data measured in steps 11–13 characterized as being MSB first or LSB first? Why?

4. Compare the serial data signals measured in each of the three formats, TTL, RS-232, and RS-422, in steps 17–19. What makes the RS-232 signal superior to the TTL signal? What makes the RS-422 signal superior to either of the other formats?

5. Determine the maximum cable run lengths for RS-232 and RS-422 by consulting the actual specifications of these standards.

DIGITAL COMMUNICATION LINK USING DELTA MODULATION CODECS

OBJECTIVES:

1. To become familiar with the process of delta modulation encoding and decoding in a data communication link.
2. To become familiar with the operating principles of the 3418 Continuously Variable Slope Delta Modulator integrated circuit.

REFERENCE:

Refer to Chapter 9 of the text.

TEST EQUIPMENT:

Dual-trace oscilloscope

Function generator

Square-wave generator

Low-voltage power supply

COMPONENTS:

3418 codec integrated circuit (2)

324 op amp integrated circuit (1)

Toggle switch

Resistors ($\frac{1}{2}$ watt): 560 Ω (3), 1 kΩ (4), 1.5 kΩ, 3.3 kΩ (9), 10 kΩ (5), 18 kΩ, 22 kΩ (4), 27 kΩ, 68 kΩ (5), 100 kΩ (4), 120 kΩ, 390 kΩ, 1.5 MΩ, 2.2 MΩ, 4.7 MΩ (4)

Capacitors: 33 pF, 75 pF, 220 pF, 470 pF (2), 820 pF (2), 1000 pF, .047 μF (8), 0.1 μF (4), 0.33 μF, 10 μF (3)

THEORY:

The 3418 is a popular codec integrated circuit that uses delta modulation, sometimes referred to as slope modulation, to encode an analog signal into a digital format. It does this by transforming each sampled segment of the analog signal into a digital instruction that states whether the analog signal's amplitude is increasing or decreas-

131

ing. Then when the 3418 is set up as a decoder, it simply reconstructs the analog signal by following the given instructions of increment or decrement. The rough edges of the reconstructed analog output signal are smoothed out by a low-pass filter. Obviously, the higher the sampling rate (clock) is, the easier it is for the low-pass filter to re-create a near-perfect replica of the original analog signal.

In this experiment, the 3418 is first set up as an encoder. In this mode the continuously variable slope delta (CVSD) encoding scheme is observed. With CVSD, the step size is increased during periods of maximum slope and reduced during periods of minimum slope. Next, a complete process of encoding digital data and subsequent decoding of this data back into its original analog form is observed. Two 3418 codecs are necessary to perform this task. Finally, a telephone voice data link is built and analyzed.

PROCEDURE:

1. Build the circuit given in Fig. 21-1. Connect a 5 V, 16 kHz positive square wave to the clock input at TP2. Close switch S1 to apply a +5 V dc level at the EN/$\overline{DE}$ control input at TP3. This sets up the codec as an encoder. In other words, it will encode the analog signal applied at TP1 into a digital format, as described in the theory section of this experiment.

2. Observe the digital output signal at TP5. Sketch the square waveform, making note of its amplitude and frequency.

3. Now apply an analog signal at TP1. Connect the function generator at TP1. Connect channel 1 of your scope to TP1 and channel 2 to TP5. Set the function generator frequency to produce a 50 Hz sine wave. Trigger on channel 1. Also invert channel 2 of your scope. Slowly increase the amplitude of the sine wave from zero to approximately 6 Vp-p. Notice that as the amplitude increases, the square wave at TP5 begins to change into a varying duty cycle pulse. You may need to slightly alter the frequency of the sine wave in order to create a stable display of the waveform of TP5 on your scope. Carefully sketch both waveforms, making note of exactly where the duty cycle of the pulse is a maximum, a minimum, and 50%.

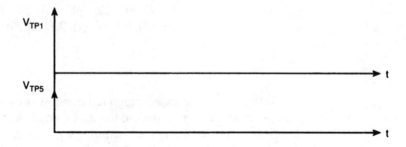

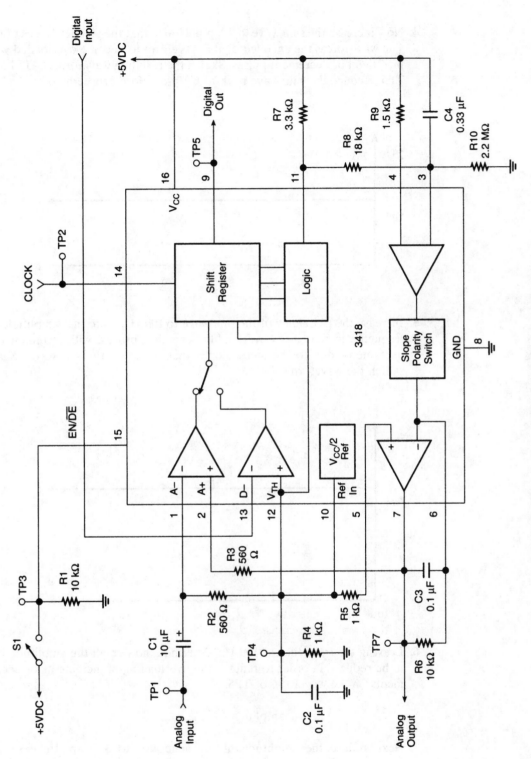

FIGURE 21-1 Encoding configuration of the 3418 codec

4. Now increase the sine wave to 7 Vp-p. Notice that the steeper slopes of the sine wave forces the encoded digital waveform to become overloaded with increment or decrement pulses. Sketch the resulting waveforms at TP1 and TP5. Reduce the sine wave back to 6 Vp-p before proceeding.

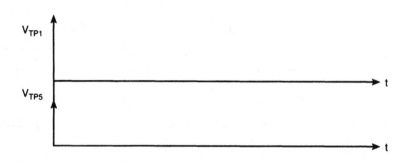

5. Increase the frequency of the sine wave to 100 Hz. Note that amplitude or frequency increases can overload the encoding process with increments or decrements due to the steep slopes that occur on the sine wave. Again sketch the waveforms.

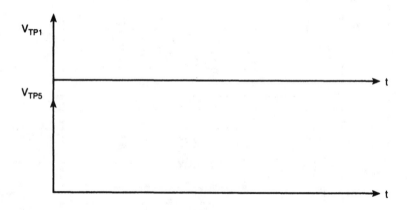

6. Leaving the frequency at 100 Hz, determine how small the amplitude has to be reduced in order to remove the overloading of increments or decrements on the waveform of TP5.

amplitude = _____

7. Next, reduce the amplitude of the generator to 3 Vp-p. Increase the frequency of the generator to 8 kHz. This time trigger on channel 2. Carefully fine-tune the frequency of the sine waves to cause the sine wave display to appear to trigger. You should see that the encoded digital display is now pretty stable at 50% duty cycle, except for an occasional toggle here and there. You have reached the Nyquist limit. The sampling of the analog input only happens once for every half-cycle of the sine wave. Thus, the encoded digital signal is no longer an accurate representation of the analog information.

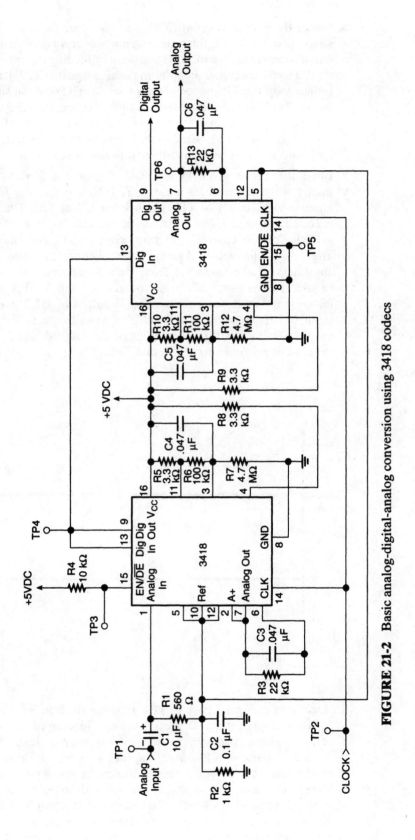

FIGURE 21-2 Basic analog-digital-analog conversion using 3418 codecs

8. Notice that by opening switch S1, we now place the codec into the decode mode. In other words, the codec can now convert encoded digital data back into its original analog form. This is done by feeding the encoded digital data into TP6 and observing the resulting analog signal at TP7. Unfortunately we cannot loop the TP5 encoded data back into TP6 of the same 3418 codec device. This is because the 3418 is a simplex codec, meaning that the internal shift registers and counters are used for either the encoding or decoding process, but not at the same time. We need two 3418 codec devices to do both operations. This is investigated in the next step.

9. Build the circuit given in Fig. 21-2. Connect a 5 V, 50 kHz positive square wave to the clock input at TP2. Notice that the EN/$\overline{\text{DE}}$ control input is high for U1 at TP3 and low for U2 at TP5. Thus, U1 is in the encode mode and U2 is in the decode mode. Apply a 500 Hz sine wave at TP1. Connect channel 1 of your scope to TP1 and channel 2 to TP4. Trigger on channel 1. Again, invert channel 2. Slowly increase the amplitude of the sine wave from zero to approximately 3 Vp-p. You should see the same effect as you did in step 3. This time, however, notice that TP4 is not only the digital output of U1 but also the digital input of U2. Reconnect channel 2 of your scope to TP6. You should see that TP6 is a fairly close replica of the original signal at TP1. Sketch the resulting waveforms at TP1 and TP6.

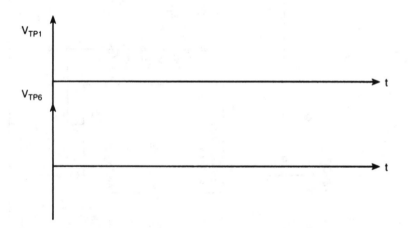

10. Reconnect channel 2 back to TP4. Increase the frequency of the sine wave to approximately 1.5 kHz. Notice the overloading of the digital encoding at the regions of maximum slope of the analog signal. Now reconnect channel 2 back to TP6. You may need to slightly alter the frequency of the sine wave to get a stable display of the digital "noise" at TP6. Note that during the digital overload time periods, the analog output shows a slope that closely resembles the steepest slopes of the input sine waves. This makes the output signal closely resemble the input signal. This is an advantage that the codec devices that have CVSD encoding have over other pulse-coded modulated devices. Sketch the resulting waveforms at TP1 and TP6.

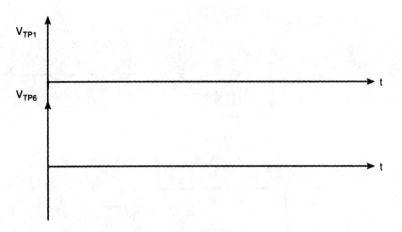

11. Sketch the resulting waveforms at TP1 and TP6 if the analog input is a 3 Vp-p, 1.5 kHz triangle wave.

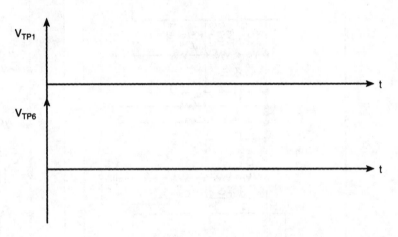

12. Repeat step 11 for a 3 Vp-p, 1.5 kHz square wave. Note that square wave slopes are very hard to reproduce.

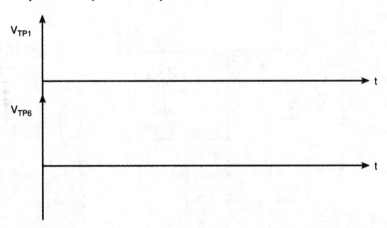

13. To make this digital communication link useful for voice communication, two improvements need to be made. First, the digital noise in the output waveform must be removed without affecting the basic waveshape of the desired signal. This can be done by the addition of an active low-pass filter in the output. Secondly, any high-frequency information in the input voice signal must not cause the encoding codec to be pushed past its Nyquist

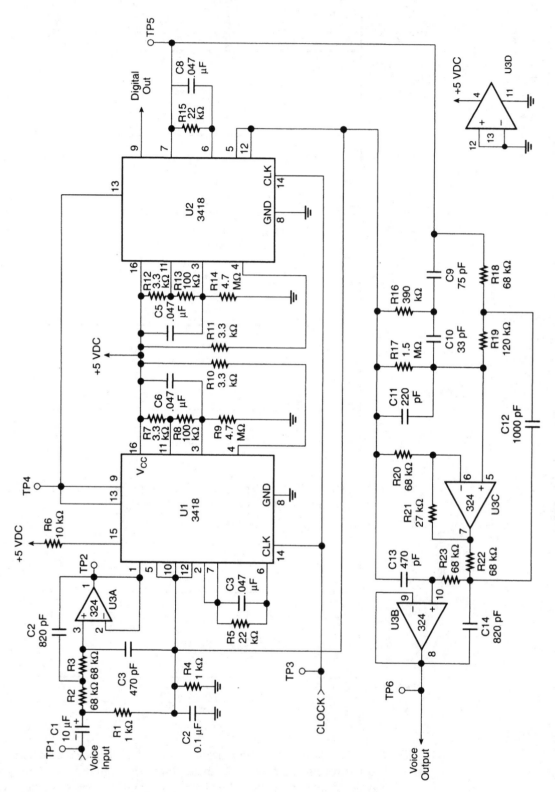

FIGURE 21-3 Telephone voice communication link using 3418 codecs

limit (half the clock rate). This is done by the addition of another active low-pass filter at the input. These two improvements are designed into the circuit of Fig. 21-3.

14. Build the test circuit of Fig. 21-3. Connect a 5 V, 50 kHz, positive square wave to the clock input at TP3. Again, U1 is set up as the encoder and U2 as the decoder. Apply a 2 Vp-p, 200 Hz sine wave at TP1. Connect channel 1 of your scope to TP1 and channel 2 to TP6. You should now see that almost all of the digital "noise" has been removed by the filter. Sketch the resulting waveforms at TP1 and TP6.

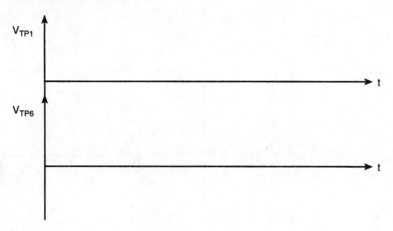

15. Repeat step 14 for a 2 Vp-p triangle wave at TP1.

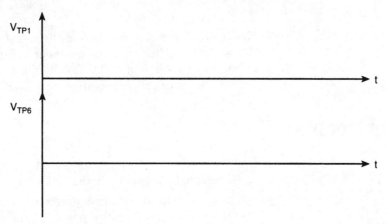

16. Repeat step 14 for a 2 Vp-p square wave at TP1.

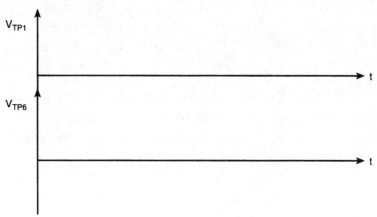

17. Reset the signal generator at TP1 as a 2 Vp-p sine wave. Determine the frequency response of this digital communication link by taking the output voltage readings at frequencies ranging from 10 Hz to 5 kHz. Calculate the output voltage in dB at each frequency, referenced to its maximum value.

18. Reset the signal generator at TP1 as a 2 Vp-p sine wave. Reduce the clock rate to 10 kHz. Notice the effect on the output waveform at TP6. Sketch the resulting waveforms.

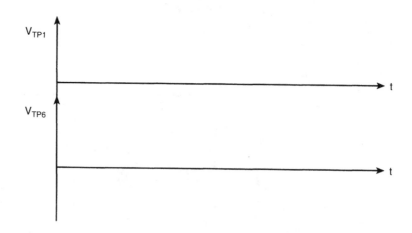

19. Notice that if the clock rate is reduced any further, the output waveform becomes unusable.

20. If time permits, try replacing the signal generator with a voice signal, by adding a microphone and amplifier. Also, amplify the output signal at TP6 in order to drive a speaker. Determine the lowest clock rate at which the recreated voice signal is still understandable.

REPORT/QUESTIONS:

1. What is meant by slope overload? In which steps did you discover this phenomena occurring in the codec?

2. Why is delta modulation preferred over basic pulse coded modulation (PCM) systems for voice signals? Explain.

3. State why in steps 12–16 the reconstructed square wave is not a close replica of the original analog input signal.

4. Why is the waveform at TP6 more distorted in step 18 than in step 14?

5. Using the results of step 17, sketch the frequency response curve in dB for the telephone voice data link.

6. What changes would be in order to have a wider frequency response than what was found in step 17?

7. What are the main advantages of delta modulation over PCM systems?

8. In what applications would PCM be preferred over delta modulation?

EWB MULTISIM—dB MEASUREMENTS IN COMMUNICATIONS

OBJECTIVES:

1. To become acquainted with the use of Electronics Workbench Multisim in simulating dB measurements in communications circuits.
2. To understand dB measurement using the EWB multimeter set to the dB mode.
3. To explore the characteristics of a T-type passive attenuator circuit.
4. To explore the procedure for setting the dB levels in a system.

VIRTUAL TEST EQUIPMENT:

AC voltage source
Multimeter
Virtual resistors, capacitors

INTRODUCTION:

This exercise introduces the reader to the techniques for making audio signal level measurements using Electronics Workbench Multisim simulations. Obtaining signal level measurements and measuring signal path performance are common maintenance, installation, and troubleshooting practices in all areas of communications. Communication networks require that proper signal levels are maintained to ensure minimum line distortion and cross-talk. This section examines the techniques for making dB (decibel) measurements on various passive circuits and setting the levels for a system.

PRELABORATORY:

1. Review Section 1-2, the dB in Communications in your text.
2. Review Chapter 11, Troubleshooting with Electronics Workbench Multisim in your text.

PROCEDURE:

You will be required to make dB level measurements on passive attenuator circuits and perform a system-wide calibration of the levels.

Part I: Measuring the Insertion Loss Provided by Passive Resistive Attenuator Circuits

1. The EWB Multisim implementation of a test circuit for making dB measurements is provided in Fig. 22-1. This circuit contains an AC signal source, a 600 Ω load, and two multimeters. The top multimeter (XMM1) is measuring the dB level and the bottom multimeter (XMM2) is being used to measure the voltage across the load. Construct the circuit and set the AC voltage source to 2.188 V at 1 kHz. Measure the voltage and dB levels across R2. Record your value. *Note*: Make sure you set the multimeter to measure AC.

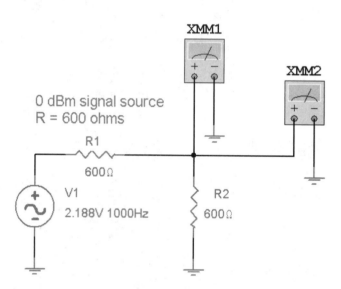

FIGURE 22-1

2. Explain how the voltage measured across R2 equates to 0 dBm (*Hint:* Refer to Section 11-13, Troubleshooting with Electronics Workbench Multisim.)

3. Construct the circuit shown in Fig. 22-2. This circuit includes a 600 Ω, 0-dBm voltage source and a 600-Ω T-type attenuator that has been inserted into the path between the signal source and the 600-Ω load resistance. Measure the amount of insertion loss provided by the T-type attenuator. This requires that you measure the dB levels at both the input and output of the attenuator. The insertion loss will be the difference in the two measurements.

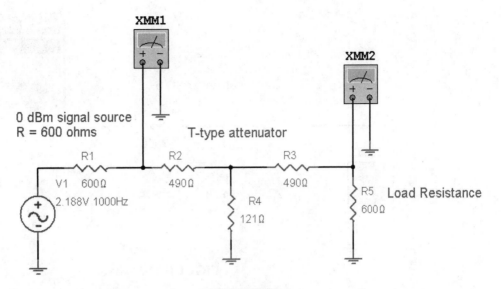

FIGURE 22-2

dB level measured at the attenuator input   _____

dB level measured at the attenuator output   _____

insertion loss (input–output)   _____

4. Repeat step 3 for the resistor values provided for the T-type attenuator in Table 22-1.

TABLE 22-1

R2	R3	R4	INPUT LEVEL (dB)	OUTPUT LEVEL (dB)	INSERTION LOSS (dB)
230	230	685			
*69	69	258			
588	588	120			
312	312	422			
563	563	38			

*Why doesn't the input level for this attenuator not equal 0 dBm?

5. Verify that the measured resistance for the T-type attenuator values listed in Table 22-1 equal 600 Ω.

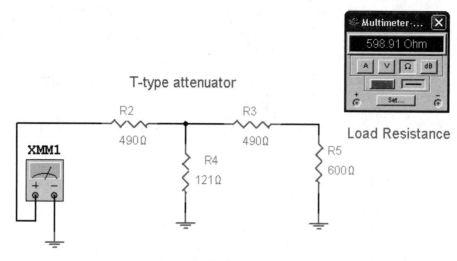

FIGURE 22-3

Repeat step 5 for the resistor values provided in Table 22-2. Make sure that the T-type attenuator is terminated with a 600-Ω resistor. The resistor R5 is the termination resistor in Fig. 22-3.

TABLE 22-2

R2	*R3*	*R4*	INPUT RESISTANCE (Ω)
230	230	685	
69	69	258	
588	588	12	
312	312	422	
563	563	38	

Part II: Setting System Wide dB Levels

This exercise demonstrates the procedure for setting dB levels in a communications system. This example is for a broadcast facility where the audio levels are being set to a predefined level throughout the system. The system is shown in Fig. 22-4. The audio outputs for a tone generator, production audio, VCR1, and a satellite feed are shown. These devices are inputted into a passive rotary switch. The output of the switch feeds the station studio transmitter link (STL). The level's input to the rotary switch (TP1) must be set to 0 dBm. The level to the STL must be set to +8 dBm (TP2). A 1-kHz tone is used to calibrate the system. Level adjustment is provided by the virtual potentiometer connected in the feedback paths for the operational amplifiers.

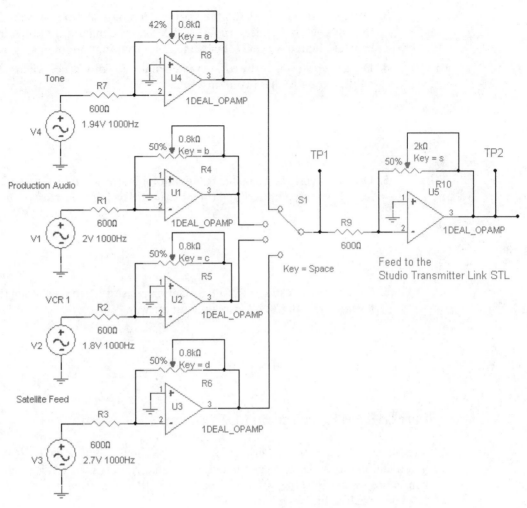

FIGURE 22-4

1. Open the file **Lab22-Part2** found on your textbook EWB CD-ROM.* This file contains the circuit shown in Fig. 22-4.

2. Set the levels according to Table 22-3.

TABLE 22-3

DEVICE	OUTPUT LEVEL	LEVEL CONTROL KEY
Tone	0 dBm	**A** increase, **a** decrease
Production Audio	0 dBm	**B** increase, **b** decrease
VCR1	0 dBm	**C** increase, **c** decrease
Satellite Feed	0 dBm	**D** increase, **d** decrease
Rotary Switch	0 dBm	None
STL	+ 8 dBm	**S** increase, **s** decrease

*The files are also available at www.prenhall.com/beasley. Click on Companion Website, then Lab 22 files.

3. Once you have completed the settings, demonstrate to your lab instructor that all levels have been properly set. Make sure that the multimeter is properly terminated with 600 Ω when making your measurements.

4. Does the position of the rotary switch make any difference when setting the audio levels? Explain your answer.

5. Do you need to add a 600-Ω termination resistor when measuring the audio level from the output of the rotary switch? Explain your answer.

6. Do you need to add a 600-Ω termination resistor when measuring the audio level feeding the STL? Explain your answer.

Part III: EWB Exercises on CD-ROM

The Multisim files for this experiment are provided on the textbook CD-ROM to give you additional experience in simulating electronic circuits with EWB. You will also gain more insight into the characteristics of the dB measurement and system-wide level settings. The file names are listed.

FILE NAME
Lab22-Fig22-1
Lab22-Fig22-2
Lab22-Fig22-3
Lab22-Part2

SMITH CHART MEASUREMENTS USING THE EWB MULTISIM NETWORK ANALYZER

OBJECTIVES:

1. To become acquainted with the use of the Electronics Workbench Network Analyzer

2. To understand impedance measurements.

3. To explore the impedance characteristics of transmission lines and impedance matching for antennas.

VIRTUAL TEST EQUIPMENT:

AC voltage source

Multimeter

Virtual resistors, capacitors, inductors

Virtual network analyzer

INTRODUCTION:

The concept of using a Smith chart has been introduced in Chapter 12 of the your text. This important impedance calculating tool is now reintroduced using Electronics Workbench Multisim simulations. Multisim provides a network analyzer instrument that contains the Smith chart analysis as one of its many features. A network analyzer is used to measure the parameters commonly used to characterize circuits or elements that operate at high frequencies. This exercise will focus on the Smith chart and the z-parameter calculations. Z-parameters are the impedance values of a network expressed in its real and imaginary components. Refer to Section 12-8 of your text for additional Smith chart examples and a more detailed examination of its function.

PRELABORATORY:

1. Review Section 12-8, the Smith Chart in your text.
2. Review Sections 12-11 and 15-12, Troubleshooting with Electronics Workbench Multisim in your text.

PROCEDURE:

You will be required to make impedance measurements on various resistive, *RC*, and *RL* circuits and transmission lines using the EWB network analyzer.

Part I: Using the Network Analyzer

1. Begin the exercise by constructing the circuit shown in Fig. 23-1. This circuit contains a 50-Ω resistor connected to port 1 (*P1*) of the network analyzer; port 2 (*P2*) is terminated with a 50-Ω resistor. The first circuit being examined by the network analyzer is a simple resistive circuit. This example provides a good starting point for understanding the setup for the network analyzer and how to read the simulation results.

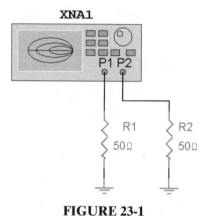

FIGURE 23-1

2. Start the simulation. The impedance calculations performed by the network analyzer are very quick and the start simulation button will quickly reset. Before you look at the test results, predict what you will to see. Based on the information you learned in Section 12-8, and the fact that you are testing a resistor, you would expect to see a purely resistive result. Double-click on the network analyzer to open the instrument. You should see a Smith chart similar to the result shown in Fig. 23-2.

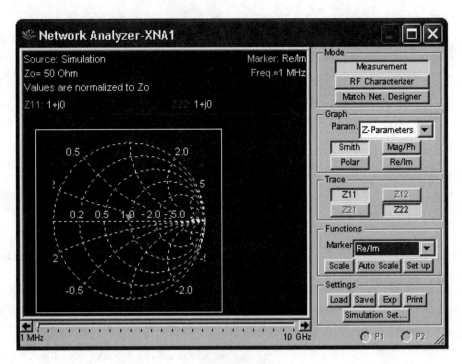

FIGURE 23-2

The Smith chart indicates the following:

$$Z_o = 50 \ \Omega$$

Values are normalized to Z_o

$$Z_{11} = 1 + j0$$

The $Z_{11} = 1 + j0$ value indicates that the input impedance for the network being analyzed is purely resistive and its normalized value is 1, which translates to 50 Ω. Recall that the values on a Smith chart are divided by the normalized resistance. Notice the marker on the Smith chart is located at 1.0 on the real axis. The 1.0 translates to 50 Ω, and this value is obtained by multiplying the Smith chart measured resistance of 1 Ω by the characteristic impedance of 50 Ω to obtain the actual resistance measured. In this case the computed resistance is 50 Ω.

The frequency at which this calculation was made is shown in the upper-right corner of the Smith chart screen. In this case a frequency of 1.0 MHz was used. The frequency range for the simulation is shown at the bottom of Fig. 23-2 and is adjusted by clicking on the left and right arrows. You can adjust the frequency to see how the impedance values can change through the frequency range. Of course an ideal resistor will not be frequency-dependent. The frequency range, used in the network analysis, is set by clicking on Simulation Set . . . at the bottom of the network analyzer screen. Click the Simulation Set . . . button to check the settings. You will notice the following:

Start Frequency	1 MHz
Stop Frequency	10 GHz
Sweep Type	Decade
Number of points per decade	25
Characteristic impedance Z_o	50 Ω

The start and stop frequency provides control of the frequency range when testing a network. The sweep type can be specified to be plotted in either a decade or linear form, but most of the time use the decade form. The number of points per decade enables the user to control the resolution of the plotted trace displayed, and the characteristic impedance Z_o provides for user control of the normalizing impedance.

3. Change the characteristic impedance of the network analyzer to 75 Ω. This will require changing the characteristic impedance setting for the network analyzer and will also require that the resistor connected to port 2 (P2) be changed to 75 Ω. Change the value of resistor R1 to 75 Ω. Restart the simulation and record the normalized value (Z_{11}); convert the normalized value to actual resistance. Repeat this for the resistor R1 values provided in Table 23-1. Record the measured normalized network impedance and calculate the actual resistance.

TABLE 23-1

R_1 (Ω)	NETWORK IMPEDANCE Z_{11}	RESISTANCE
75		
50		
100		
600		
300		

Part II: Measuring Complex Impedances with the Network Analyzer

The next two Multisim exercises provide examples of using the Multisim network analyzer to compute the impedances of RC and RL networks. These exercises will help you better understand the Smith chart results when analyzing complex impedances.

1. Construct the circuit shown in Fig. 23-3. This is a simple RC network of R = 25 Ω and C = 6.4 nF. The network analyzer is set to analyze the frequencies from 1 MHz to 100 MHz. The results of the simulation are shown in Fig. 23-4. At 1 MHz, the normalized input impedance to the RC network shows that $Z_{11} = .5 - j.497$. Multiplying these values by the normalized impedance of 50 Ω yields approximately a Z of 25 − j25, which is the expected value for this RC network at 1 MHz. Verify this by calculating the capacitive reactance (X_c) of the 6.4 nF capacitor at 1 MHz. Record your result.

$$X_c = \underline{\qquad}$$

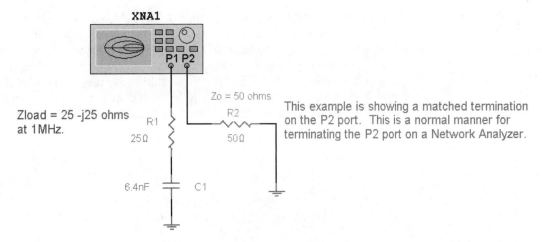

Zload = 25 -j25 ohms
at 1MHz.

Zo = 50 ohms

This example is showing a matched termination on the P2 port. This is a normal manner for terminating the P2 port on a Network Analyzer.

R1
25Ω

R2
50Ω

6.4nF   C1

FIGURE 23-3

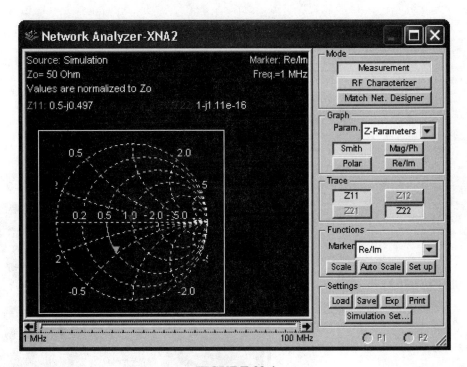

FIGURE 23-4

2. Repeat step 1 to measure the impedance of the *RC* networks specified in Table 23-2 for the frequencies listed.

TABLE 23-2

R_1 (Ω)	C_1	FREQUENCY	NETWORK IMPEDANCE, Z_{11}
25	6.4 nF	100 kHz	
25	6.4 nF	100 MHz	
10	10 pF	1 MHz	
10	10 pF	100 MHz	
50	50 nF	100 kHz	

3. Next construct the circuit shown in Fig. 23-5. This circuit contains a simple *RL* network of $R = 25\ \Omega$ and $L = 4.0\ \mu\text{H}$. The network analyzer is set to analyze the frequencies from 1 MHz to 10 GHz. The results of the simulation are shown in Fig. 23-6. At 1 MHz, the normalized input impedance to the *RL* network shows that $Z_{11} = 0.5 + j0.5$. Multiplying these values by the normalized impedance of 50 Ω yields $Z = 25 + j25$, which is the expected value for this *RL* network at 1 MHz. Calculate X_L of the 4-μH inductor at 1 MHz. Record your value.

$$X_L = \underline{\hspace{2cm}}$$

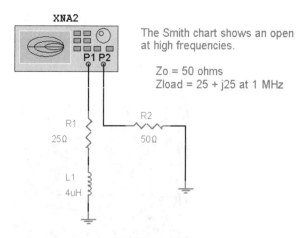

FIGURE 23-5

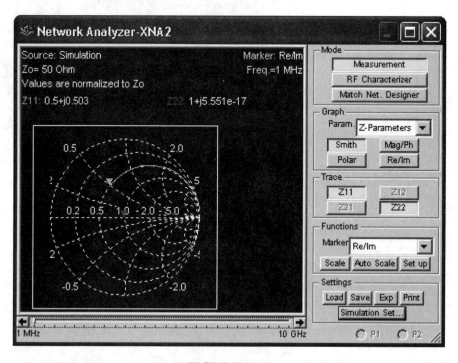

FIGURE 23-6

4. Measure the impedance of the *RL* networks specified in Table 23-3 for the frequencies listed.

TABLE 23-3

R_1 (Ω)	L_1	FREQUENCY	NETWORK IMPEDANCE, Z_{11}
25	4 µH	10 kHz	
25	4 µH	100 kHz	
25	4 µH	10 MHz	
25	4 µH	100 MHz	
25	4 µH	1 GHz	

5. What observation can you make about the *RL* network of step 4 at low and high frequencies? (*Hint:* Express your observation in terms of the changes in the inductance value).

Low frequencies:

High frequencies:

Part III: Transmission and Waveguide Impedance Measurements

1. Part III provides the student with the opportunity to explore the properties of a low-loss waveguide. Begin by opening the **Lab23-Part3** file on your textbook CD-ROM.[*] This circuit contains a sample waveguide attached to the network analyzer. The circuit is shown in Fig. 23-7. Both ends of the waveguide are connected to the ports of the network analyzer.

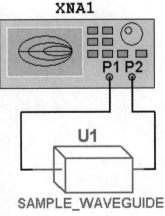

L-band 1 - 2G
56 ohms

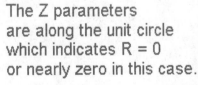

The Z parameters
are along the unit circle
which indicates R = 0
or nearly zero in this case.

FIGURE 23-7

[*]The files are also available at www.prenhall.com/beasley. Click on Companion Website, then Lab23 files.

Before starting the simulation change, click on the network analyzer. Set the characteristic impedance of the network analyzer to 50 Ω and change the number of points per decade to 200, the start frequency to 1 GHz, and the stop frequency to 2 GHz. This change provides a smoother plot of the simulation results and a realistic frequency range for the waveguide. Start the simulation and view the results on the network analyzer. You should see a result similar to the one shown in Fig. 23-8.

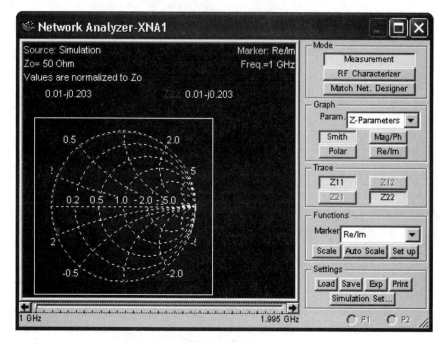

FIGURE 23-8

The plot of the data is along the outside perimeter of the Smith chart. This indicates that the line is low loss. Move the frequency marker on the network analyzer to 1 GHz. The input impedance of the waveguide at 1 GHz is $Z_{11} = 0.01 - j0.203$ or very little resistive loss.

2. Begin step 2 by opening the **Lab23-Part3_6** file on your textbook CD-ROM. This circuit contains several waveguides and a network analyzer. Connect each waveguide to the network analyzer and record the normalized resistance of the waveguide. This is the point where the resistance and reactance terms for Z_{11} are minimum. Also record the frequency at which the measurement was made. For example, the minimum resistance and reactance, Z_{11}, for waveguide A is $0.01 - j0.0081$ at 1.122 GHz. Complete this information for the

waveguides listed in Table 23-3 and provided in the file **Lab23-Part3_6.** Limit the frequency sweep of the waveguide to 500 MHz to 2 GHz.

WAVEGUIDE SECTION	NETWORK IMPEDANCE, Z_{11}	FREQUENCY
A		
B		
C		
E		
F		
G		

3. Based on the measurements made in step 2, which waveguide has the greatest loss? Record your answer.

_____

4. Based on your measurements made in step 2, what frequency are the waveguide sections A-E designed to carry? Record your answer.

_____

5. Based on your measurements, what frequency is waveguide G designed to carry? Record your answer. (*Hint:* This is the frequency of the waveguide when the resistance and reactance are at a minimum.)

_____

6. Connect the network analyzer to waveguide H and sweep it from 1 GHz to 2 GHz. Explain what the Smith Chart is showing when the frequency is set to 1.12 GHz.

Part IV: EWB Exercises on CD-ROM

The Multisim files for this experiment are provided on your textbook EWB CD-ROM so you can gain additional experience simulating electronic circuits with EWB and gain more insight into the characteristics of impedance measurements using the network analyzer.

FILE NAME
Lab23-Fig23-1
Lab23-Fig23-3
Lab23-Fig23-5
Lab23-Fig23-7
Lab23-Part1_Step3_75_ohms
Lab23-Part3_6

TONE DECODER

OBJECTIVES:

1. To investigate frequency sensing and detection.
2. To investigate the bandwidth and sensitivity of a tone decoder.
3. To build and test a tone decoder circuit.

REFERENCE:

An application of tone sensing can be found in Chapter 11 of the text.

TEST EQUIPMENT:

Dual-trace oscilloscope
Sinusoidal function generator
Low-voltage power supply
Prototype board

COMPONENTS:

Integrated circuit: 567
Capacitors: 10 μF, 4.7 μF, 1 μF, 0.22 μF, 0.01 μF
Resistors: 1 kΩ
Potentiometers: (10-turn trim) 25 kΩ

PRELABORATORY:

Calculate the f_o, and bandwidth for the circuit shown in Fig. 24-1. Use the equations shown in the data sheet for the LM567 Tone Decoder included in the Appendix (assume the potentiometer is 10 kΩ).

$f_o = $ _____

$BW = $ _____

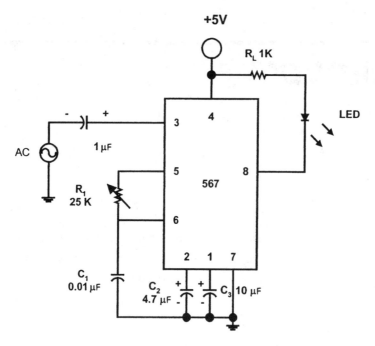

FIGURE 24-1

PROCEDURE:

1. Build the tone decoder circuit shown in Fig. 24-1.

2. Apply the DC power to the circuit, but do not apply a signal from the function generator yet.

3. Connect channel 1 of your oscilloscope to pin 5 of the LM567 and adjust R_1 until the free-running frequency of the decoder is the same as your calculated f_o.

4. Set the function generator frequency to equal the calculated f_o and set the output voltage level to 1.0-V rms. Connect the function generator to the input of the circuit.

5. Reduce the function generator voltage level to the minimum voltage that will still have the LED lit.

6. Slowly increase the frequency of the generator until the LED turns off. This is the upper-lock frequency.

   Record the frequency: _____

7. Slowly decrease the frequency of the generator until the LED turns back on. Continue to decrease the frequency until the LED turns off. This is the lower-lock frequency.

   Record the frequency: _____

8. Subtract the lower-lock frequency from the upper-lock frequency to get the bandwidth, and divide the bandwidth by f_o to get the Bandwidth %.

$$\text{Bandwidth \%} = \frac{\text{bandwidth}}{f_0} * 100$$

Bandwidth % = _____

9. Increase the voltage level of the function generator, and repeat steps 6 through 8.

Bandwidth % = _____

10. Demonstrate to your instructor that the tone decoder will detect the presence of a signal by lighting an LED when the function generator frequency is adjusted to within the bandwidth of the tone detector and there is sufficient voltage level.

REPORT/QUESTIONS:

1. What is the Bandwidth % of the tone decoder at the minimum input voltage level?
2. What happened to the bandwidth when the input voltage was increased?

NAME _____

DIGITAL COMMUNICATION USING FREQUENCY-SHIFT KEYING

OBJECTIVES:

1. To become familiar with modems.
2. To analyze the XR2206 function generator and observe how it can be used to encode digital information into an FSK signal.
3. To analyze the XR2211 FSK decoder and observe how it can be used to convert an FSK signal back into digital data.

REFERENCE:

Refer to Chapter 10 of the text.

TEST EQUIPMENT:

Dual-trace oscilloscope
Function generator
Frequency counter
Low-voltage power supply (2)

COMPONENTS:

Integrated circuits: XR2206, XR2211
Capacitors: 0.0047 μF, 0.022 μF, 0.033 μF, 0.1 μF (5), 1 μF (2), 10 μF
Resistors: 220 Ω, 3.9 kΩ, 4.7 kΩ (4), 10 kΩ (2), 18 kΩ, 100 kΩ, 220 kΩ, 470 kΩ (2)
Potentiometers: (10-turn trim) 1 kΩ (2), 10 kΩ, 50 kΩ

THEORY:

One form of modulation of digital signals onto an RF carrier uses a technique called "frequency-shift keying" (FSK). Early forms of radio-teletype used such a form of modulation. Today, this digital modulation technique is quite obsolete in its basic form, although the general principles of FSK are used in more advanced data encoding techniques.

In FSK, the two digital logic states, "1" and "0," are converted to a constant-amplitude sine wave that is shifted between two possible frequencies. These two frequencies are referred to as the "mark" and "space" frequencies. These frequencies are usually in the audio-frequency spectrum. Popular mark/space frequency pairs are 1070/1270 Hz, 2025/2225 Hz, and 2125/2290 Hz. For example, the mark frequency of 2025 Hz might represent the binary "1" and the space frequency of 2225 Hz might represent the binary "0." In a radio transmitter, if an FSK signal is fed into the microphone input and single-sideband modulation is used, the RF carrier at the output of the transmitter will then be shifted between the corresponding RF mark and space frequencies. Refer to Fig. 25-1.

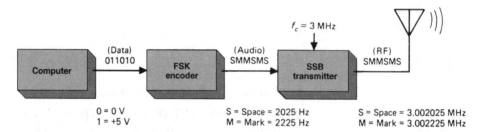

Figure 25-1 FSK encoder and transmitter

Note that the resulting SSB output signal ends up being an elementary form of FM, since only a single sine wave actually modulates the RF carrier at a time. Being FM makes it quite immune to noise interference. This is the main advantage of using FSK in a digital communications system. As seen in Fig. 25-2, the SSB receiver would detect the original audio frequencies and the FSK decoder would then convert them back to the original digital format.

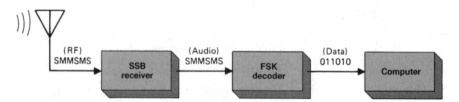

Figure 25-2 Digital communication receiver and decoder

In some digital communications systems, regardless if radio or telephone wires are being used to send the FSK signals, there is a need for two-way communications to occur simultaneously. Thus the FSK encoding and decoding are needed at both ends of the communications link. If this is being done, "modems" are used (Fig. 25-3). "Modem" is an acronym for a device that contains an FSK encoder or modulator and an FSK decoder or demodulator. In this laboratory exercise, the XR-2206 FSK modulator and the XR-2211 FSK demodulator are investigated.

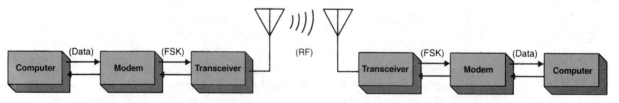

Figure 25-3 Digital communications link using modems

PRELABORATORY:

1. Using the data sheets for the XR2206 and the given mark and space frequencies of 2025 and 2225 Hz, respectively, determine the theoretical values for R_1 and R_2 in the FSK encoder of Fig. 25-4.

2. Using the design notes for the XR2211 found in the Appendix and the given mark and space frequencies used above, determine the required values for R_9, C_{11}, R_8, C_9, and C_8 for the FSK decoder of Fig. 25-5. These are listed as R_0, C_0, R_1, C_1, and C_f, respectively, on the XR2211 data sheets.

PROCEDURE:

1. Build the circuit given in Fig. 25-4. Apply 10 V dc to the circuit. Place a jumper between TP_1 and ground. Monitor the output voltage at TP_3 with an oscilloscope. Adjust R_4 for a 1 V_{p-p} amplitude.

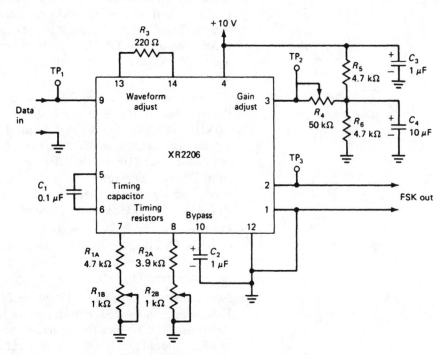

Figure 25-4   XR2206 FSK modulator stage

2. Measure the frequency of the signal at TP_3 with a frequency counter. Adjust R_{2B} for a frequency of exactly 2225 Hz.

3. Open the jumper between TP_1 and ground. Adjust R_{1B} for a frequency of exactly 2025 Hz at TP_3.

4. Apply a 2-V dc voltage at TP_1. Decrease the voltage at TP_1 until you notice the frequency of V_O at TP_3 changing to the alternate value. Determine the critical voltage at TP_1 where the frequency switching occurs.

5. Apply a 3 V positive square wave at TP_1. Adjust the frequency for approximately 2 Hz. Notice the output waveform switching between the mark and space frequencies at TP_3. The square-wave generator is simulating a

computer sending digital data. The voltage at TP_3 is the encoded FSK signal. Do not disassemble this circuit before proceeding.

6. Build the circuit given in Fig. 25-5. Notice that the XR2211 is actually a phase-locked loop similar to the 565 used in Experiments 16 and 17, except that it is specifically designed for use in decoding FSK signals. This is seen in the fact that the VCO output signal is fed back through two separate loops. The standard loop that is used in all PLLs is the phase detector followed by a loop low-pass filter. In addition to this, the XR2211 has a separate phase detector that drives a lock filter. Both loop and lock filters drive separate comparators that cause output signals at pins 5, 6, and 7 to be binary in nature. The data output at pin 7 is the result of converting the FSK mark and space frequencies back into digital form. The lock-detect outputs switch into their active states each time the XR2211 senses mark and space frequencies at its input.

7. Apply 10 V dc to the circuit. Apply a 100-mV_{p-p} sine wave at a frequency of 2125 Hz at TP_4. Observe the waveform at TP_6 with channel A of the oscilloscope. Observe the dc voltage at TP_7 with channel B. You should notice that when the XR2211 locks up, the waveform at TP_6 should stabilize and the dc voltage at TP_7 jumps to a digital "low state." Then when the sine wave is removed at TP_4, the waveform at TP_6 should free-run and the voltage at TP_7 should jump up to 10 V. TP_8 is just the complement logic state of TP_7.

8. Now observe the waveform at TP_5 with channel B. Note what happens when the frequency of the generator is slowly increased and decreased between 2.0 kHz and 2.3 kHz. Measure the critical frequency at which the voltage at TP_5 changes states. It should be approximately 2.125 kHz. If not, adjust R_{10} to make it 2.125 kHz. You will find that there is a small amount of hysteresis at TP_5. In other words, the frequency at which a low to high transition occurs is different from that at which a high to low transition occurs. Adjust R_{10} so that 2.125 kHz is the average of these two critical frequencies.

9. Decrease the amplitude of the function generator. Determine the sensitivity of the FSK decoder. In other words, determine the minimum amplitude at TP_4 that guarantees that successful decoding takes place. Remove the generator from TP_4 before proceeding.

10. Now connect a jumper from TP_3 of Fig. 25-4 to TP_4 of Fig. 25-5. The output FSK signal from the FSK encoder is now being applied to the FSK decoder. Apply a 20-Hz, 3-V positive square wave at TP_1. Observe the FSK output at TP_3. It should slowly switch back and forth between the mark and space frequencies. If it doesn't switch states, you may need to fine-tune R_{10} to make it happen.

11. Observe the original digital data (square wave) at TP_1 with channel A and the re-created digital data at TP_5 with channel B. Sketch the resulting waveforms.

12. Increase the baud rate of the digital data by increasing the frequency of the function generator. Determine the highest square-wave frequency that can be sent through this simple digital communications link. Again, fine-tuning R_{10} may make this upper frequency limit a bit larger.

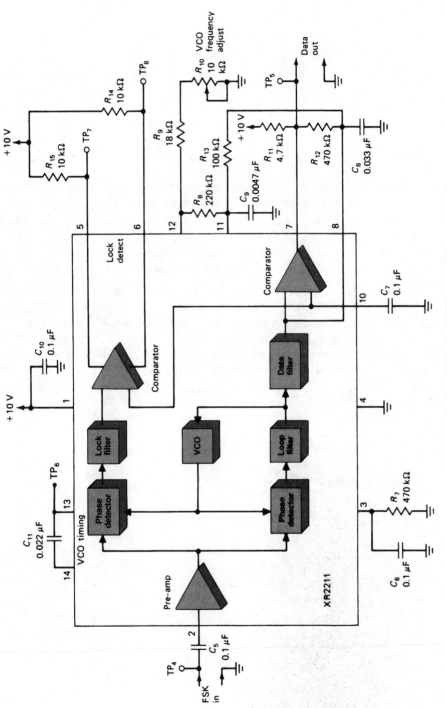

Figure 25-5 XR2211 FSK demodulator stage

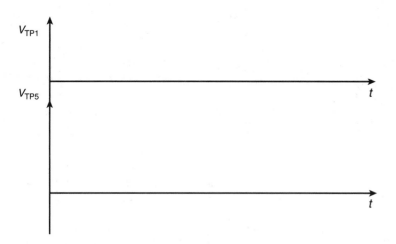

REPORT/QUESTIONS:

1. Calculate the maximum baud rate that the digital communications link could handle without any errors resulting.

2. In your own words, write a brief theory of operation section for the digital communications link analyzed in this experiment. Assume that this system is being sold as a product to the technical public and that this theory section is to be incorporated as part of a manual for the product.

3. For the manual described in question 2, write a technical test procedure for calibration of the FSK encoder for generation of 2025/2225 Hz FSK. Also, write a test procedure for calibration of the FSK detector for successful detection of 2025/2225 Hz FSK. Assume that this procedure is to be read and followed by a bench technician who has a function generator, oscilloscope, and power supply at the test bench.

DTMF SIGNALING TECHNIQUES

OBJECTIVES:

1. To become familiar with dual-tone multifrequency signals used in telephone systems that use touch-tone keypads.

2. To build and test a DTMF encoder using a TP5089 integrated circuit and a matrix keyboard.

3. To build and test two types of DTMF decoder circuits.

REFERENCE:

Refer to Section 11-2 of the text.

TEST EQUIPMENT:

Dual-trace oscilloscope

Function generator (2)

Low-voltage power supply

COMPONENTS:

Matrix keypad (3 by 4 or 4 by 4)

Integrated circuits: TP5089, 567 (2), 145436, CA4051

Resistors: 330 Ω, 470 Ω (8), 4.7 kΩ, 8.2 kΩ, 18 kΩ (2), 1 MΩ

Capacitors: 0.1 μF (4), 0.33 μF, 0.47 μF, 1 μF (2), 4.7 μF, 10 μF

Potentiometers: 5 kΩ (2)

Crystal: 3.579545 MHz, (TV color-burst, parallel resonant, C = 18 pF) (2)

LEDs (7)

THEORY:

Most telephones today use a keypad rather than a rotary dial for entering the phone number of the telephone to be called. This system uses twelve pushbuttons configured in a 4 row by 3 column matrix to represent the numbers "0" through "9" and the symbols "*" (star or asterisk) and "#" (pound or number). An alternate 4 row by 4 column matrix can also be used in some systems to add four additional push-buttons: "A", "B", "C", and "D."

The keypad is interconnected to a system of seven or eight audio oscillators, of which two (one row oscillator and one column oscillator) are enabled when a given key is depressed. These frequencies, which are intentionally harmonically unrelated, are shown in Figure 26-1. Thus, the output signal of this DTMF (dual-tone multifrequency) encoder is a composite signal containing each of these two frequencies. The TP5089 is a popular integrated circuit that creates each of these eight possible frequencies and adds the proper row and column frequencies together when a particular key is depressed. It creates these frequencies from a master oscillator that uses a popular 3.579545 MHz TV color-burst crystal, which must be connected to two of its pins in order for it to work properly. In Part I of this experiment, this encoder circuit is built and tested.

The central office or local PBX of the telephone network must be able to decode this DTMF signal in order to provide proper connections to the called party. Two such DTMF decoder circuits are built and tested in parts II and III of this experiment. One design uses a popular 567 phase-locked loop tone decoder integrated circuit. By following a cookbook design procedure, the proper resistor and capacitor values are determined to make the 567 respond to a pair of DTMF frequencies. All other combinations of frequency pairs are ignored by the 567. The drawback to this design is the excessive number of 567 integrated circuits needed to decode each of the twelve or sixteen keys on a keypad. The other design uses a 145436 integrated circuit that, like the TP5089, utilizes a master oscillator that uses a 3.579545 MHz TV color-burst crystal to tune itself to each of the possible combinations of DTMF frequency pairs that could be present for successful decoding to take place. The advantage of this design over the 567 design is that only one 145436 integrated circuit is needed.

PROCEDURE:

Part I: The TP5089 Encoder

1. Build the DTMF encoder circuit given in Figure 26-1. Make sure that the keypad that you are using is of the proper design. When a key is depressed, the row and column pins relating to that particular key should not only be

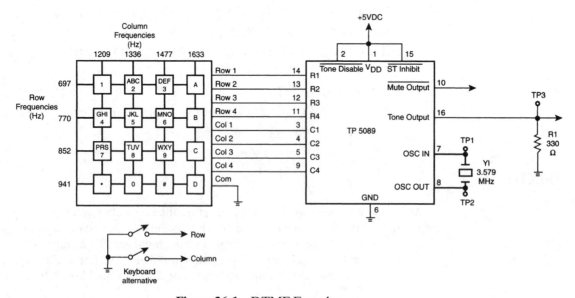

Figure 26-1 DTMF Encoder

connected together but also be connected to the common pin of the keypad. If this doesn't happen, you have the wrong type of keypad and may need to use the alternate switch arrangement shown in Figure 26-1 in order to produce the same results.

2. Make sure that pins 1, 2, and 15 of the TP5089 are tied to +5 V dc. Energize the circuit and connect the oscilloscope to TP1. Depress any key on the keypad. Measure the amplitude of the 3.579 MHz signal being produced by the crystal. Repeat for the signal at TP2. Note that the oscillator is enabled only when a key is depressed.

3. Depress the "1" key on the keypad. This should cause the signals at pins 3 and 14 of the 5089 to switch to a digital low state. Verify this. Now observe the waveform at TP3. It should be a complex waveform that is impossible for your scope to trigger on. This is because the waveform is the summation of two sine waves: one at 697 Hz (row frequency) and one at 1209 Hz (column frequency). These two frequencies are not harmonically related. Measure the amplitude of this signal and attempt to sketch the waveform as accurately as possible. (If a storage scope is available, use it to capture a few cycles of this waveform.)

4. Notice that it is possible to observe each of the row and column sinusoidal signals separately by depressing two keys in the same row or in the same column. Now the encoder will produce a single sinusoidal signal pertaining to the row or column that both depressed keys are in. This will make it possible for the scope to trigger on the measured sine wave. Verify that the DTMF encoder is putting out the proper frequencies by measuring each row and column frequency in this manner. Sketch the waveform that results when the "1" and "4" keys are depressed at the same time.

First row frequency   = _____   First column frequency   = _____

Second row frequency = _____   Second column frequency = _____

Third row frequency   = _____   Third column frequency   = _____

Fourth row frequency = _____   Fourth column frequency = _____

5. Set this circuit aside for now. It will be used as a DTMF signal source in later steps of the procedure.

Part II: DTMF Decoder Using 567 PLL Tone Decoders

6. Build the DTMF decoder circuit given in Figure 26-2. Energize the circuit and apply a 560 mVp-p sine wave at TP4. Slowly adjust the generator frequency from 500 Hz to 2 kHz and observe the dc voltage level at TP5. It should start out in the digital high-state but switch to an active low-state at frequencies near 852 Hz. Then above 852 Hz, it should switch back to a high-state. If it doesn't do this, adjust the potentiometer, R3, until 852 Hz is in the middle of this narrow band of frequencies where TP5 is in the low-state.

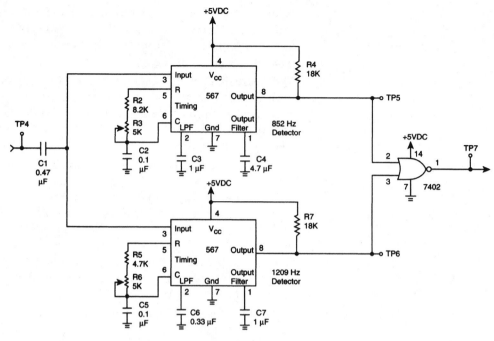

Figure 26-2 DTMF Decoder

7. Measure the final two frequencies at which the dc voltage at TP5 switches states. Also, check and make sure that it doesn't false trigger at other frequencies.

Frequency (low end of band) = _____

Frequency (high end of band) = _____

8. Now observe the dc voltage at TP6. Again, carefully adjust the generator frequency from 500 Hz to 2 kHz. You should again see switching action near the column frequency of 1209 Hz. Adjust the potentiometer, R6, until 1209 Hz is in the middle of a narrow band of frequencies where TP6 is low.

9. Measure the final two frequencies at which the dc voltage at TP6 switches states. Again, check and make sure that it doesn't false trigger at other frequencies.

Frequency (low end of band) = _____

Frequency (high end of band) = _____

10. The decoder circuit is now calibrated in order to respond to the DTMF tones produced when a "7" key is depressed. Disconnect the generator from TP4. Connect the output of the encoder circuit of Figure 26-2 at TP3 to the input of this decoder circuit at TP4. Monitor the output of the decoder at TP7. The dc level of the output voltage at TP7 should switch from a low state to a high state when the "7" key is depressed. Any other keys being depressed should have no effect on the voltage at TP7. Determine if the decoder is working. Once the circuit has been verified to be working properly, disconnect the encoder output at TP3 from the decoder input at TP4 before continuing on to the next step.

11. Build the DTMF decoder circuit given in Figure 26-3. Energize the circuit and connect the scope to TP9. Measure the amplitude of the 3.579 MHz signal being produced by the crystal.

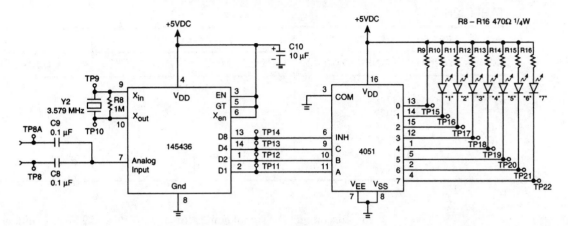

Figure 26-3   DTMF Decoder

12. Check the logic levels of each of the outputs of the decoder at TP11, TP12, TP13, and TP14. They all should be at a logic low state. Check the "0" output of the demultiplexer at TP15. It also should be low, while all the outputs, TP16-TP22, remain high. Verify this.

13. Set one generator to produce a 560 mVp-p 697 Hz sine wave. Set a second generator to produce a 560 mVp-p 1209 Hz sine wave. Connect the first generator to TP8 and the other one to TP8A. Observe the waveform at pin 7 of the decoder. It should be the complex waveform that was hard to trigger, observed in step 3. It should cause the decoder to respond as if a "1" key was depressed on a matrix keyboard connected to a DTMF encoder. Thus, the "1" LED should light. Verify this.

14. Determine the exact upper and lower frequency limits of each of the generators that cause the "1" LED to light.

Generator #1: minimum frequency = _____   maximum frequency = _____

Generator #2: minimum frequency = _____   maximum frequency = _____

15. Set the frequency of each of the two generators to produce each of the pairs listed in Table 26-1. For each of these settings determine the logic states of the voltages at TP11, TP12, TP13, and TP14 and enter the results in Table 26-1. Also, list in the table which LED lights up.

TABLE 26-1

FREQUENCY (Hz)		LOGIC LEVEL				
GENERATOR #1	GENERATOR #2	TP11	TP12	TP13	TP14	ENABLED LED
697	1209					
697	1336					
697	1477					
770	1209					
770	1336					
770	1447					
852	1209					

16. Disconnect the signal generators from TP8 and TP8A. Connect the output of the encoder built in Part I of this experiment to the input of this decoder. Specifically, connect TP3 of the encoder to TP8 of the decoder. Try depressing each of the keys of the keypad, one at a time. Verify that each of the keys 1, 2, 3, 4, 5, 6, and 7 causes its respective LED to light. Also, verify that keys 8, 9, 0, *, #, and A, B, C, and D (if you have a 4 × 4 matrix keypad) do not cause false lighting of the seven LEDs.

REPORT/QUESTIONS:

1. Refer to the 567 design section of this experiment. Follow the design procedure in order to design a DTMF decoder using two 567 integrated circuits that detects when a "5" key is being depressed. Show all final values for R1, C1, C2, and C3 for each of the tone detectors.

2. How many 567 tone detector integrated circuits would be required to detect every one of the sixteen keys on a 4 × 4 matrix keypad? How many 7402 NOR-gates would be required?

3. Add some additional circuitry (another 4051 and an inverter) to the 145436 decoder to enable an LED to light when each of all sixteen keys on a 4 × 4 matrix is depressed. In other words, include LEDs for the 8, 9, 0, *, #, A, B, C, and D keys, in addition to the seven LEDs, already designed in Figure 26-3. Include a schematic drawing of your final design.

Part IV: Tone Decoding

To make the 567 PLL respond to one of the eight possible oscillator frequencies used in DTMF encoding, three capacitors (C1, C2, and C3) and one resistor (R1) must be determined. These externally connected components are shown with the 567 in Figure 26-4. R1 and C1 are the timing components that determine the natural frequency of oscillation, f_o, for the PLL:

$$f_0 = \frac{1.1}{R1\,C1} \qquad (26\text{-}1)$$

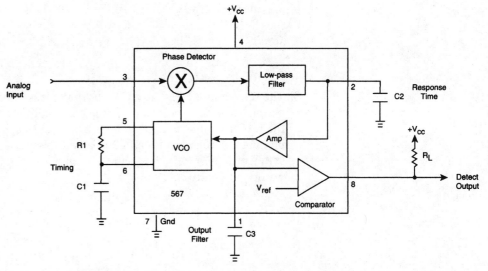

Figure 26-4 Tone Detector Using 567

This frequency, f_o, is the frequency that the 567's VCO will run at in the absence of an input signal. This is also the frequency applied at pin 3 that will cause the PLL to go into the locked state. In this application, this would be one of the eight possible DTMF frequencies. The bandwidth, BW, is the range of frequencies centered about f_o that will cause a lock-up state to result. It is found by:

$$BW = 1070\sqrt{\frac{(Vin)}{f_o C2}} \qquad (26\text{-}2)$$

where Vin<200 mV rms and C2 is in μF.

The following design shows a 567 as an 852 Hz tone detector:

1. The center frequency of 852 Hz is due to the proper selection of R1 and C1. R1 should be between 2 kΩ and 20 kΩ. If C1 is picked to be 0.1 μF, then:

$$R1 = \frac{1.1}{(852)(0.1\mu F)} = 12.9 \text{ k}\Omega, \text{ which is in the desired range.}$$

2. A study of the spacing between adjacent DTMF frequency values reveals that the BW needs to be less than 2% of f_o to avoid overlap of the BW ranges, which could result in false triggering. In other words, the PLL could detect an adjacent DTMF frequency if the BW was set too high. Thus, to provide a little safety margin, the BW is picked to be 2% of f_o. The input voltage is assumed to be 200 mV rms:

$$2\% f_o = 1070\sqrt{\frac{Vin}{f_o C2}}$$

$$.02(852)f_o = 1070\sqrt{\frac{0.2}{852\ C2}}$$

Solving for C2, we get:

$$C2 = 1.080 \ \mu F$$

So let C2 = 1 μF.

3. The selection of C3 is easy. C3 must be at least twice C2 to avoid false triggering from frequencies outside the bandwidth of the PLL. If C3 is too large, the response time will be too high, producing sluggish operation. C3 should be kept less than ten times C2 to avoid this. So let C3 = 4.7 μF. These resulting values of R1, C1, C2, and C3 can be seen in the 852 Hz tone detector portion of Figure 26-2.

TIME-DOMAIN REFLECTOMETRY

OBJECTIVES:

1. To determine the delay time of a delay line.
2. To observe the effects of open, shorted, and matched terminations on the input and output ends of a transmission line.
3. To design a minimum-loss unbalanced matching pad.

REFERENCE:

Refer to Chapter 12 of the text.

TEST EQUIPMENT:

Dual-trace oscilloscope
Pulse generator

COMPONENTS:

Inductors: 1.0 mH (21)
Capacitors: 0.01 μF (20), 0.001 μF
Resistors: selected design values from Figs. 27-7 and 27-8 for R_1 and R_2 of the minimum loss pad

THEORY:

Any transmission line can be analyzed by drawing equivalent circuits using "lumped" values of impedances, which represent sections of the line. If we consider the entire transmission line to have an infinite number of these sections, we approach an accurate representation of the transmission-line characteristics. Each section of lumped impedances, in its simplest form, consists of a set of two series inductances of equal value and a shunt capacitance, such as shown in Fig. 27-1.

As a voltage is applied at the input of this section, the inductive and capacitive properties act to delay the arrival of this signal to the output end of this section. The amount of delay time is dependent on the values of the series inductance and shunt capacitance and, for a real piece of transmission line, its total length or number of sections. The input impedance of a given transmission line consisting of an infinite

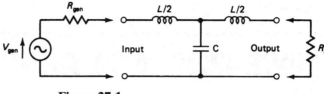

Figure 27-1

number of equivalent, cascaded sections is what is known as the characteristic impedance of the transmission line, Z_o. If the internal impedance of a signal generator connected to the input of the transmission line happens to be exactly equal to the characteristic impedance of the line, the result is maximum transfer of power from the generator to the line, certainly a desired phenomenon. Similarly, the other end of the line should be terminated with a purely resistive load exactly equal to the characteristic impedance for there to be maximum power transfer between the line and the load.

If we do not have $R_L = Z_o$, the result is what is called a mismatch. The characteristics of the inductances and capacitances of the transmission line act to restore some of the electrical energy to appear to reflect the signal back in the direction of the generator. Similarly, if we do not have $R_{gen} = Z_o$, any reflected energy returning to the generator end of the line can be partially reflected back in the direction of the load. Thus, when the line is not matched to the generator or the load, we can end up with multiple reflections or a "Ping-Pong game" effect. Fortunately, these multiple reflections eventually die away, since the series inductors always have a small amount of resistance in them, which causes the signal to be attenuated each time it travels down the transmission line.

If the load end of the transmission line is either open- or short-circuited, we find that all the energy is reflected by the load and none of it is dissipated as heat at the load. If R_L does not equal Z_o, and the load is not a short or an open circuit, we find that some of the energy is dissipated as heat by the load, and some of the energy is reflected back toward the generator. Finally, if the line and load are matched and the load is purely resistive, we find that all the energy is dissipated by the load as heat.

An interesting technique that illustrates this principle of reflection is to send a pulse of very short duration down the transmission line. Of course, its pulse width must be smaller than the delay time of the line. Otherwise, it is difficult to distinguish between the original incident pulse and the reflected pulse, since they algebraically add together if they occur at the same time. If the time duration of the pulse is kept small enough, separate incident and reflected pulses can be observed on the oscilloscope. This visual technique is known as time-domain reflectometry. In this experiment you will use this technique to study what kind of reflections actually occur in various mismatched and matched states. Waveforms will be observed as shown in Fig. 27-2.

The transmission line to be investigated in this experiment is not really a transmission line but rather a circuit that behaves like an actual transmission line (Fig. 27-3). The only difference is that the values for L and C in each lumped section are much larger than those that really exist in an actual transmission line. Also, in actual transmission lines these inductances and capacitances are evenly distributed throughout the cable rather than being lumped together in specific locations. This is designed intentionally so that the resulting time delays are large enough to measure on standard bench oscilloscopes. Also, it keeps the effects of line attenuation to a minimum. Other than these

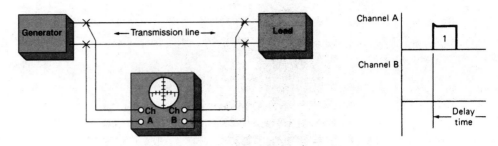

Pulse 1: Incident (original) pulse at the transmission line input.
Pulse 2: Reflected pulse at the transmission line input.
Pulse 3: Incident plus reflected pulse at the transmission line output.
They occur at the same time, so they add together.

Figure 27-2   Time-domain reflectometry test setup

differences, it is important to realize that the behavior of the simulator circuit of Fig. 27-3 is very close to that of an actual transmission line.

Time-domain reflectometry is used as a standard test procedure in detecting faults in transmission lines. It is especially useful in cases where it is difficult to inspect the transmission line visually for obvious physical defects. This includes troubleshooting buried cables or cables that are wound on a spool. Not only can the type of fault be determined, but the approximate location of the fault can be determined.

PRELABORATORY:

Refer to the test circuit diagram in Fig. 27-6 and the design notes for the matching networks of Figs. 27-7 and 27-8. Design a minimal loss pad that will match the internal impedance of the function generator to the simulated transmission line that has a characteristic impedance of 300 Ω.

PROCEDURE:

1. Show your prerequisite homework design values to the lab instructor for verification that they are correct before you proceed.

2. Assemble the test configuration as shown in Fig. 27-6. The function generator should be set up to produce a 2-V positive-going pulse of 50-μs duration with a period of 1 ms.

3. Using channel A of the oscilloscope, observe the pulse at TP_2 and adjust the load resistor, R_L, to produce a waveform that has a flat, zero volt baseline, such as that shown in Fig. 27-4. You are matching the load to the transmission line.

4. With the oscilloscope in the dual-trace mode, observe the voltage at TP_2 with channel A and the voltage at TP_3 with channel B. Sketch the two pulses as accurately as possible, making sure that their amplitudes, periods, and relative positions with respect to each other are clearly visible. (Do not forget to record the delay time between incident and reflected pulses).

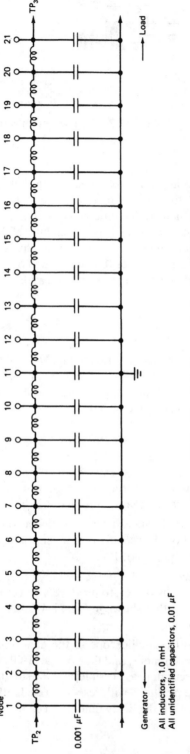

Figure 27-3 Simulated transmission line (delay line)

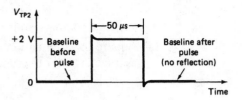

Figure 27-4

5. Disconnect the load resistor from the end of the line. Set the scope so that the incident (original) voltage and the reflected voltage at TP_2 can be observed. Accurately sketch these pulses along with the pulse at TP_3. Again, make sure that all amplitudes, periods, and relative positions are clearly visible, such as those shown in Figure 27-2.

6. Short circuit the load resistor and repeat step 5.

7. Remove the matching pad and try driving the delay line directly with the generator. Repeat steps 3 through 6 and note exactly when multiple reflections occur. Use a larger horizontal time/div setting so that you may detect these multiple reflections. Make sure that you measure the amplitudes of the first three pulses that occur at both ends if there are multiple reflections.

8. Reconnect the matching pad and increase the pulse width to 250 μs.

9. Repeat step 5 using an open-line termination and a 250-μs pulse width.

10. Return the pulse width back to 50 μs. Now attach a 600-Ω load to the transmission line and record the resulting waveforms at TP_2 and TP_3.

REPORT/QUESTIONS:

1. What is the measured value of the simulated transmission line's delay time for a one-way trip? Use any of your sketched waveforms to determine this.

2. Calculate the theoretical value for the delay time determined in question 1 using the nominal values of L and C in each lumped section and the following equation:

$$t_d = N\sqrt{LC}$$

where N is the number of cascaded lumped sections in the simulated transmission line. Compare the theoretical and measured values.

3. Briefly explain each of the sketches drawn in steps 4–6.

4. Calculate the attenuation factor for the transmission line. Remember that the waveform observed at TP_3 is the sum of the amplitudes of the incident and reflected pulses, since they occur at the same time.

$$\alpha = \text{attenuation} = \frac{E_{\text{incident}} \text{ at } TP_3}{E_{\text{incident}} \text{ at } TP_2}$$

5. Calculate the amount of attenuation that a signal would undergo for a two-way trip, that is, down and back again, assuming a total reflection at the load

end of the line. Use both of the following methods in determining this quantity and then compare your results.

(a) Square the attenuation for a one-way trip. This is valid since the signal is traveling over the same line on the return trip.

$$\text{Two-way attenuation} = \alpha^2$$

(b) Using the open termination data gathered in step 5, calculate the ratio of the reflected voltage at TP_2 to the incident voltage at TP_2.

$$\text{Two-way attenuation} = \frac{E_{\text{reflected}} \text{ at } TP_2}{E_{\text{reflected}} \text{ at } TP_2}$$

6. Calculate Γ_{load}, the reflection coefficient at the load end of the line for each of the loaded conditions in steps 4–6.

7. Calculate the theoretical value for Γ_{gen}, the reflection coefficient at the generator end of the transmission line in step 7 when the load was short-circuited. Use the following equation:

$$\Gamma_{\text{gen}} = \frac{R_{\text{gen}} - Z_o}{R_{\text{gen}} + Z_o}$$

8. Verify that the measured amplitude of the first reflected pulse at TP_3 in step 10 is correct by use of the following procedure.

(a) Calculate the reflection coefficient at the load end of the line using

$$\Gamma_{\text{load}} = \frac{R_L - Z_o}{R_L + Z_o}$$

(b) Determine the incident signal arriving at TP_3 after making a one-way trip through the line by multiplying the original signal at TP_2 by the attenuation factor of the line:

$$E_{\text{inc}} = \alpha E_{TP_2}$$

(c) Determine the reflected signal leaving TP_3 by multiplying the incident signal by the reflection coefficient of the load end of the line:

$$E_{\text{ref}} = \Gamma_{\text{load}} E_{\text{inc}}$$

(d) Add the incident and reflected voltages at TP_3 together to determine the resulting pulse amplitude at TP_3. Since the incident and reflected pulses are happening (arriving and departing) simultaneously, what you see on the scope is the sum.

$$E_{TP3} = E_{\text{inc}} + E_{\text{ref}}$$

9. Calculate the expected amplitude of the first reflected pulse observed at the generator end of the transmission line in step 7 when the load was open circuited using the procedure given below. Compare your calculated value to the actual measured value. Why do they or don't they agree?

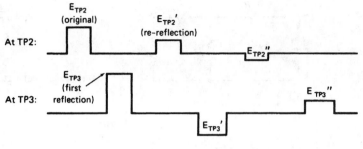

At TP2:
At TP3:

Figure 27-5

(a) Calculate the incident signal arriving back at TP$_2$ after making a two-way trip through the line by multiplying the original signal by the two-way attenuation factor:

$$E_{\text{inc}} = \alpha^2 E_{\text{TP}_2}$$

(b) Calculate the re-reflected signal leaving TP$_2$ for another trip by multiplying the incident signal by the reflection coefficient of the generator end of the line.

$$E_{\text{ref}} = \Gamma_{\text{gen}} E_{\text{inc}}$$

(c) Add the incident and re-reflected voltages at TP$_2$ together. Since they are happening (arriving and departing) simultaneously, what you see on the scope is their sum.

$$E'_{\text{TP}_2} = E_{\text{inc}} + E_{\text{ref}}$$

10. Explain the strange waveforms that resulted in step 9 when the pulse width was increased to 250 μs.

11. Write a technical test procedure using time-domain reflectometry that a lab technician could follow successfully in order to determine the value of an unknown load resistance on the simulated transmission line used in this experiment. (*Hint:* Review the procedures used in answering question 8 and make the necessary modifications.)

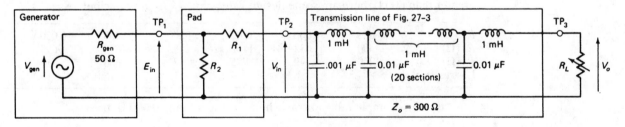

Figure 27-6 Transmission-line test circuit

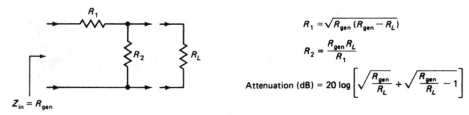

Figure 27-7

Designing Minimum-Loss Pads

For matching two impedances where R_{gen} is greater than R_L, use the circuit and design equations shown in Fig. 27-7. For example, when R_{gen} is 600 Ω and R_L is less than 600 Ω, here are some design values for R_1, R_2, and the resulting decibel loss:

R_L (Ω)	500	300	200	100	75	50	30	25
R_1 (Ω)	245	424	490	548	561	575	585	587
R_2 (Ω)	1225	425	24	110	80.2	52.2	30.8	25.6
Loss (dB)	3.8	7.6	10.0	13.4	14.8	16.6	18.9	19.7

For the preceding example table, when R_L is less than 25 Ω, use

$$R_1 = 600 - \frac{R_{gen}}{R_L}$$

$$R_2 = R_L$$

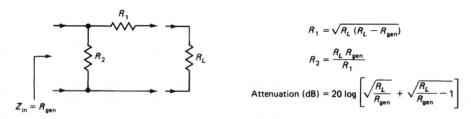

Figure 27-8

For matching two impedances where R_L is greater than R_{gen} use the circuit and the design equations shown in Fig. 27-8. For example, when R_{gen} is 600 Ω and R_L is greater than 600 Ω, here are some design values for R_1, R_2, and decibel loss:

R_L (Ω)	800	1000	1500	3000	5000	10,000
R_1 (Ω)	400	632	1162	2683	4690	9695
R_2 (Ω)	1200	949	775	671	638	619
Loss (dB)	4.8	6.5	9.0	12.5	15.0	18.1

For the example table above, when R_L is less than 10,000 Ω, use

$$R_1 = R_L - 300$$

$$R_2 = 600 \ \Omega$$

NAME _____

STANDING-WAVE MEASUREMENTS OF A DELAY LINE

OBJECTIVES:

1. To take data for plots of the voltage standing-wave patterns of a simulated transmission line terminated with open, shorted, and matched loads.
2. To use data and a Smith chart to determine the value of an unknown impedance.
3. To use ac circuit theory to determine the value of an unknown impedance.

REFERENCE:

Refer to Section 12-8 of the text.

TEST EQUIPMENT:

Dual-trace oscilloscope
Function generator
Volt-ohmmeter
Pulse generator
Impedance bridge

COMPONENTS:

Resistors: 10 Ω, 22 Ω, 180 Ω, 270 Ω, and other values needed to obtain a proper match between generator and transmission line (explained in the procedure)
Potentiometer: 500 Ω
Capacitors: 0.001 μF, 0.01 μF (20), 0.1 μF, 0.22 μF
Inductors: 1 mH (20)

THEORY:

If an impedance is given, the VSWR of a system can be obtained from a specialized graph called the Smith chart. The radius from the center of the Smith chart to the impedance (written and plotted in normalized form) can be used to determine the VSWR on the standing-wave ratio axis of the Smith chart. Conversely, if the VSWR is known, a circle can be drawn on the Smith chart that passes through the coordinates of all possible normalized impedances. Refer to Fig. 28-1.

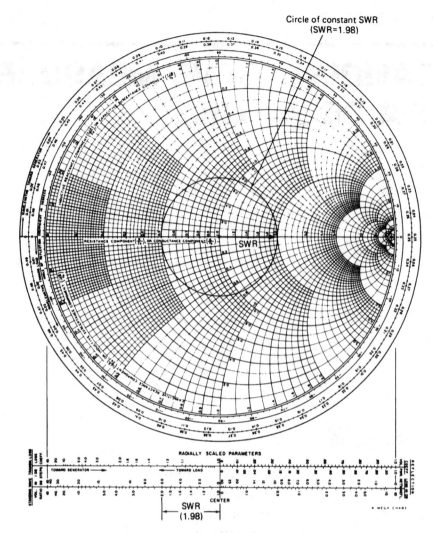

Figure 28-1   Example SWR circle

The problem then remains of determining which point on the circle represents the actual impedance. The problem can be solved by comparing the location on the transmission line of a voltage minimum referred to as a null caused by the unknown load with the location of a voltage null caused by a short.

Any mismatch on the transmission line will cause a standing wave to result. In other words, sinusoidal voltages measured at various points on the line will have different amplitudes. The change in amplitude with respect to location will exhibit a regular pattern. This pattern, which has the same shape as that of a full-wave rectifier's waveform, is referred to as the standing-wave pattern, as shown in Fig. 28-2. Keep in mind, however, that this is a plot of the magnitude of the sine wave versus distance away from the load, not versus time. Thus do not think of this pattern as a waveform itself.

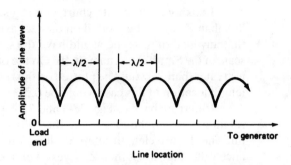

Figure 28-2 Standing wave when $R_L > Z_o$

A short circuit will cause a voltage null to occur at the load end of the line and additional nulls every $\lambda/2$ distance away from the load toward the generator. "λ" represents the wavelength of the signal being transmitted down the transmission line. It can be computed by dividing the velocity of the signal passing through the line by its frequency ($v = f\lambda$). Any purely resistive load with an ohmic value less than the characteristic impedance of the line, Z_o, will cause voltage nulls and peaks to occur at the same locations as those observed when the load is a short, but with different amplitudes. Refer to Fig. 28-3.

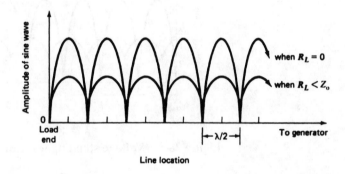

Figure 28-3 Voltage standing wave of a short when $R_L < Z_o$

An open circuit will cause a voltage peak at the load end of the line, a null at a location $\lambda/4$ down the line, and additional nulls at locations every $\lambda/2$ after that. Any purely resistive load with an ohmic value greater than Z_o will cause voltage nulls and peaks to occur at the same locations as those observed when the load is an open circuit, but with different amplitudes. Refer to Fig. 28-4.

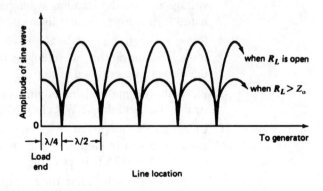

Figure 28-4 Voltage standing wave when $R_L > Z_o$

Looking at the Smith chart, you can see that any resistive load quantity that is less than Z_o would be plotted on the real axis between zero and 1.0. Also, you can see that any load of this type would have the same reference location on the wavelength scale of the Smith chart, namely, 0λ. On the other hand, any resistive load quantity that is greater than Z_o would be plotted on the real axis between infinity and 1.0. The wavelength location in this case would be 0.25λ.

Referring back to Figs. 28-3 and 28-4, you can observe that there is a direct correlation between the location of the voltage standing-wave nulls on the transmission line and the wavelength reference locations on the Smith chart. Any resistive impedance value that is less than Z_o gives a voltage null at location zero on the transmission line and a reference location 0λ on the Smith chart. Any resistive impedance value that is greater than Z_o yields a voltage null at a location $\lambda/4$ away from point zero on the line and is also plotted at the reference location 0.25λ on the Smith chart. Thus, it can be deduced that any complex impedance, either inductive or capacitive, will cause voltage nulls to fall at line locations other than 0λ or $\lambda/4$ with respect to nulls caused by a short. Refer to Fig. 28-5. The location of these voltage nulls and the VSWR can be used to determine the actual value of an unknown load impedance. The following steps below summarize the proper procedure to be used.

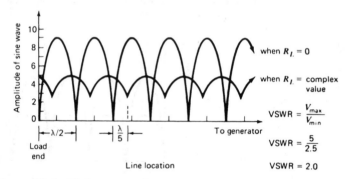

Figure 28-5   Voltage standing wave of a short and complex load

1. From the maximum and minimum voltage amplitude data, calculate the VSWR. (VSWR = 2.0 in example)

2. Draw the VSWR circle on the Smith chart using the calculated value of the VSWR to set the radius of the circle.

3. Find the location of a voltage null using the standing wave pattern for the transmission line terminated with the unknown load. Determine the distance in wavelengths toward the load (clockwise) to the first encountered voltage null on the standing wave pattern for the transmission line terminated with a short circuit. (distance = $\lambda/5$ in example)

4. Draw a straight line on the Smith chart from the center of the chart to the same location on the "wavelengths toward the load" scale that represents the distance determined in step 3.

5. Read the normalized impedance of the load at the intersection of the straight line and the VSWR circle. ($Z_{Ln} = 1.55 - j0.65$ in example)

6. Calculate the unknown load impedance by multiplying the normalized value obtained in step 5 by the characteristic impedance of the line, Z_o. ($Z_L = 77.5 - j32.5\ \Omega$ in example)

Refer to Fig. 28-6 for the solution to the example problem given in Fig. 28-5.

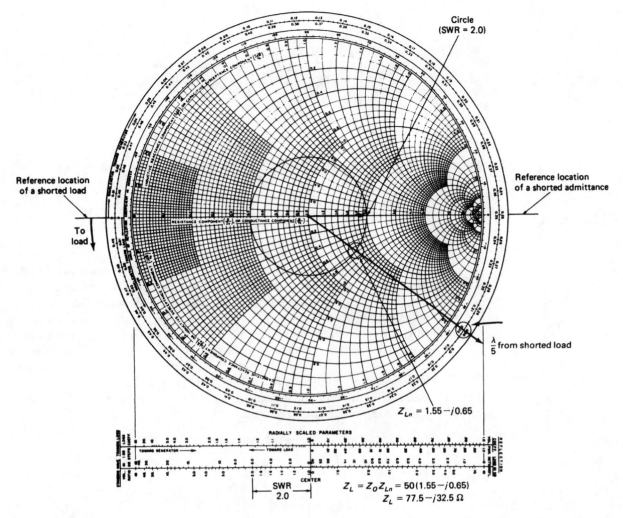

Figure 28-6 Use of the Smith chart in determining the value of R_L with $Z_o = 50\ \Omega$ using complex voltage

The distance between voltage nulls also provides a convenient way to calibrate the transmission line in units. The distance between voltage nulls must always be $\lambda/2$. Matching of a variable resistive load can also be achieved by using the short circuit data if the transmission line has negligible losses. The location of a short-circuit peak and null voltage can be monitored simultaneously. The load can then be adjusted until the voltage at the peak location is equal to the voltage at the null location. Under these conditions, the voltage everywhere on the line must be equal and the line must be matched. The sinusoidal voltage observed at the peak and null locations is $\lambda/4$ or 90° out of phase. Line loss will introduce some error into these impedance matching and measuring techniques. The effect of line loss can be observed in this experiment, but it will have minimal effect on the impedance measurement. Refer to Fig. 28-7 to observe the effects of line loss on the standing-wave pattern.

PRELABORATORY:

Using the given values of the components within the complex load of Fig. 28-8 and the given frequency of 10 kHz, calculate the theoretical value for the complex impedance and express in rectangular coordinates.

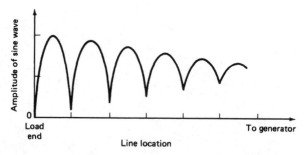

Figure 28-7 Voltage standing wave of a short on a line containing a high resistive loss

PROCEDURE:

1. Construct the test configuration as shown in Fig. 28-9. Short the load end of the transmission line. Apply a 50-μs pulse width, 2.0-ms period pulse to the input of the transmission line. Adjust the matching network as necessary to obtain a match at the generator end of the transmission line. This is done by adjusting resistors R_1, R_2, and R_3 to eliminate any re-reflections from occurring at the generator end of the line. Measure and record the values of R needed for a matched impedance condition. The values of R_1, R_2, and R_3 given in Fig. 25-8 are for a 50-Ω generator. If your generator has other than a 50-Ω impedance, other values of R_1, R_2, and R_3 will have to be used. In all cases, the pulse generator and function generator must have the same internal impedance for this procedure to be useful.

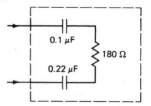

Figure 28-8 Complex load

2. Disconnect the pulse generator and connect the function generator in its place. Apply a 10-kHz sinusoidal voltage of convenient amplitude to the circuit of Fig. 28-9. Check the frequency with a frequency counter. Connect the potentiometer to the load end of the line and match it to the line. Do this by following the procedure given at the end of the theory section of this experiment. With the load matched, measure each of the 21 node voltages and record all measured values in Table 28-1. Also, disconnect the potentiometer and measure its matched resistance value. It should be close to 300 Ω.

3. Open the load end of the transmission line. Measure the amplitudes of each of the sinusoidal waveforms observed at the 21 nodes on the line and record in Table 28-1.

4. Short the load end of the line. Again, measure each of the 21 node voltages and record in Table 28-1.

5. Connect the complex load (Fig. 28-8) to the transmission line. Again, measure each of the 21 node voltages and record values in Table 28-1.

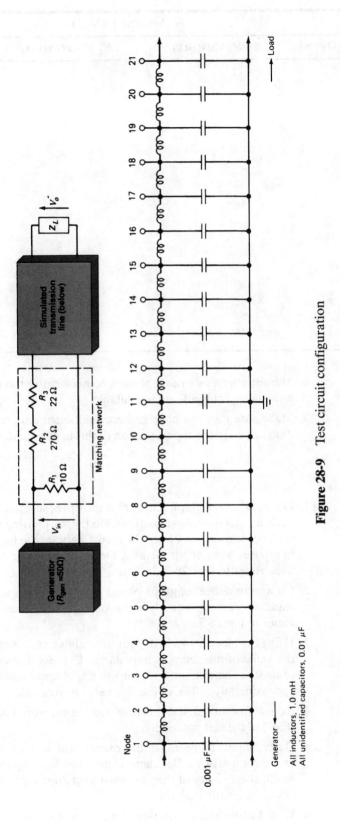

Figure 28-9 Test circuit configuration

TABLE 28-1

| NODE | VOLTAGE (V_{P-P}) | | | |
	R_L (OPEN)	R_L (SHORT)	R_L (MATCHED)	R_L (UNKNOWN)
1				
2				
3				
4				
5				
6				
7				
8				
9				
10				
11				
12				
13				
14				
15				
16				
17				
18				
19				
20				
21				

6. Measure the exact value of each component in the complex load using an impedance bridge if one is available.

7. Make sure that you have gathered sufficient data to complete each of the items given in the report section of this experiment.

REPORT/QUESTIONS:

1. On one sheet of graph paper, plot the open, shorted, and matched-load-impedance standing-wave patterns. Do this by plotting voltage amplitude versus line location away from the load. Calibrate the location axis in nodes and in quarter-wavelength sections. Draw all three patterns on the same set of axes. Refer to Figs. 28-2, 28-3, and 28-4.

2. On a second sheet of graph paper, plot the shorted and complex impedance standing-wave patterns. Use the same plotting technique as in question 1. Refer to Fig. 28-5.

3. Using the plots drawn in the question above and the Smith chart, determine the value of the complex impedance. Express this value in rectangular coordinates. The characteristic impedance of the simulated transmission line is approximately 300 Ω, so use this value in your calculations.

4. Repeat question 3, but this time use the measured Z_o value determined in step 2 of the test procedure.

5. Describe fully all results of this experiment, with particular attention to relevant comparisons. Be numerically specific. In particular, compare your Smith chart values of the unknown load impedance to the calculated value based on circuit analysis.

6. What factors in this experiment may have led to discrepancies in the resulting values of the complex load impedance? Which value do you feel is the most accurate? Why?

FIBER OPTICS COMMUNICATION LINK

OBJECTIVES:

1. To become familiar with fiber optic mounting and fabrication procedures.

2. To build and test a fiber optic transmitter and receiver.

3. To evaluate three types of fiber optic emitters.

4. To evaluate two types of fiber optic detectors.

REFERENCE:

Refer to Chapter 18 of the text.

TEST EQUIPMENT:

Dual-trace oscilloscope

Function generator

Low-voltage power supply (2)

Hot cutting knife

COMPONENTS:

Integrated circuits: LM5534 (2), LM386-3 (2)

Resistors: 10 Ω (2), 12 Ω (2), 27 Ω, 100 Ω—1 W (2), 1 kΩ, 2.2 kΩ, 10 kΩ, 100 kΩ, 220 kΩ

Potentiometer (10-turn trim): 1 kΩ (2), 20 kΩ

Capacitors: 500 pF, 0.001 μF, 0.01 μF, 0.1 μF (4), 10 μF (4), 470 μF (4)

Microphone: Panasonic P-9930 miniature cartridge

Speaker: 8 Ω

Fiber optic emitters: Motorola MLED-71, MFOE-71, low-profile red LED (e.g., HLMP-3200)

Fiber optic detectors: Motorola MRD-721, MFOD-71 (2)

Fiber optic cable (plastic): three 8-ft lengths of AMP 501232

Fiber optic plugs: AMP 228087-1 (3)

Fiber optic device mounts: AMP 228040-1, AMP 228709-1 (2)

Prepare three 8-ft fiber optic cables with terminations as shown in Fig. 29-1.

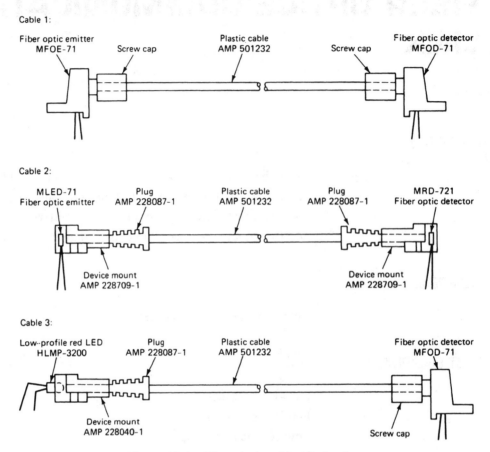

Figure 29-1 Fiber optic cable fabrication

For minimal attenuation between transmitter and receiver it is essential that each of the cable ends be severed with a smooth surface. Thus a simple cut with a pair of diagonal cutters is undesired. A recommended procedure to follow is to use a hot cutting knife. A plastic cutting knife such as the Weller SP23 HK performs this task quite well. If this is not available, simply heat up an X-Acto knife with a lighted match. Apply a straight perpendicular cut with a firm downward force. Do not allow excessive burning or melting of the dark coating of the cable to occur.

To fabricate cable 1, first remove the caps of the MFOD-71 and MFOE-71. Cut the cable at both ends as described above. Mount the screw caps onto the cable. Insert the cable into the mating mount as far as it will go. Then screw the caps into the mount. This should apply a sufficient press fit onto the cable as the caps are hand-tightened.

To fabricate cable 2, the cable is cut at both ends as described above. The black jacket of the plastic cable must be stripped away from the last ½ in. of the cable at both ends. When using wire strippers to do this, be careful not to score the outer surface of the clear plastic fiber in the center of the cable. Now, insert the cable into the plugs. Needle-nosed pliers should be used to force the cable into the plugs. The ends of the cable should line up at the edge of the plug at approximately the same position where the cable cannot be inserted any farther into the plug. Now insert the plugs at each end of the cable into their respective mating device mounts. At the transmitter end of the cable, insert the MLED-71 emitter into the rectangular opening at the base of the device mount. Make sure that the LED is pointed toward the cable (white backing away from the cable). At the receiver end of the cable, insert the MRD-721 detector into the rectangular opening at the base of the device mount. Again, make sure that the detector is pointed toward the cable (blue backing away from the cable).

To fabricate cable 3, at the transmitter end of the cable follow the same procedure as in cable 2 for mounting the cable to the plug and mounting the plug to the device mount. The device mount is a different part number since it has a TO-92 opening rather than a rectangular opening for the LED. Carefully glue the LED into the TO-92 opening. Use 5-minute epoxy. Do this by carefully applying a small amount of epoxy to the sides of the LED, being careful not to smudge the top surface of the LED's lens. Insert the LED into the TO-92 opening. Let the bottom surface of the LED stick out approximately 0.1 in. away from the bottom surface of the device mount. This will ensure that the plug does not push into the top of the LED when inserted into the mount. At the receiver end of the cable, follow the same procedure that was used in cable 1 for mounting the detector to the cable.

PROCEDURE:

1. Build the fiber optic transmitter given in Figure 29-2. Use the MFOE-71 of cable 1 as the fiber optic emitter. Make sure that C_2 is placed as close to pins 4 and 6 of the 386-3 as possible. This will make sure that the amplifier does not break into oscillations. Apply ± 10 V dc to the circuit. Adjust R_6 for approximately 100 mA of current through the MFOE-71.

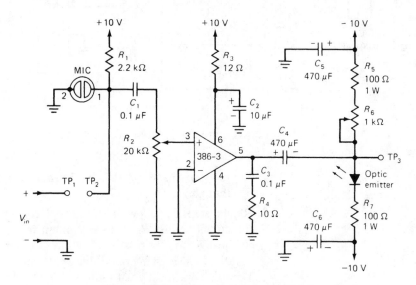

Figure 29-2 Fiber optic transmitter*

2. Connect a jumper between TP_1 and TP_2. Apply a 60-mV$_{p-p}$, 5-kHz sine wave as V_{in}. Adjust R_2 to maximize the amplitude of the voltage at TP_3. Measure

*Original design and testing of these fiber optic circuits were made by William J. Mooney and Scott Smith of the Optical Engineering Technology Department at Monroe Community College.

the voltage at TP_3. Calculate the decibel voltage gain of the 386-3 transmitter amplifier using

$$A_{V(dB)} = 20 \log \frac{V_{TP3}}{V_{TP1}}$$

3. Determine the 3-dB bandwidth of the amplifier stage by finding the frequencies at which V_{TP3} drops 3 dB below its maximum value. The bandwidth should be quite large. Do not disassemble this circuit before proceeding.

4. Build the fiber optic receiver given in Fig. 29-3. Again, make sure that capacitors C_7, C_9, C_{12}, and C_{13} are placed as close to the integrated circuits as possible so that oscillations will not occur. Use the MFOD-71 as the fiber optic detector. Leave TP_6 disconnected from TP_7.

5. The transmitter and receiver stages should be linked together with cable 1. Apply a 60-mV$_{p-p}$, 5-kHz sine wave as V_{in} to the transmitter. Measure V_{TP6}. Adjust R_{12} so as to force V_{TP6} to be exactly 6 V$_{p-p}$.

6. Record V_{TP1}, V_{TP3}, V_{TP4}, V_{TP5}, and V_{TP6}.

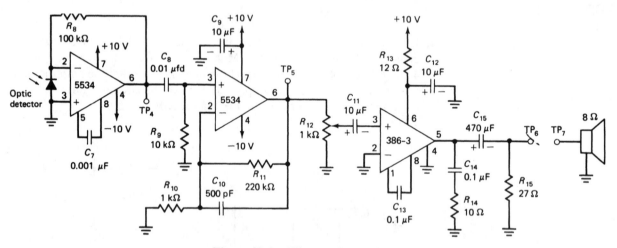

Figure 29-3 Fiber optic receiver*

7. Repeat steps 5 and 6 with cable 1 replaced with cable 2. (Don't readjust for 6 V though.)

8. Repeat steps 5 and 6 with cable 2 replaced with cable 3. Before reenergizing the circuit, adjust R_6 to its maximum resistance. Energize the circuit and readjust R_6 for approximately 20 mA of current through the LED. Notice that the LED is giving off visible light, whereas the emitters used with cables 1 and 2 work mostly with infrared light, which is not visible. After making these measurements, temporarily remove the screw cap from the receiver end of the cable. You should be able to see visible light at the receiver end of the cable. Restore the screw cap on the end of the cable before proceeding.

*Original design and testing of these fiber optic circuits were made by William J. Mooney and Scott Smith of the Optical Engineering Technology Department at Monroe Community College.

9. Replace cable 3 with cable 1. Readjust R_6 for 100 mA of current flow. Disconnect the function generator from the transmitter by disconnecting the jumper between TP_1 and TP_2. Repeat steps 5 and 6. Connect a jumper between TP_6 and TP_7 of the receiver in order to connect the speaker at the output. Whistle into the microphone and you should be able to hear the whistle in the speaker. Adjust R_{12} so that the output signal is at a comfortable listening level. Too large an output voltage may cause feedback to occur between the speaker and the microphone, resulting in an undesirable squeal produced by the speaker.

10. Try each of the three fiber optic cables. Determine by listening to the speaker which of the cables offers minimal attenuation between transmitter and receiver. Do not forget that R_6 needs to be increased to limit the current to 20 mA in the LED of cable 3.

OPTIONAL PROJECTS:

1. Determine how long each of the three cables can be made before there is so much attenuation between the transmitter and receiver that the output signal at the speaker is too small to be heard very well.

2. Place this fiber optic link between the modulator and demodulator stages used in the digital communication systems investigated in:

   (a) Experiment 18: Pulse-amplitude modulation and time-division multiplexing

   (b) Experiment 19: Pulse-width modulation and detection

   (c) Experiment 25: Digital communication using frequency-shift keying

   Determine how well the total digital communication system works with the fiber optics added. List any modifications that are needed to produce satisfactory results.

REPORT/QUESTIONS:

1. Determine the decibel voltage gain between the output of the receiver at TP_6 and the input of the transmitter at TP_1 for each of the three cables used in steps 5–8.

2. Based on the calculated results in question 1 and observed results in steps 9 and 10, which of the fiber optic cables produced the minimal amount of attenuation? Which produced the largest attenuation? Since each of the cables are of the same type and length, what must be the main reason for differences in the amount of attenuation? Refer to the data sheets of the fiber optic emitters and detectors for clues.

3. Give a brief description of how the applied signal at TP_1 is processed through the fiber optic link by each stage as it proceeds to the speaker at TP_7.

FIBER OPTIC CABLE SPLICING

OBJECTIVES:

1. To become familiar with the proper use of strippers and cleavers in the preparation of fiber optic cable for fusion splicing or mechanical splicing.
2. To become familiar with the proper cleaning techniques in the preparation of fiber for fusion splicing or mechanical splicing.
3. To become familiar with fusion splicing and/or mechanical splicing techniques.

REFERENCE:

Refer to Chapter 18 of the text.

TEST EQUIPMENT:

Fiber optic inspection microscope—FIS model F1-0111-E

SUPPLIES:

Black bench mat tube—Anixter model 167438
Safety glasses
Alcohol bottle—FIS model F1-000728
Fiber optic disposal container—FIS model F1-8328
KIM wipes—FIS model 34-155
Canned air—FIS model F1-1007
Tweezers
Jacket strippers—FIS model F1-0016
Miller strippers—FIS model WO-1224
Clauss strippers—FIS model WO-1225
Clauss No-Nik strippers—FIS models NN175 and NN203
Microstrip tool kit—FIS model MSFOK1
Diagonal pliers
Regular scissors
Kevlar scissors—FIS model F1-0020
Pocket cleaver—FIS model WO-2220
Fitel cleaver—FIS model F1-0010
Thomas & Betts cleaver—FIS model 92208

Alcoa Fujikura cleaver—FIS model CT07

Single-mode bare fiber optic cable—FIS model SMF-28

Multimode simplex fiber optic cable—FIS model 601-IN-SFS-62PFD/s

Paper towels

Toothpick

Scotch tape

Stick-on labels

OPTIONAL SUPPLIES:

3M Fibrlok mechanical splice—FIS model 2529

3M Fibrlok mechanical splice assembly tool—FIS model 2501

Siecor camsplice—FIS model 95-000-04

Siecor camsplice assembly fixture—FIS model 2104040-01

AMP corelink—FIS model F04590

AMP corelink assembly fixture—Anixter model

Fusion splicer

SAFETY RULES:

It is mandatory that safety and cleanliness procedures be followed when working with glass fibers. Following these procedures will ensure that stray fibers do not end up in your skin and eyes. It will also ensure that stray fibers and dirt do not end up in your tools and fiber optic hardware. The following rules should be followed when working with a fiber optic installation:

1. Always work on a dark bench surface with adequate lighting so that stray fibers can always be located.

2. Always dispose of fiber scraps in a special disposable container that can be disposed of separately from other waste products.

3. Always wear safety glasses when working with fiber. Never rub your eyes when working with fiber installations.

4. Always wash your hands before and after working with a fiber installation project.

5. Check your clothing and workbench for fiber scraps after completing a fiber installation.

6. Always clean connectors and glass fibers with alcohol and KIM wipes when preparing them for connectorization.

7. Always blow out all connectors and tool openings with canned air.

8. Keep your workbench neat and organized throughout the fiber installation project. Keep all small parts in a special area so that they will not be misplaced when you need them.

9. Never look into a fiber connector when you are not sure if it is active. The light produced by an active connector is usually invisible but could be harmful to your eyes.

10. Try not to breathe the fumes produced by the chemicals (alcohol, epoxy, solvents, etc.) used in the fiber installation process.

PROCEDURE:

1. To set up your bench for this experiment, follow these procedures:

   a. Wash your hands to ensure that all body oils have been removed.

   b. Retrieve each of the following items for your bench top. These items make up your fiber optic workbench supply kit, which is also referred to in Experiment 31.

      (1) Black bench mat tube

      (2) Safety glasses

      (3) Alcohol bottle

      (4) Fiber optic disposal container

      (5) KIM wipes

      (6) Canned air

      (7) Tweezers

   c. Make sure that you are working in a well-lit area. You may need to use a portable light fixture at your workbench.

   d. Open the mat and place it in the center of the work area.

   e. Put on your safety glasses and wear them whenever glass fiber scraps may be in the work area.

   f. Identify an area on your workbench where all small parts will be kept for easy retrieval.

2. Obtain a few 2-foot lengths of single-mode bare fiber optic cable and multimode simplex fiber optic cable. In the next few steps, you will be preparing these fibers for mechanical or fusion splicing. Also, obtain as many of the handtools found in the supply list as your laboratory can offer. To view the end of your fiber under your inspection scope, the fiber should be stripped and cleaved as shown in Fig. 30-1(a) so that it will fit in the recommended test fixture given in Fig. 30-2. To prepare your fiber for actual splicing, it should be stripped and cleaved as shown in Fig. 30-1(b). The next few steps will show you how to strip and cleave your fiber.

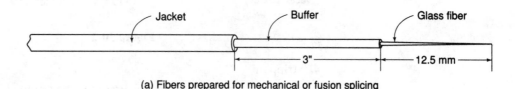

(a) Fibers prepared for mechanical or fusion splicing

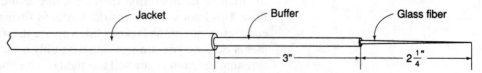

(b) Fibers being viewed on fiber inspection scope like the one shown in Fig. 27-2

Figure 30-1

3. The test fixture given in Fig. 30-2 should be constructed to allow for viewing of the end of the fiber with the inspection microscope. The microscope is inserted into its plastic base, which is held in place on the bottom layer of vector board, using doublesided tape. Also, a splice protection sleeve (FIS model F1-1002) is secured to the five layers of vector boards by two ³/₁₆-inch cable clamps, each held in place by a small screw and locknut. Notice that the end of the protection sleeve is pushed under the slide clip of the microscope. The fiber being viewed is then carefully inserted and slid through the protection sleeve until it appears in view. Scotch tape can then be used to hold the fiber in place. This fixture works better with the 900-micron diameter buffered multimode fiber. If a fusion splicer is available in your laboratory, then this test fixture may not be needed because the fusion splicer may then be used to view the end of the fiber.

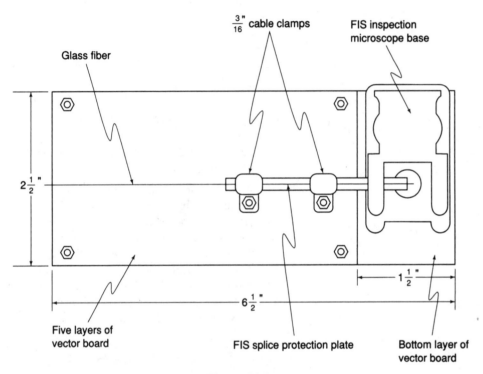

Figure 30-2

4. Stripping

    a. If your fiber has a jacket, it first must be stripped back. Use the jacket strippers to do this. Use the #16AWG size stripping hole in ¼-inch increments to ensure that the glass does not break inside. Strip the jacket back at least 8 inches to give you plenty of buffered glass to work with. With the multimode fiber, first tie a knot in the middle of the length of the fiber. This knot will prevent the buffered fiber from accidently slipping completely out of the jacket while you are stripping off the ¼-inch increments because you are working with only a short length of fiber. Make sure that the knot is not tied too tight or else you could break the fiber!

    b. Blow on the end of your stripped cable to force all the Kevlar yarns to move to one side of the buffered fiber. Then carefully twist the Kevlar yarns into a tight strand as it comes out of the jacket to make them easy to cut. Use the special Kevlar scissors to cut the Kevlar strands so that they are completely removed at the jacket.

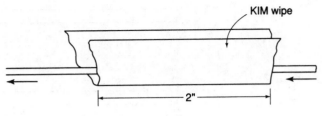

Figure 30-3

   c. Review the manufacturer's instruction sheets for each of the Miller, Clauss, and Microstrip stripping tools. Practice stripping the buffer of the bare glass using each of the following stripping tools. Remember to strip the buffer back at least 6 inches to be able to produce the final dimensions given in Fig. 30-1.

      (1) Jenson red-handled jacket strippers (These will not work very well!)

      (2) Miller yellow-handled buffer strippers

      (3) Clauss yellow-handled jacket/buffer strippers (Use the small hole for stripping the buffer.)

      (4) Clauss No-Nik mustard-handled buffer strippers (175-micron opening for 250-micron diameter buffered single-mode fiber.)

      (5) Clauss No-Nik red-handled buffer strippers (203-micron opening for 900-micron diameter buffered multimode fiber.)

      (6) Microstrip tool kit buffer strippers (with proper insert guide and blades)

5. Cleaning

   a. Once your buffer has been stripped off, the bare fiber will most likely have small pieces of leftover fiber on it. These must be cleaned off with a KIM wipe and alcohol.

   b. Take one KIM wipe tissue and fold it along its crease into a square. Then fold it down the middle two more times, once horizontally and once vertically, to form a small square. Then fold it one more time down the middle to form a V-groove, such as the one shown in Fig. 30-3. Hold the KIM wipe tissue between the index finger and thumb of your left hand. Open up the V-groove a little and drop four to five drops of alcohol into the V-groove. Then lay the stripped fiber into the V-groove and carefully pull the fiber out of the KIM wipe while slightly pinching the KIM wipe against the fiber.

   c. Repeat this cleaning procedure with another few drops of alcohol. You should note that once the glass is clean, it squeaks just like when you are washing a window with an alcohol cleaner.

   d. The KIM wipe can be used again for other fibers after unfolding your three folds and refolding to use another part of the KIM wipe tissue. (This keeps your bench free of used KIM wipe tissues.)

6. Cleaving

   a. The objective here is to produce a clean cut of the end of the fiber so that the end is perfectly perpendicular to the center-line axis of the fiber, as shown in Fig. 30-4. To do this, the cleaving tool first scratches the outside surface of the glass where the cut is to be made. Then a perpendicular force is firmly applied at the exact location of the scratch. The glass should then break cleanly at the desired location.

Figure 30-4

b. Review the manufacturer's instruction sheets for the following cleavers that are available in your laboratory. Practice the cleaving process with your fibers using each of the following tools:

  (1) Diagonal pliers—will crush the glass, producing unacceptable results.

  (2) Regular scissors—will crush the glass, producing unacceptable results.

  (3) Pocket cleaver

  (4) Fitel cleaver

  (5) Thomas & Betts cleaver

  (6) Alcoa Fujikura cleaver

c. Using the inspection microscope and fixture, observe the end of the fiber after using each of the cleaving tools. Carefully rotate the fiber once you have its cleaved end clearly in view. If it is cleaved properly, it should appear as the good cleave shown in Fig. 30-4 when viewed at all angles. Make a sketch of your findings for each of the cleavers used in step 6b. You should find that the more expensive cleavers will produce acceptable results more consistently.

d. It is absolutely necessary that an acceptable cleave be produced before any splicing can be successful. An unacceptable cleave will not allow the two fibers to be spliced and to be butted next to each other in perfect alignment. Thus, the splice will end up with a high loss. Therefore, do not proceed to step 7 or 8 until acceptable cleaves have been produced.

e. The fibers being prepared for splicing in either step 7 or 8 must have their other ends terminated with an appropriate connector that can be connected to a light source and power meter or to an optical time domain reflectometer (OTDR). This termination is necessary for measuring the resulting power loss of the splice. For the use of an OTDR, it is recommended that two 100- to 150-meter rolls of fiber be used in steps 7 and 8, as shown in Fig. 30-5. Most OTDRs cannot make accurate measurements on the first 10-80 meters of fiber that are connected to their laser output connectors due to saturation of their detectors by the high amounts of scattered light energy. This is referred to as the OTDR's dead zone.

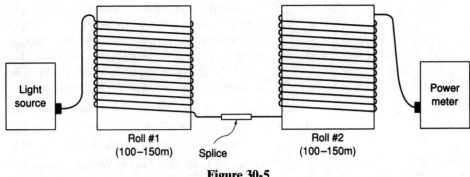

Light
source

Power
meter

Roll #1
(100–150m)

Splice

Roll #2
(100–150m)

Figure 30-5

7. Mechanical Splicing (Optional)

a. A mechanical splice is a small plastic device in which prepared fiber ends are inserted from either end into V-grooves. The center of the V-groove, where the ends butt together, is filled with index matching gel to help create the lowest loss possible.

b. Review the manufacturer's instructions for the mechanical splices listed below that are available in your laboratory. Note that each of the mechanical splices requires the use of a special test fixture to assemble a low-loss splice properly. Some of the splices can be attempted repeatedly until acceptable results are achieved. This may require the use of special small handtools to reopen the mechanical splice. Eventually, however, the index matching gel inside the V-groove will become contaminated with dirt and broken pieces of fiber. Then acceptable results can never be achieved. At that point, repreparation of the fibers and a new mechanical splice will be necessary.

(1) 3M Fibrlok

(2) Siecor Camsplice

(3) AMP Corelink

c. To minimize the power loss in the mechanical splice, a light source is placed at one end of the fiber and the power meter is placed at the other end. The fibers are adjusted in the mechanical slice until minimal loss is achieved. (The procedure for measuring actual loss is covered in more detail in Experiment 31.)

d. You should notice that the more expensive mechanical splices are easier to use. They make more consistent low-loss splices and are more capable of being readjusted for optimum results.

8. Fusion Splicing (Optional)

a. A fusion splice is created by heating the glass fibers with a predetermined arc of electrical energy. As the glass melts, the fibers are carefully pushed together in near-perfect alignment so that, when the glass solidifies, it becomes a continuous piece of fiber.

b. If your laboratory is equipped with a fusion splicer, review the manufacturer's instructions for its use. The earlier models require the operator to align carefully the two fibers being fused by manual adjustments along both horizontal and vertical axes of view. The later models perform this alignment automatically by computer.

c. Practice these procedures with your cleaved fibers until you feel that you have mastered these skills. Again, to check your resulting splice loss, a light source and power meter are attached at opposite ends of the fiber being tested and measurements are made. The alternate procedure is to "shoot" one end of the fiber with an OTDR laser source and observe the resulting pattern on the OTDR display.

9. Cleanup—It is important to follow these procedures for your protection and to prolong the functional life of your tools and fiber installation project.

   a. Cover all fiber optic connectors that are not in use with their safety caps.

   b. Clean all tools that may have been exposed to pieces of glass fiber and stripped buffer. Use canned air, piano wire, cleaning brushes, KIM wipes, and alcohol to remove all scraps, dirt, and debris from all tools and your work area.

   c. Carefully inspect the workbench and your clothing for broken fibers. Pick up any stray fibers with tweezers and drop them into a disposable container.

   d. Return all tools and working mat back to their original containers and store them in their proper location.

   e. Wash your hands and face with soap and water.

REPORT/QUESTIONS:

1. Why is it so important to follow safety and cleanliness procedures when working with fiber optic cable? List the most important procedures that must always be followed.

2. Why did you need to tie a small knot in the 2-foot fiber optic cable before stripping off the jacket?

3. Which of the buffer stripping tools worked best for you? List any problems you have with any of the buffer strippers.

4. What is the advantage of folding the KIM wipe as detailed in step 5?

5. Which of the cleavers produced ideal results most consistently when viewed under the fiber inspection microscope? List any problems you had with any of the cleavers listed in step 6.

6. Which of the mechanical splices in step 7 seemed to produce low loss most consistently? Which was the easiest to readjust for optimum results?

FIBER OPTIC CABLE CONNECTORIZATION

OBJECTIVES:

1. To become familiar with the procedure used to connectorize multimode fiber optic cable with ST connectors.

2. To become familiar with the procedure used to measure the loss of connectors on a patch cord using a light source and power meter.

3. To recognize the differences between ST connectors that have stainless steel ferrules and those that have ceramic ferrules.

REFERENCE:

Refer to Chapter 18 of the text.

TEST EQUIPMENT:

Fiber optic light source—FIS model 9050-0000

Fiber optic power meter—FIS model F1-8513HH with universal adapter

Fiber optic inspection microscope—FIS model F1-0111-E

SUPPLIES:

Stainless-steel ferrule connectors—FIS model F1-0066 (2)

Ceramic ferrule connectors—FIS model F1-0061 (2)

Four-foot piece of simplex multimode fiber optic cable—FIS model 601-IN-SFS-62PFD/S

ST mating sleeve—FIS model F1-8102

Fiber optic workbench supply kit (detailed in Experiment 30)

Fiber optic jacket strippers—FIS model F1-0016

Fiber optic buffer strippers—any of those detailed in Experiment 30

Kevlar scissors—FIS model F1-0020

Scribe—FIS model F0-90C

Piano wire—FIS model F1-8265

Epoxy—FIS model F1-7070

Glass polishing plate—FIS model F1-9111A

Soft pad—FIS model 05-00053

Polishing disk—FIS model F1-6928

Polishing paper—FIS models F1-0109-5, F1-0109-1, and F1-0109-3

Connector heat oven (optional)—FIS model F1-9772 with 9451S curing stand

Paper towels

Toothpick

Scotch tape

Stick-on labels

PROCEDURE:

1. The consumable supplies for this experiment are the ST connectors, multi-mode simplex fiber optic cable, and polishing paper. The ST connectors consist of three parts: the back shell, the rubber boot, and the safety cap, as shown in Fig. 31-1. Two of the ST connectors have a stainless steel ferrule (silver) and two have a ceramic ferrule (white). The ceramic ferrule connector is less forgiving in the connection procedures because the ceramic material will not polish like the stainless steel will. Carefully store these parts in a safe location because they are small enough to get lost easily. There are three types of polishing paper: coarse (5-micron grit) black paper, medium (1-micron grit) green paper, and fine (0.3-micron) white paper.

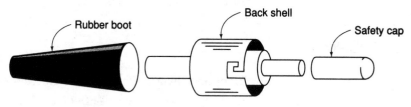

Figure 31-1

2. Setup—Set up your lab bench following the procedures outlined in Experiment 30. You will also need the following tools at your lab bench:

   a. Fiber jacket strippers

   b. Fiber buffer strippers

   c. Kevlar scissors

   d. Scribe

   e. Epoxy—Tube A and Tube B

   f. Fiber optic crimping tool

   g. Glass polishing plate

   h. Soft pad

   i. Polishing disk (puck)

   j. Paper towel

   k. Toothpick

   l. Scotch tape

   m. Small stick-on labels

3. *Initial Preparation*—To prepare for two patch cords, you need to cut your 3- to 4-foot fiber optic cable into two equal lengths. Cut your cable in the middle of its length with the cutting blade of your jacket strippers. This will cut the jacket and the glass fiber. Then you will need to cut through the Kevlar yarns with the special Kevlar scissors. Next, tie a knot in the middle of each of your two patch cords. Do not make the knot so tight that you break the fiber. It does have to be tight enough to prevent jacket slippage when you strip the buffer in step 4. Also, place sticker labels on each of your patch cords. Label with your last name followed by SS or CM, for stainless steel or ceramic ferrules, respectively. This labeling will prevent your patch cord from being mixed up with another student's cable when it is placed in the oven or being tested for loss with the power meter.

4. *Stripping*—The ends of each of the two patch cords need to be stripped back so that they match Fig. 31-2. To produce these results, the following procedure should be followed for each of the four cable ends:

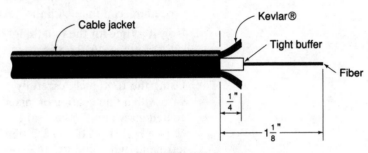

Figure 31-2

a. Strip back the jacket approximately 1⅛ inch to expose the buffer and Kevlar yarn. Use your jacket strippers. Use the #16AWG size stripper hole and strip in ¼-inch increments to ensure that the glass does not break inside.

b. Blow on the end of your stripped fiber to force all the Kevlar yarns to move to one side of the buffered fiber. Then carefully twist the Kevlar yarns into a tight strand as they come out of the jacket to make them easy to cut. Use the special Kevlar scissors to cut the Kevlar strands so that they all stick out away from the jacket approximately ¼ inch.

c. Next, use the buffer stripping tool to strip the buffer away from the glass. The glass (core and cladding) should stick out approximately ⅞ inch beyond the end of the buffer to conform to the dimensions given in Fig. 31-2. Stripping the buffer should be done in short increments if you are using a Clauss or Miller hand stripper. If you are using the No Nik or Microstrip strippers, you should be able to strip ⅝ inch in one complete pass. Don't forget to clean your stripping tool with canned air when you are finished stripping the fiber.

5. *Cleaning*—Carefully clean the glass with alcohol and KIM wipes. Refer to step 5 of Experiment 30 for detailed instructions on the cleaning procedures.

6. *Epoxying/Crimping*—Two patch cables will be fabricated. The first one will use the stainless-steel ferrule connectors and the second will use the ceramic ferrule connectors. The following procedure should be used for the stainless-steel ferrule connectors:

a. Carefully slide on the rubber boot. Make sure that the wide end of the boot faces the back end of the back shell. It must be installed now because it won't slide over your connector later!

b. Look through the back end of the back shell of the ST connector using an eye loupe and back light to make sure that there is no dirt or obstruction inside the passageway that would interfere with the glass fiber's ability to be pushed through the connector. If there is an obstruction, blow it out with canned air or push it out with piano wire.

c. Try a "dry run." Carefully feed your glass fiber into the back end of the back shell of the ST connector and see if it will slide out of the tip of the ferrule. Be very careful not to trap the fiber and subsequently break it by making it bend against itself. This procedure is very delicate like threading a needle, and practice is necessary. You should notice that you will be able to continue feeding the glass fiber through the connector until the buffer jams into the back end of the ferrule. It will not be narrow enough to enter the ferrule opening. Once you feel that you have mastered this procedure, continue with the next step.

d. Now it is time for the real thing! Mix a small blob of epoxy by using a 2:1 ratio of epoxy "A" and epoxy "B." Do this on a scrap of paper or paper towel. Stir the two quantities together with a toothpick or equivalent.

e. Using the toothpick, carefully paint the glass fiber, buffer, and Kevlar yarn with a thin coating of mixed epoxy. Then carefully feed your epoxy coated glass into the back of the main shell just as you practiced in step 6c (see Fig. 31-3). If you are right-handed, hold the connector with your left hand and guide the fiber into the connector with your right hand. Continue pushing until the buffer hits the back entrance of the ferrule. The glass fiber should protrude out of the tip of the ferrule. Be very careful not to let the glass break as it protrudes out of the opening of the ferrule. If it does break, you will have to start the whole preparation process again! If you are left-handed, hold the connector with your right hand and guide the glass fiber in with your left hand.

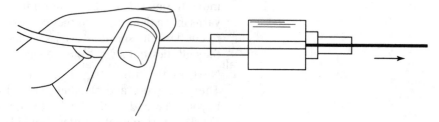

Figure 31-3

f. As the jacket gets closer to the back of the connector, the epoxy covered Kevlar should fold back against the jacket and tuck in between the jacket and the side of the back shell of the connector. This will allow for sealing the fiber cable into the connector after the crimp is made and the glue has hardened. With the fiber fully inserted into the back of the connector, very carefully position the side of the back shell against the .128 die of the crimping tool. Start closing the crimping tool (see Fig. 31-4). Once the crimp is started, it is impossible for the crimpers to be reopened until the crimpers have been fully closed, completing the crimp. You will need to push hard with both hands to complete the crimp. While you are pushing hard with the crimpers, be careful to position the front of the connector so that the protruding glass fiber does not hit something and break off.

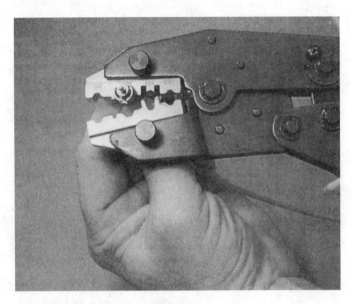

Figure 31-4

g. Carefully remove the connector assembly from the die of the crimpers without breaking the glass. Now the epoxy must be left to harden, which takes twenty-four hours at room temperature or thirty minutes at 100 degrees C. If you are using an oven to speed the curing process, carefully guide the connector into the circular opening of the oven so you do not break off the fiber strand that is sticking out of the tip of the ferrule. Using Scotch tape, secure the jacket of the fiber to the side of the curing stand. Don't forget to clean any leftover epoxy from the die of the crimping tool.

h. Obviously, steps 6a through 6g must be completed for the opposite end of the patch cord to make a complete patch cord.

i. For the second patch cord, which is made of the ceramic ferrule connectors, the procedure is the same except that, after completing step 6f, place a small bead of epoxy on the tip of the ferrule, around the protruding glass fiber. This ensures that the fiber does not accidentally break below the ferrule tip surface during the polishing process. The ceramic material will not polish like the stainless-steel material does. The cured epoxy will polish away.

7. *Scribing*—Carefully remove the crimped connector assembly from the oven or storage area when your epoxy has completely cured. Be very careful at this point because if the glass fiber strand is accidentally bumped, it could easily break off, making it difficult if not impossible to polish. Be aware that the back shell may need five minutes to cool enough so that you can handle it. Carefully slide the rubber boot over the crimped part of the connector to give

the fiber more strain relief where it comes out of the back of the connector. Now complete the scribing procedure as outlined below (see Fig. 31-5):

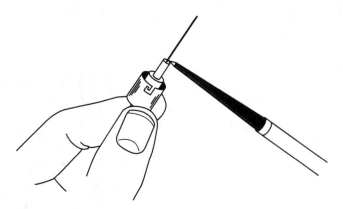

Figure 31-5

a. Using a scribe, carefully scratch the side of the glass fiber close to but just above the ferrule tip surface. One scratch should be sufficient. If you scratch several times, the glass will most likely break off in an unpredictable location.

b. Next, carefully pinch the end of the glass fiber with your thumb and index finger and pull the glass away from the tip of the ferrule. The glass strand should break away fairly easily at the exact location of the scratch. This should leave you with a small nub of glass sticking out of the end of the connector. Dispose of the glass fiber scrap properly. Make sure that the nub is not too long. If it is too long, the next step of air polishing may take an excessive amount of time or result with a broken glass nub. Thus, you may want to try to rescribe a bit closer to the ferrule tip surface if it is possible.

8. *Air Polishing*—The objective of this procedure is to polish away the glass nub so that the glass fiber ends right at the ferrule tip surface. Refer to Fig. 31-6 and complete the steps outlined below:

a. Take a sheet of black polishing paper (5-micron grit) and hold it at one end so that the other end is hanging in free space with the coarse (dull) side facing up. You need to bend the polishing paper in a slight U-shape to give it a little stress support.

b. Push the glass nub down very lightly against the opposite end of the polishing paper on its rough side. Begin rubbing the tip of the nub against the polishing paper in a small circular motion. Carefully repeat this polishing action without pushing too hard. You should notice that a fine circular scratch is being drawn on the polishing paper.

c. After a dozen circular traces have been completed, view the end of the ferrule using your eye loupe. You should notice that the length of the nub has decreased due to the polishing action.

d. Carefully continue this polishing process until the nub has been completely polished away. When this happens, if you are using the stainless-steel

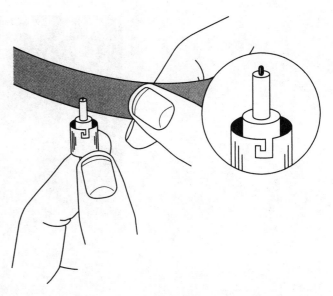

Figure 31-6

ferrule connectors, you will find that the circular traces on the polishing paper will take on a much thicker appearance because you are now starting to polish the stainless-steel surface along with the end of the fiber. If you are using the ceramic ferrule connectors, you will notice that the scratching sound will disappear and you will not be polishing anything anymore because the ceramic ferrule cannot be polished by the polishing paper.

e. Be careful that you don't push too hard or the nub will break off unpredictably and you may have to start the entire procedure again!

f. When the nub has been completely polished away, you are now past the critical point in the connection process. The connector assembly is now not as delicate and prone to breakage due to mishandling. You can now proceed to the polishing procedure.

9. *Polishing*—Now the objective is to make the end of the glass fiber at the tip of the ferrule as smooth and clear as possible. This mirror surface is achieved by the use of three polishing papers, a glass polishing table, and a polish disk (puck). The following procedures are followed with the stainless-steel ferrule connectors:

a. Using the fiber inspection microscope, view the tip of the ferrule. It should look like a fuzzy black dot on a scratchy silver or white background. Check for any evidence of broken glass pieces. It is hard to determine if the fiber is broken at this time.

b. Place the black (0.5-micron grit) polishing paper on the glass polishing table so that the rough side faces up. Clean the polishing puck and place it on top of the polishing paper. Now insert the ferrule of the connector into the polishing puck and push it down until it cannot be inserted any further. Lightly push the puck/connector assembly around in a figure-eight pattern so that all edges are polished equally. Approximately eight to ten figure-eight strokes should suffice. See Fig. 31-7.

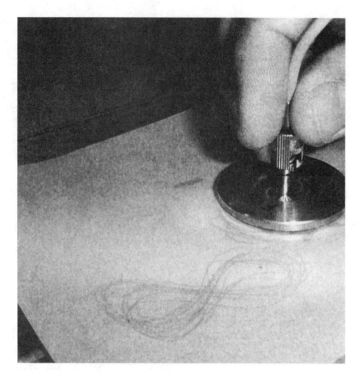

Figure 31-7

c. Remove the connector from the puck. Clean the end of the ferrule with alcohol and a KIM wipe. When it is squeaky clean, let the alcohol evaporate and then view the tip of the ferrule with the inspection microscope. It should now look more like a gray dot against a silver or white background. If not, repeat step 9b until it does. There may be a few scratches and epoxy spots on the glass at this point. The use of finer grit polishing paper should take care of this. What you do not want to see is a dark black dot representing the glass surface. If this happens, you most likely have broken the fiber inside the ferrule and you will have to start the whole process again!

d. Clean the bottom of the polishing puck with alcohol and a KIM wipe. This keeps the new polishing paper from being contaminated with the scraps from the old polishing paper. Now use the green polishing paper (1-micron grit) and repeat step 9b. This should remove more of the remaining scratches and epoxy spots from the surface of the glass fiber. Again, eight to ten figure-eight strokes should suffice.

e. Again, remove the connector from the polishing puck. Clean the end of the ferrule with alcohol and a KIM wipe. You should again hear the ferrule squeak clean when it has been totally cleaned off. Once the alcohol has evaporated, view the tip of the ferrule with the inspection microscope. It should have less scratching and spotting, and the glass should be a light gray color. If not, repeat step 9d until it does.

f. Clean the bottom of the polishing puck with alcohol and a KIM wipe to avoid contaminating the new polishing paper. Now use the white polishing paper (0.3-micron grit) and repeat step 9b. This should remove most of the remaining scratches and epoxy spots from the surface of the glass fiber. This time, you should complete twenty to twenty-five figure-eight strokes.

g. Again, remove the connector from the puck. Clean the end of the ferrule with alcohol and a KIM wipe until it is squeaky clean. Once the alcohol has evaporated, view the tip of the ferrule with the inspection microscope. It should be almost completely free of scratches and epoxy spots. The glass should be a clear, light gray circle. If not, then complete steps 9h and 9i. If it is clear, then skip steps 9h and 9i and proceed to step 9j.

h. Clean the bottom of the polishing puck with alcohol and a KIM wipe. This time, place a few drops of alcohol on the white polishing paper in an unused area. Continue polishing the connector by pushing the puck/connector assembly through the alcohol drops as they evaporate from the surface of the polishing paper. This should remove the last few specks of epoxy from the glass surface.

i. Make one last inspection of the ferrule with the inspection microscope. It should now look perfectly clear of epoxy specks and scratches. You do not want to see any of the last five displays shown in Fig. 31-8. If you do, repeat step 9h until the ferrule is perfectly clear.

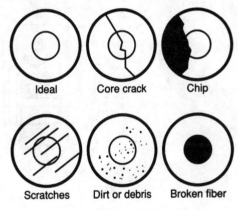

Figure 31-8

j. Obviously, steps 9a through 9i must be completed for the opposite end of the patch cable.

k. For the patch cable using the ceramic ferrule connectors, the polishing procedure is identical, except that you should place a soft pad (sheet of rubber) between the polishing paper and glass. Also steps 9d and 9f may take more than twenty to twenty-five figure-eight strokes. The soft pad allows you to remove extra epoxy deposits that may become trapped on the glass below the surface of the ferrule tip. Remember that it is impossible for the polishing paper to polish away any of the ceramic material.

10. *Protection*—To protect the ferrules from dirt and scratches, make sure that you install the protective caps on the ferrules when they are not in use.

11. *Connector Loss Measurement*—To determine the loss of each of the connectors at either end of the fabricated patch cord, the following procedure must be completed. Make sure that all connectors on the test equipment and the patch cords are thoroughly cleaned before any measurements are

made. Also, the fiber optic cables must be kept as straight as possible to produce the smallest loss. Follow this procedure for each connector of both of the patch cords that were fabricated in steps 1 to 10. Enter the resulting measurements in Table 31-1.

TABLE 31-1

PATCH CABLE	CONNECTOR	REFERENCE POWER LEVEL	POWER LEVEL	dB LOSS
Stainless-steel	A			
ferrule	B			
Ceramic	A			
ferrule	B			

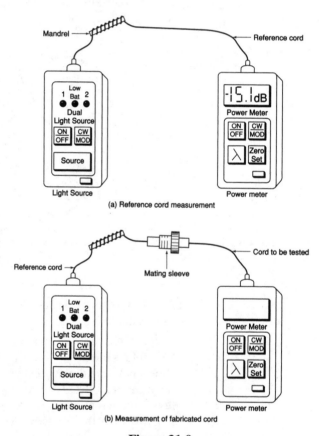

(a) Reference cord measurement

(b) Measurement of fabricated cord

Figure 31-9

a. Place a low-loss reference patch cord between the LED light source and power meter, as shown in Fig. 31-9(a). The end of the patch cord nearest the light source should be wound five to six times around a mandrel to ensure that the higher ordered modes of light are not propagated through the fiber, thus causing erroneous readings to be made by the power meter. The mandrel is nothing more than a pencil or pen or wooden doweling of the same diameter. You may need to hold the fibers in place with Scotch tape.

b. Turn on the light source and power meter. Make sure that the light source and power meter are designed for the same wavelength (850 nm). Measure the power level displayed by the power meter in dBm. It should be approximately −17 dBm. Turn off the light source before proceeding.

c. Disconnect the power meter from the sample patch cord. Add an ST mating sleeve and add your constructed patch cord between the reference patch cord and the power meter, as shown in Fig. 31-9(b).

d. Turn on the light source. Again, measure the power level displayed by the power meter in dBm. The difference between this reading and the reading made in step 11b is the loss of the connector of your patch cord that is connected to the mating sleeve. The other connector that is plugged into the power meter is not being tested because the power meter's mating connector allows for almost all of the light to shine on the detector surface, even if the connector has substantial loss.

e. Reverse your constructed patch cable end for end to measure the loss of the other connector of your cable. If your results in steps 11c and 11d are less than 0.5 dB, your patch cable is acceptable. If the loss is larger than 0.5 dB, you may need to reconnect one or both of your connectors!

12. Clean your workbench following the procedures outlined in Experiment 30.

REPORT/QUESTIONS:

1. State the differences in preparing a fiber optic cable for connection in comparison to preparing a fiber optic cable for splicing.

2. Why is it so important to clean your tools, connectors, and work area constantly when fabricating patch cords?

3. Which steps did you find had the greatest chance of breaking a glass fiber? What extra precautionary steps could be taken to avoid breaking the glass?

4. List the differences in the connection of stainless-steel ferrule connectors versus ceramic ferrule connectors? How do you account for the differences?

5. What is a mandrel used for?

6. Which connector is being tested by the test procedures followed in step 11? Why aren't the other connectors shown in Fig. 31-9(b) being tested?

USING A SPECTRUM ANALYZER

OBJECTIVES:

To learn how to use a spectrum analyzer and apply this to the analysis of the frequency components of a square wave.

REFERENCE:

Refer to Section 2-7 of the text.

TEST EQUIPMENT:

Function generator
Spectrum analyzer, with a resolution bandwidth of 3 kHz or less

COMPONENTS:

Mini-Circuits®, Model HAT-20, 20 dB attenuator

THEORY:

When an electronics student first learns about AC electronics, the primary instrument used to demonstrate various principles is the oscilloscope. This provides a graph of voltage on the vertical scale and time on the horizontal scale. In electronic communications, the primary instrument used to measure signals is the spectrum analyzer. Three reasons why a spectrum analyzer is preferred instead of an oscilloscope are:

1. Frequency range. An inexpensive oscilloscope (< $1,000) will work from DC to about 50 MHz. Note that some expensive oscilloscopes (> $30,000) will work as high as 1 GHz. An inexpensive spectrum analyzer (< $4,000) does not work at DC. It works between about 30 kHz and 1 GHz. Expensive spectrum analyzers (> $100,000) can work between 1 Hz and 100 GHz.

2. On-screen dynamic range. This is the ability to measure weak signals when strong signals are present. A spectrum analyzer has more than 100 dB of dynamic range, compared to about 20 dB for an oscilloscope.

3. Simplicity of the display. A sine wave on an oscilloscope appears as a single trace. If more than two frequencies are present, it may be difficult to measure the voltage at each frequency. With a spectrum analyzer, this is very easy because each frequency is displayed separately.

The following are some comparisons between spectrum analyzer and oscilloscope displays:

1. The vertical scale for a spectrum analyzer is power instead of voltage. A spectrum analyzer is normally terminated with a 50-Ω load. The input to an oscilloscope is normally a high impedance.

2. The vertical scale on a spectrum analyzer is usually set for 10 dB/division. The vertical scale on an oscilloscope is linear and the volts/division is adjusted according to the size of the signal.

3. The reference level for an oscilloscope is normally a horizontal line across the middle of the screen. This is zero volts. The reference level for a spectrum analyzer is usually set at 1 line below the top division. For example, if the reference level is set at −10 dBm, then a signal 3 divisions below this reference would have a power of −40 dBm.

4. The horizontal display for a spectrum analyzer is frequency. The horizontal reference line is usually the center of the screen and this is called the "center frequency."

5. The width of horizontal scale for a spectrum analyzer is adjustable and this is called the "span." For example, the center frequency could be set for 10 MHz with the span set at 10 kHz per division; a signal one division to the right side of the center frequency would be 10.01 MHz. A signal one division to the left side of the center frequency would be 9.99 MHz.

Spectrum Analyzer Tests: Sine Wave Signal Input

The following tests are used to demonstrate what the output spectrum will be for a sine wave or a square wave signal input. Follow the procedure provided and record the measurements as directed.

Procedure (Sine Wave):

1. Connect a 20 dB attenuator to the RF input of the spectrum analyzer to prevent damage. Use a BNC cable to connect the output of the function generator to the 20 dB attenuator. Set the function generator to produce a sine wave at 1 MHz. Adjust the output of the function generator for fully CW (clockwise) for maximum power.

2. Use the spectrum analyzer to make sure that the frequency is 1 MHz. Change the span and center frequency controls to get the best accuracy.

3. Using the spectrum analyzer, measure the power output of the function generator. You may need to adjust the "Ref Level" on the spectrum analyzer to get the best accuracy. The power should be about +18 dBm.

Note: A power in excess of +30 dBm can damage the spectrum analyzer.

4. Adjust the output of the function generator fully CCW (counter-clockwise) for minimum power.

5. Repeat steps 1-3 using the function generator. The power output range should be about +2 dBm to −55 dBm.

6. Adjust the function generator power level so that the power measured on the spectrum analyzer is −10 dBm. Record your measurement in Table 32-1.

7. Without changing the frequency of the function generator, change the spectrum analyzer center frequency to 2 MHz. Record the power in Table 32-1. Do the same for the other frequencies listed in the table.

TABLE 32-1   Sine Wave Harmonics

FREQUENCY	1 MHz	2 MHz	3 MHz	4 MHz	5 MHz	6 MHz	7 MHz
Power							

Procedure (Square Wave):

8. Change the function generator so that a square wave is produced.

9. Measure the power of the odd harmonic frequencies shown in Table 32-2, and record the results in Table 32-2.

10. Measure the power of the even harmonic frequencies shown in Table 32-3, and record the results in Table 32-3.

11. Compute the measured power, relative to that at 1 MHz and record the results (measure of harmonics) in Table 32-2 and Table 32-3.

12. Compute the theoretical power, relative to that at 1 MHz, and record the results in Table 32-2 and Table 32-3. Use the following equation:

$$dB = 20 \log \left(\frac{V_2}{V_1} \right)$$

Note that the theoretical voltage at 3 MHz is 1/3 that at 1 MHz. The voltage at 5 MHz is 1/5 that at 1 MHz, etc.

TABLE 32-2   Odd Square Wave Harmonics

FREQUENCY	1 MHz	3 MHz	5 MHz	7 MHz	9 MHz	11 MHz
Measured dBm						
dB, Relative to 1 MHz						
Calculate dB, Relative to 1 MHz						

TABLE 32-3   Even Square Wave Harmonics

FREQUENCY	2 MHz	4 MHz	6 MHz	8 MHz	10 MHz	12 MHz
Measured dBm						
dB, Relative to 1 MHz						
Calculate dB, Relative to 1 MHz						

NAME _____

USING CAPACITORS FOR IMPEDANCE MATCHING

OBJECTIVE:

   To use a Smith chart for determining the size and location of series and shunt impedance-matching capacitors.

REFERENCE:

   Refer to Section 12-8 of the text.

TEST EQUIPMENT:

   None required

COMPONENTS:

   Smith charts (11"×17")
   Compass
   Ruler

THEORY:

   A Smith chart is input impedance mapped on to the plane of reflection coefficient. Moving clockwise along a circle (centered in the middle of the Smith chart) shows how the input impedance is "transformed" to different values corresponding to moving towards the RF generator.

PROCEDURE:

   Part I: Circuit Description

   A coaxial transmission line has the following parameters: $Z_o = 50\ \Omega$, $E_r = 1$, and frequency = 1 GHz. Compute wavelength in centimeters. $C = 3 \times 10^{10}$ cm per second.

   _____ (a)

Part II: Measuring Z_{in} along the Transmission Line

1. On the Smith chart, mark the location for $Z_L = 25\ \Omega$.
2. Using a compass, draw a circle, centered at $Z_{in} = 50\ \Omega$ and passing through the point noted in step 1.
3. What is the VSWR for all points on this circle?

   _____(b)
4. Compute the number of wavelengths for a distance of 3 cm.

   _____(c)
5. From the point identified in step 1, move along the circle (towards the generator) by 3 cm. Using a ruler measure $Z_{in}/Z_o =$ _____ + j _____ (d)
6. Calculate $Z_{in} =$ _____ + j _____ (e)

Part III: Impedance Matching with a Series Capacitor

1. Find the number of wavelengths from the point in Part II, step 1 required to make $R_e\ (Z_{in}) = 50\ \Omega$.   _____(f)
2. Compute the corresponding physical distance.   _____(g)
3. What is the magnitude and phase of the reflection coefficient at this point?

   _____(h)
4. What is $I_m\ (Z_{in})$?   _____(i)
5. Compute the amount of series capacitive reactance required to make the reflection coefficient equal to zero.   _____(j)
6. Compute the corresponding capacitance for step 5.   _____(k)

Part IV: Impedance Matching with a Shunt Capacitor

1. Start at a point defined in $Z_L = 15 + j15\ \Omega$.
2. Draw a circle through this point.
3. In order to convert from Z_{in} / Z_o to Y_{in} / Y_o: Draw a line through this point and the center of the Smith chart and extend it through the other side of the circle. Y_{in} / Y_o is located at the other intersection.
4. Note the wavelength towards the generator and record.

   _____(l)
5. Follow the circle further towards the generator until $R_e\ (Y_{in} / Y_o) = 1 + j0$.
6. Note the wavelength towards the generator and record.

   _____(m)
7. What is the physical distance between step 4 and step 6? _____(n)
8. What is $I_m\ (Y_{in} / Y_o)$?   _____(o)
9. Compute Y_{in}.   _____ −j _____(p)
10. Compute the value of shunt capacitance required to make the reflection coefficient equal to zero.   _____(q)
11. Compute the value of inductance for a reactance of 15 Ω at 1 GHz.

   _____(r)

Part V: Change Frequency and Dielectric Constant

1. For the series-matching capacitor:

   a. Calculate the capacitance if the frequency is changed from 1 GHz to 100 MHz: _____(s)

   b. Calculate the length for the matching element if the dielectric constant is 2.1 (teflon): _____(t)

   HINT: See equation 12-21. The velocity equals "c" divided by the square root of the dielectric constant.

2. For the parallel-matching circuit, start with a 25-Ω load. Change the frequency to 100 MHz and the dielectric constant to 2.1.

   a. Calculate the capacitive reactance and capacitance:

      (i) $X_C =$ _____(u)

      (ii) $C =$ _____(v)

   b. Calculate the # wavelengths and length for the matching element:

      (i) Wavelengths _____(w)

      (ii) Length _____(x)

NAME _____

ELECTRONICS WORKBENCH MULTISIM—IMPEDANCE MATCHING

OBJECTIVE:

To use the network analyzer in Multisim to demonstrate impedance matching.

REFERENCE:

Refer to Section 12-8 of the text.

THEORY:

A network analyzer measures the reflection coefficient and input impedance of an electronic circuit. S parameters are measurements that fully define the transfer characteristics of a microwave circuit. The reflection coefficient for the input port of a circuit is called **S11**. It has both magnitude and phase information.

PROCEDURE:

Part I: Setting Up the Series Capacitor Circuit

1. Make all of the connections shown in Fig. 34-1. The lossless transmission line, "W1," is located in the MISC file. This circuit uses a transmission line and a series capacitor to compensate for the mismatch that is caused by a 25-Ω load positioned at the end of a transmission line having a characteristic impedance of 50 Ω.

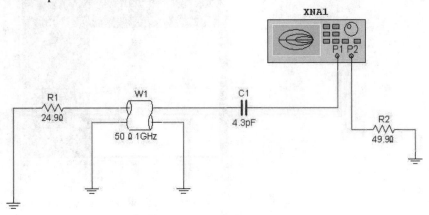

FIGURE 34-1 Initial connections

2. Click on the transmission line and make the following settings, as shown in Fig. 34-2:

Nominal Impedance: 50 Ohm

Frequency: 1 GHz

Normalized Electrical Length: 0.15

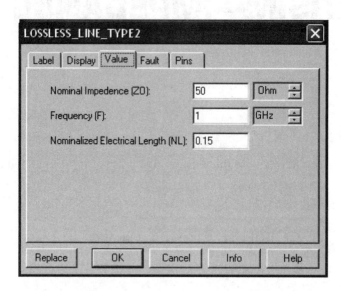

FIGURE 34-2   Transmission line settings

3. Click on the network analyzer and make the following settings, as shown in Fig. 34-3:

MODE: Measurement

Param: S-Parameters and Smith

Marker: Re/Im

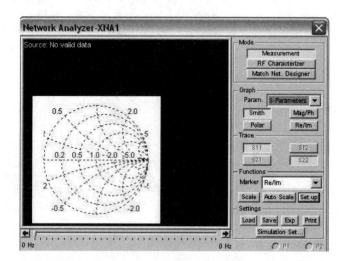

FIGURE 34-3   Network analyzer settings

4. Set the frequency limits by clicking on **Simulation Set**, . . . shown in Fig. 34-3. Then make the following settings, as shown on Fig. 34-4:

Start frequency: 500 MHz

Stop frequency: 1.5 GHz

Sweep type: Decade

Number of points per decade: 500

Characteristic Impedance (Zo): 50 Ohm

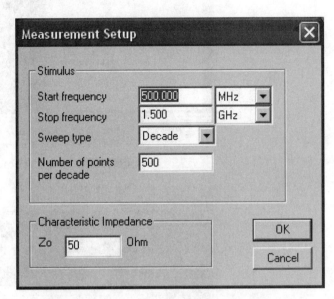

FIGURE 34-4   Frequency range settings

Part II: Measurements for the Series Capacitor

1. Start the Multisim simulation of the circuit shown in Fig. 34-1. Click on the network analyzer. A display similar to that shown in Fig. 34-5 should be provided. Move the slider across the screen to change frequency. Note the movement of the marker (red triangle) on the Smith chart. Make the following measurements:

a. Record the minimum value of S11. _____

b. Record the frequency when S11 is minimum. _____

Ask your instructor to initial your results. _____

2. On the Graph section of the Network Analyzer screen, select Mag/Ph instead of Smith. Under Functions section, select Auto Scale. A display should appear on the Network Analyzer screen that is similar to that shown in Fig. 34-6.

3. Move the frequency slide to the left and record the frequency when S11 has the following values: .05, .10, and .15. Use Table 34-1.

4. Move the frequency slide to the right of that for minimum S11 and record the frequency when S11 has the following values: .05, .10, and .15.

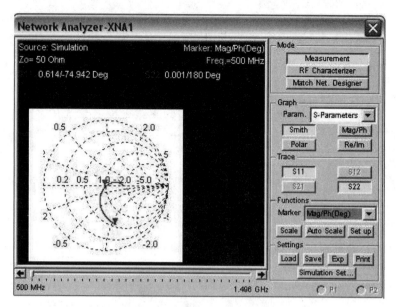

FIGURE 34-5 S11 measurements

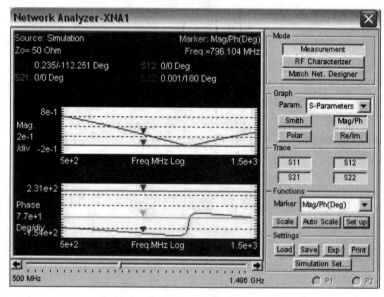

FIGURE 34-6 S11 measurements

TABLE 34-1 Results for Series Capacitor Circuit

#	S11	FREQUENCY (MHz)	VSWR	BANDWIDTH (MHz)
1	Min:			
2	.05L			
3	.10L			
4	.15L			
5	.05R			
6	.10R			
7	.15R			

5. Use equation 12-27 in your textbook to calculate the VSWR for each of the values of S11 recorded in Table 34-1. Note that the magnitude of "S11" is the same as "Γ".

6. Calculate the bandwidth for each value of VSWR.

Part III: Measurements for a Shunt Capacitor

1. Change the circuit as shown in Fig. 34-7.

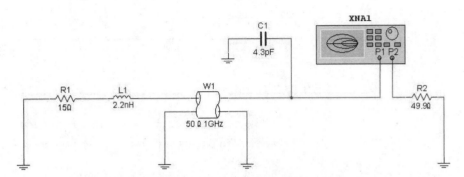

FIGURE 34-7   Connections for a shunt capacitor

2. For the transmission line remember to change the wavelengths from .15 to .026.

3. Start the Multisim simulation. A display should appear on the Network Analyzer screen that is similar to that shown in Fig. 34-8. Move the slide across the screen to change frequency and record the following:

Minimum S11 _____

Frequency at Min S11 _____

Instructor's initials _____

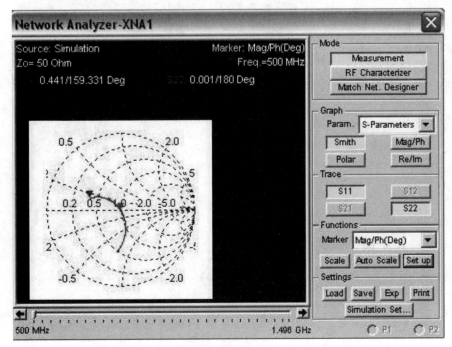

FIGURE 34-8   Measurements for a shunt capacitor

4. Make the measurements for the shunt capacitor circuit in a manner similar to that used for the series capacitor. Enter the results in Table 34-2.

TABLE 34-2 Results for Shunt Capacitor Circuit

#	S11	FREQUENCY (MHz)	VSWR	BANDWIDTH (MHz)
1	Min:			
2	.15L			
3	.15R			

Part IV: Changing Frequency and Dielectric Constant

1. Make the following changes to the shunt circuit.

   a. For the circuit, shown in Fig. 34-7, change R1 and L1 to a single 24.9 Ω resistor.

   b. For the circuit, shown in Fig. 34-7, change C1 to 22 pF.

   c. For the transmission line, shown in Fig. 34-2, change the frequency from 1 GHz to 100 MHz and the wavelengths from .15 to .10.

   d. For the Frequency Range, shown in Fig. 34-4, change the Start frequency to 50 MHz and the Stop frequency to 150 MHz.

   e. Click on the "I/0" switch shown in Fig. 34-7. A display should appear on the Network Analyzer screen that is similar to that shown in Fig. 34-9.

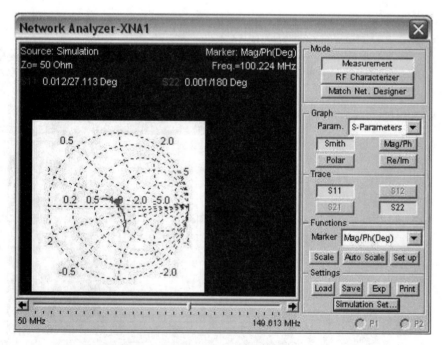

FIGURE 34-9 100 MHz Smith chart for a shunt capacitor

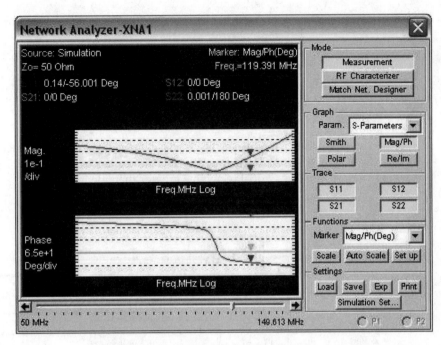

FIGURE 34-10 100 MHz S11 for a shunt capacitor

Part V: Optimizing the Shunt Capacitor Design

1. Replace C1, shown in Fig. 34-7, with a virtual capacitor.

2. With the frequency set for minimum S11, record the results in Table 34-3, for the different values of C1. An example of the set-up for the EWB Network Analyzer is shown in Fig. 34-10.

3. Note the large phase change when C1 is changed from 23.0 pF to 23.5 pF. This shows that the capacitance has been changed from a value that is less than ideal to a value greater than ideal.

4. Similarly, note the large phase change when C1 is changed from 23.1 pF to 23.2 pF.

TABLE 34-3   Optimizing Results for Shunt Capacitor Circuit

#	C1	S11	PHASE (DEGREES)	FREQUENCY (MHz)
1	22.0 pF			
2	22.5 pF			
3	23.0 pF			
4	23.5 pF			
5	24.0 pF			
6	23.1 pF			
7	23.2 pF			
8	23.3 pF			

Part VI: Optimizing the Series Capacitor Design

1. Refer to Fig. 34-1. Change C1 to a 43.7 pF capacitor.

2. Refer to Fig. 34-2. Set the Normalized Electrical Length at .15. Set the frequency at 100 kHz.

3. With the frequency set for minimum S11, record the results in Table 34-4, for the different values of C1.

4. Note the large phase change when C1 is changed from 50 pF to 51 pF. This shows that the capacitance has been changed from a value that is less than ideal to a value greater than ideal.

5. Similarly, note the large phase change when C1 is changed from 44.0 pF to 44.1 pF.

TABLE 34-4   Optimizing Results for Series Capacitor Circuit

#	C1	S11	PHASE (DEGREES)	FREQUENCY (MHz)
1	43.7 pF			
2	43.8 pF			
3	43.9 pF			
4	44.0 pF			
5	44.1 pF			
6	44.2 pF			
7	44.3 pF			
8	44.4 pF			

This EWB laboratory has demonstrated the steps for impedance matching. The textbook EWB CD-ROM* contains the simulation files for this exercise.

File Name

34-1

34-7

34-Part-IV

---

* The files are also available at www.prenhall.com/beasley. Click on Companion Website, then Lab 34 files.

NAME _____

AM GENERATION USING AN ELECTRONIC ATTENUATOR

OBJECTIVE:

To use an electronic attenuator to develop an amplitude-modulated signal.

REFERENCE:

Refer to Chapter 2 of the text.

TEST EQUIPMENT:

Function generator (2)
Spectrum analyzer, with a resolution bandwidth of 3 kHz or less

COMPONENTS:

Mini-Circuits®, Model ZAS-3, electronic attenuator

THEORY:

A ZAS-3 is a PIN diode attenuator (called an "electronic attenuator"). Changing the DC current bias on the control port of ZAS-3 from 0 mA to 10 mA changes the attenuation between the input and output ports from about 60 dB to 1 dB. In order to produce an amplitude-modulated signal, a carrier signal is applied to the input port and the modulating signal is applied to the control port. The amplitude modulated signal is then produced at the output port.

PROCEDURE:

1. Use a coaxial cable to connect the output of the function generator directly to the spectrum analyzer. Set the frequency of the function generator to 2 MHz, and set the power level of the function generator to the maximum level. Record the power:

_____

2. Refer to Fig. 35-1. Connect one end of the coaxial cable to the function generator and the other end to the IN port of the ZAS-3 electronic attenuator. Connect one end of another coaxial cable to the spectrum analyzer and the other end to the OUT port of the ZAS-3 electronic attenuator. Record the power:

_____

3. Apply a DC bias to the CON port, increasing the bias current to 10 mA. Refer to the connection diagram provided in Fig. 35-1. Record the power at 2 MHz out of the OUT port.

_____

4. Set the frequency of signal generator #2 (modulation) to 10 kHz. Increase the signal level until the spectrum analyzer shows sidebands that are 6 dB below the carrier. This corresponds to 100% modulation. You can also use an oscilloscope to verify 100% modulation if needed.

5. Connect a coaxial tee to the OUT port of the electronic attenuator. Connect one output of the tee to the spectrum analyzer and the other to an oscilloscope. Verify that the oscilloscope display looks like Fig. 2-3(b) of the textbook.

6. Find the upper and lower limits of measurable carrier frequency for the spectrum analyzer and the oscilloscope. Enter the results into Table 35-1. Find the upper and lower limits of measurable modulation frequency for the spectrum analyzer and the oscilloscope. Enter the results in Table 35-1. Estimate the lowest measurable modulation level (dynamic range) and enter the results in Table 35-1.

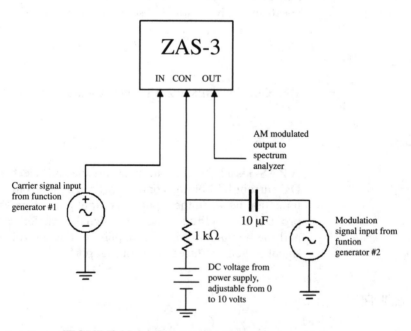

FIGURE 35-1   Electronic attenuator connections

TABLE 35-1   Measurement Limits

	OSCILLOSCOPE	SPECTRUM ANALYZER
Minimum Carrier Frequency		
Maximum Carrier Frequency		
Minimum Modulation Frequency		
Maximum Modulation Frequency		

NAME _____

GENERATING FM FROM A VCO

OBJECTIVE:

To use a voltage-controlled oscillator (VCO) to develop a frequency-modulated signal.

REFERENCE:

Refer to Section 5-2 of the text.

TEST EQUIPMENT:

Function generator
Spectrum analyzer, with a resolution bandwidth of 3 kHz or less
DC power supplies (2)

COMPONENTS:

Mini-Circuits®, Model ZX95-100, voltage-controlled oscillator
Mini-Circuits®, Model SM-BF50, adapter, SMA-M to BNC-F
50 Ω coaxial cable, 3 foot, BNC-M connectors (2)
Banana connector jumper cables, 3 foot (6)

PRELABORATORY:

The instructor may choose to provide the entire assembly (shown in Fig. 36-1) on a metal mounting plate. See Figs. 36-2 and 36-3. Components are:

Metal plate, 8" × 12"
10 kΩ potentiometer, 10 turns
10 μF capacitor
BNC-F panel mount connector
Banana panel mount connectors (6)

THEORY:

A ZX95-100 is a voltage-controlled oscillator that can be tuned from 50 MHz to 100 MHz by changing the voltage at the Vtune port from 0.5 V to 17 V. Another DC voltage of +12 V must be applied to the Vcc. The modulating signal is provided via a 10 kΩ (10 turn) potentiometer and a 10 μF capacitor.

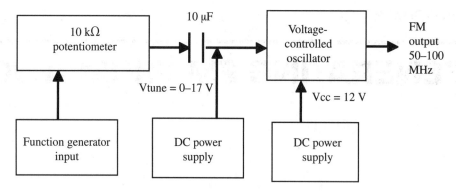

FIGURE 36-1 VCO assembly

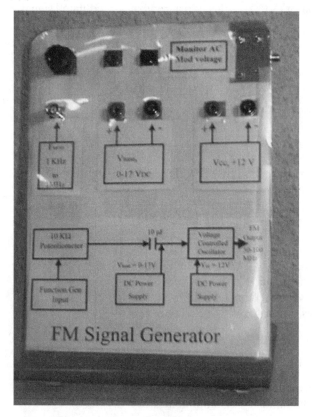

FIGURE 36-2 FM signal generator assembly, front view

FIGURE 36-3 FM signal generator assembly, side/back view

PROCEDURE:

1. Connect the RF OUT port of the VCO to the spectrum analyzer via a coaxial cable with BNC connectors.
2. Set the spectrum analyzer center frequency to 50 MHz and set the span at 10 MHz/div.
3. Apply +12 V dc to the Vcc port of the VCO.
4. Observe the signal on the spectrum analyzer at 50 MHz.

5. Connect another DC power supply to the Vtune port of the VCO. Observe the spectrum analyzer display while increasing the DC voltage until the center frequency is 70 MHz.

6. Adjust the spectrum analyzer center frequency so that the signal is in the middle of the screen.

7. Connect the function generator to the Vtune port via a 10 kΩ potentiometer and a 10 μF capacitor, as shown in Fig. 36-1.

8. Adjust the frequency of the function generator to 100 kHz and the amplitude to the maximum value.

9. Observe the spectrum analyzer display and start reducing the span until you reach 100 kHz/div.

10. Adjust the amplitude of the function generator from a very low value to a very high value and note that the frequency spectrum varies as illustrated in Fig. 5-4 of the textbook. Further note that the amplitude value of the carrier and each sideband follow the Bessel Functions shown in Table 5-2 in the textbook.

11. Measure and record the rms voltage of the 100 kHz signal at the first J_0 null ($x = 2.4$). _____

12. Measure and record the rms voltage of the 100 kHz signal at the 2nd J_0 null ($x = 5.5$). _____

13. Record the AC voltage of the 100 kHz modulating input voltage when $x = 5.5$. _____

14. Calculate the AC voltage for $x = 5.0$. _____

15. Set the AC voltage at the level calculated in step 14.

16. Look at section 5-3, Table 5-2 of the textbook and record the J_0–J_8 in Table 36-1.

17. Convert J_0–J_8 to dB and record in Table 36-1.

18. With x set at 5 for the VCO, measure and record the values of J_0–J_8 in Table 36-1.

Note that J_0–J_5 can be measured directly off the spectrum analyzer. In order to measure J_6–J_8, the center frequency can be changed from 70.01 MHz to 70.31 MHz.

TABLE 36-1   Calculations and Measurements for VCO Project.
(Bessel Functions for $x = 5$)

	J #	VALUE FROM TABLE 5-2 IN THE TEXTBOOK	CALCULATED (dB)	MEASURED (dBm)	MEASURED MINUS CALCULATED VALUES
1	J_0		0 ref	Ref:	———
2	J_1				
3	J_2				
4	J_3				
5	J_4				
6	J_5				
7	J_6				
8	J_7				
9	J_8				

UPCONVERSION AND DOWNCONVERSION

OBJECTIVE:

To gain an appreciation of signal levels and frequencies for various orders of mixing products, used in upconverters and downconverters.

REFERENCE:

Refer to Sections 2-2 and 7-2 of the text.

TEST EQUIPMENT:

Function generator (2)
Spectrum analyzer, with a resolution bandwidth of 3 kHz or less

COMPONENTS:

Mini-Circuits® Model ZP-3 double-balanced mixer
Mini-Circuits® Model ZFL-500HLN 20 dB amplifier

THEORY:

An **upconverter** uses a mixer and a local oscillator. The "order" of the mixing product is defined at $m + n$, where m and n are integers used in the following equation:

$$F_0 = mF_1 \pm nF_2$$

Where F_0 = output frequency
  F_1 = input frequency
  F_2 = local oscillator frequency

In order to minimize the conversion loss between frequencies, the local oscillator power usually needs to be at least +5 dBm. In order to avoid distortion, the input frequency needs to be equal to or less than -20 dBm. The theoretical conversion loss when $m + n = 2$ is 6 dB. Consider the following example:

1. $F_2 = 20$ MHz and $F_1 = 1$ MHz.

2. When $n = 1$ and $m = 1$, the "upconverted" frequency would be 21 MHz. If the power at F_1 is -20 dBm, then the theoretical power at 21 MHz would be -26 dBm.

3. Note that the local oscillator power does not affect the output power, as long as it exceeds the minimum level of $+5$ dBm. For local oscillator power less than $+5$ dBm, the mixer is "starved" and the conversion loss increases.

4. Note than $m - n$ produces a frequency of -19 MHz. A negative frequency is 180° out of phase with a positive frequency. A typical spectrum analyzer has no way of distinguishing a positive frequency from a negative frequency, so a signal will appear at $+19$ MHz. The power level will be also about -26 dBm.

The same circuit can be used as a **downconverter**. Consider the following example:

1. $F_2 = 20$ MHz and $F_1 = 21$ MHz. (Note: You can set $R = 3$ MHz and $L = 2$ MHz if your function generators won't go to 20 and 21 MHz.)

2. When $n = 1$ and $m = 1$, the "downconverted" frequency would be 1 MHz. If the power at F_1 is -20 dBm, then the theoretical power at 1 MHz would be -26 dBm.

3. The frequencies for "higher order" modes ($m + n > 2$) can be calculated. These are calculated for all orders, up to 5 and shown in Table 37-1.

4. The approximate conversion loss increases at a rate of 6 dB per order. For example, a 3rd order product has 12 dB conversion loss and a 4th order product has 18 dB conversion loss. Note that these approximations are for a single mixer. The Mini-Circuit ZP-3 is a "double-balanced mixer" so the actual power level of the mixing products may be substantially different from the 6 dB per order approximation.

PROCEDURE:

1. Use a coaxial cable to connect one function generator to the mixer. Set the frequency at 21 MHz and the power level at -2 dBm.

2. Remove the coaxial cable from the spectrum analyzer and reconnect it to the mixer port labeled "R". Refer to Fig. 37-1 for the location of "R".

3. Use a coaxial cable to connect the other function generator to the mixer. Set the frequency at 20 MHz and note the maximum power level available. If this power level is less than $+5$ dBm, use the 20 dB amplifier and set the power level to $+10$ dBm.

4. Remove the coaxial cable from the spectrum analyzer and reconnect it to the mixer port labeled "L".

5. Use another coaxial cable to connect the spectrum analyzer to the mixer port labeled "X".

6. Measure the power level at all of the frequencies shown in Table 37-1.

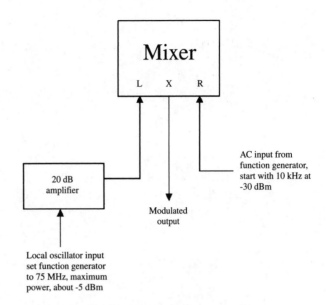

FIGURE 37-1   Mixer connections

TABLE 37-1   Calculated Signals Generated for a Mixer, When $F_2 = 20$ MHz and $F_1 = 21$ MHz $(-20$ dBm).

#	FREQUENCY (MHz)	m	n	ORDER	* CALC. POWER (dBm)	MEASURED WITH −20 dBm INPUT AT 21 MHz
1	1	1	1	2	−26	
2	2	2	2	4	−38	
3	20	1	0	1	—	
4	21	0	1	1	−20	
5	22	1	2	3	−32	
6	23	2	3	5	−44	
7	19	2	1	3	−32	
8	39	3	1	4	−38	
9	41	1	1	2	−26	
10	43	1	3	4	−38	
11	59	4	1	5	−44	
12	61	2	1	3	−32	
13	62	1	2	3	−32	
14	64	1	4	5	−44	
15	81	3	1	4	−38	
16	82	2	2	4	−38	
17	83	1	3	4	−38	
18	101	4	1	5	−44	
19	102	3	2	5	−44	
20	104	1	4	5	−44	

PART II

ELECTRONICS WORKBENCH (EWB) MULTISIM EXPERIMENTS

SIMULATION OF ACTIVE FILTER NETWORKS

OBJECTIVES:

1. To become acquainted with the use of Electronics Workbench Multisim for simulating Active Filter Networks.
2. To understand frequency response measurements using the EWB Bode plotter.
3. To measure the frequency response and phase of an active filter and compare the simulation results with the experimental results.

VIRTUAL TEST EQUIPMENT:

Dual-trace oscilloscope
AC voltage source
Bode plotter
Power supplies
Virtual resistors, capacitors
Simulation model of the 741 operational amplifier

THEORY:

The computer simulation of electronics circuits provides a way to verify the functionality and performance characteristics of your circuit prior to its fabrication and assembly. Many simulation models of both active (e.g., op-amps and transistors) and passive (resistors, capacitors, and inductors) devices are available and provide accurate simulation results. For example, in this experiment the 741 operational amplifier is specified for the operational amplifier. EWB Multisim includes a model of the 741 available for use in the simulation. Using this model, instead of an ideal operational amplifier, enables you to obtain computer simulation results that more closely resemble your experimental results.

PRELABORATORY:

Review your experimental results from Experiment 1–Active Filter Networks.

In this exercise you will be required to build an active filter circuit using Electronics Workbench Multisim. You will be asked to make both frequency and phase measurements on the circuit, and to compare your simulation results with those obtained in the experimental portion of Experiment 1.

Note: Do not forget to apply basic troubleshooting techniques when performing a computer simulation. Visual checks and power supply voltage verification are still a must.

Part I: Plotting the Output Magnitude for the Butterworth Second Order Low-Pass Active Filter

1. The EWB Multisim implementation of a Butterworth Second Order Low-Pass Filter is provided in Fig. E1-1. Your first step is to construct this filter in EWB Multisim. After completing your design, start the simulation and use the oscilloscope to verify that the circuit is functioning properly. To do this, connect the oscilloscope to both the input and output of the filter and verify that the output sine wave is 180° out of phase with respect to the input sine wave. The output signal will also be attenuated with respect to the input (gain less than 1). The connections you need to make to the oscilloscope are shown in Fig. E1-1.

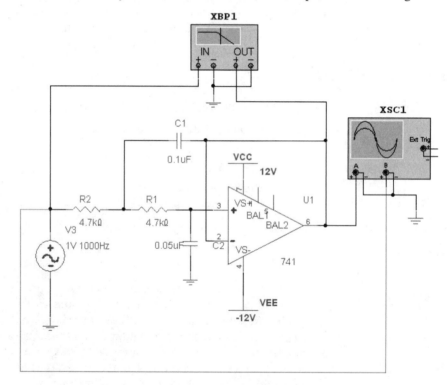

FIGURE E1-1

2. EWB Multisim provides a Bode plotter for making frequency and phase response measurements. A Bode plot graphs the output magnitude and phase as a function of the input signal frequency. Connect the Bode plotter as shown in Fig. E1-1. To use the Bode plotter, double-click on it and set the sweep frequency to an initial value (**I**) of 1 Hz and a final value (**F**) of 20 kHz. Both the Vertical and Horizontal settings should be set to Log. Adjust the initial and final level settings for the vertical axis as needed to display the data results. For this exercise click on the Magnitude button. The settings for the Bode plotter and the expected simulation results are shown in Fig. E1-2.

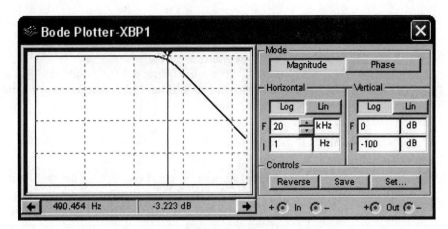

FIGURE E1-2

3. Use the slider on the Bode plotter to measure the critical frequency. This will be the point where the plot drops from 0 dB to −3 dB. Record the value.

_____

4. Move the slider to 1 kHz and record the dB level.   _____

5. Move the cursor to 10 kHz and record the dB level.   _____

6. Calculate and record the dB level difference for 1 kHz and 10 kHz.

_____

This is the amount in dB that the magnitude of the output voltage is being attenuated. Is this value correct? For assistance, refer back to Part I of Experiment 1 (the experimental lab) for a discussion on the Butterworth Second Order Filter.

Part II: Plotting the Output Phase for the Butterworth Second Order Low-Pass Active Filter

1. In this exercise you will make phase measurements on the Butterworth Second Order Low-Pass Filter provided in Fig. E1-1 using the EWB Bode plotter. Your first step is to double-click on the Bode plotter and select on the phase mode by clicking on the Phase button. Change the initial (**I**) phase setting to −360° and final (**F**) phase setting to 180°. The settings for the Bode plotter and the expected simulation results are shown in Fig. E1-3. You may need to rerun the simulation after making the changes.

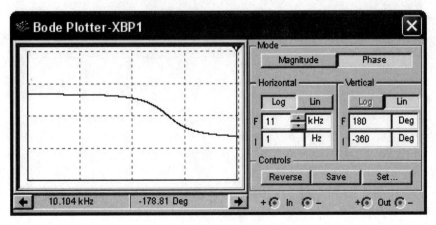

FIGURE E1-3

2. Move the slider to the critical frequency you measured in step 3 of Part I. Record the amount of phase shift, in degrees. _____

> At the critical (or 3-dB frequency) the phase shift of a second order Butterworth filter should be $-90°$. Verify that your measurement is close to this value.

3. Move the cursor to 10 kHz and record the phase. _____

> Your answer should be close to $-180°$. This is the amount of phase shift for an ideal second order Butterworth active filter. Double-click on the Bode plotter and change the final frequency to 500 kHz. Restart the simulation and you will see that the phase shift actually exceeds 180°. This is due to the additional contributions in phase shift provided from the LM741. The simulation for the LM741 models its behavior on real operating characteristics including the non-ideal characteristics such as the increased phase shift.

Part III: Additional EWB Exercises on CD-ROM

The active filter networks for EWB Experiment 1 are provided on the textbook CD-ROM[*] so that you can gain additional experience simulating electronic circuits with EWB and gain more insight into the characteristics of active filter networks. The file names and their corresponding figure numbers are listed.

FILE NAME	FIGURE NUMBER
Lab1-low_pass_filter	Figure 1-1
Lab1-high_pass_filter	Figure 1-2
Lab1-bandpass_filter	Figure 1-3
Lab1-notch_filter	Figure 1-4

[*] The files are also available at www.prenhall.com/beasley. Click on Companion Website, then Lab 1 files.

USING THE SPECTRUM ANALYZER IN THE SIMULATION AND ANALYSIS OF COMPLEX WAVEFORMS

OBJECTIVES:

1. To become acquainted with the use of Electronics Workbench Multisim in analyzing complex waveforms.

2. To understand the operation of the EWB spectrum analyzer.

3. To measure and analyze the spectral content of a sinusoid, a triangle wave, and a square wave.

VIRTUAL TEST EQUIPMENT:

Dual-trace oscilloscope

Function generator

Spectrum analyzer

Virtual resistor

THEORY:

The method of analyzing complex repetitive waveforms is known as Fourier analysis. It permits any complex wave to be represented by a series of sine or cosine waves. The application of Fourier analysis, when using digital sampling oscilloscopes and spectrum analyzers, is provided through the use of the fast Fourier transform (FFT). Examples of using the FFT are presented through the use of the EWB Multisim virtual spectrum analyzer. Refer back to the theory section in Experiment 2, which mathematically defines the relationships for the sine wave, the square wave, and the triangle wave.

PRELABORATORY:

1. Review Section 1-6, Information and Bandwidth, in your text.

2. Review your experimental procedures and results from Experiment 2, Frequency Spectra of Popular Waveforms.

You will be asked to use the EWB Multisim virtual spectrum analyzer to examine simple and complex waveforms. You will gain a better understanding of the use of the spectrum analyzer, including the setup and configuration of the instrument.

Note: Make sure you have Section 1-6 from your text available for reference when working on this laboratory.

Part I: Using the EWB Spectrum Analyzer and the Spectral Analysis of a Sine Wave

1. The first step is to assemble the test setup shown in Fig. E2-1. The setup includes an oscilloscope and a spectrum analyzer, each connected to the output of a function generator. The function generator will be used to generate the waveforms being analyzed.

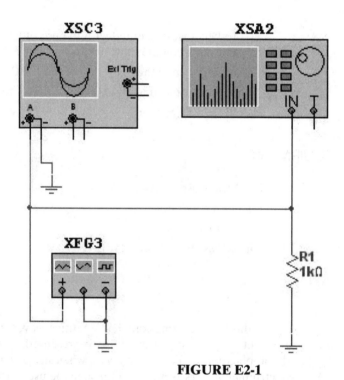

FIGURE E2-1

2. Double-click on the function generator. You should see the panel display of the function generator as shown in Fig. E2-2.

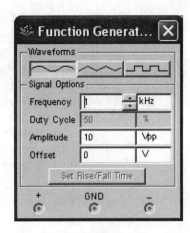

FIGURE E2-2

3. Notice that you have the choice of outputting either a sine wave, a triangle wave, or a square wave. Click on the sine wave. Set the amplitude to 10 V and the offset to 0 V.

4. Start the simulation. Double-click on the oscilloscope and verify that you are outputting a 1-kHz sine wave. Next, double-click on the spectrum analyzer. You should see a display similar to the one shown in Fig. E2-3. Change the settings to match those shown in the figure using the guidelines outlined in the following steps.

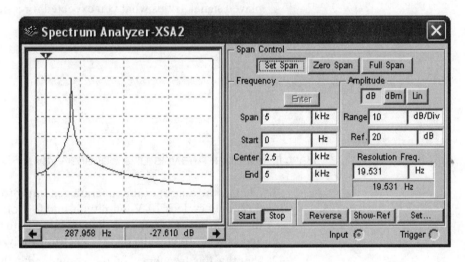

FIGURE E2-3

5. Click on the Set Span button. This provides the user with the ability to set the Span and Frequency controls. Set the Span to 5 kHz, the Start frequency to 0 Hz, the End to 5 kHz, then click on Enter. The center frequency is automatically calculated. The span setting controls the frequency range displayed by the spectrum analyzer.

6. Set the amplitude to dB and the range to 10 dB/Div.

7. Set the Resolution Frequency to 19.531 Hz. This setting defines how many data points are to be used by the spectrum analyzer to display the frequency spectra. The smaller the resolution frequency, the better and more accurate the display. The smallest resolution frequency for the EWB spectrum analyzer is determined by the equation $f\_end/1024$ $(2^{10} = 1024)$ where f_end is the end frequency of the span. In this example, the end frequency equals 5 kHz; therefore, the minimum frequency resolution is 4.8828 Hz. (*Important:* The resolution frequency specified must be a multiple of the minimum resolution frequency such as two times, four times, six times, etc. The 19.531-Hz resolution frequency specified is four times 4.8828 Hz.)

8. Calculate the minimum resolution frequency for the following, given an end frequency of:

END FREQUENCY	MINIMUM RESOLUTION FREQUENCY
100 kHz	
455 kHz	
108 MHz	
433 MHz	

9. How many frequency components are displayed by the spectrum analyzer? One, two, three, . . . ? Use the slider to determine the frequency of the displayed signal. Is this what you expected for a sine wave? Record and justify your answer.

_____

Part II: Spectral Analysis of a Square Wave

1. Change the output of the function generator to a 1-kHz square wave with a 50% duty cycle, a 10-V amplitude, and 0-V offset voltage. Double-click on the spectrum analyzer and change the span to 15 kHz. Start the simulation and observe the results on the spectrum analyzer.

2. Identify the frequency of the components (harmonics) being displayed, their dB levels, and harmonic number. Do this by placing the slider directly over each harmonic. You should be seeing the 1st through 13th harmonics. The 1st harmonic is 1 kHz, the 3rd harmonic is 3 kHz, and so on.

frequency   ____   ____   ____   ____   ____   ____   ____

harmonic   ____   ____   ____   ____   ____   ____   ____

dB value   ____   ____   ____   ____   ____   ____   ____

3. Which harmonic frequency has the greatest amplitude (dB value)?

_____

4. How many dB down is the 7th harmonic relative to the 1st harmonic?

_____

5. How many dB down is the 3rd harmonic relative to the 1st harmonic?

_____

Part III: Spectral Analysis of a Triangle Wave

1. Change the output of the function generator to a 1-kHz triangle wave with a 50% duty cycle, a 10-V amplitude, and 0-V offset voltage. Double-click on the spectrum analyzer and change the span to 15 kHz. Start the simulation and observe the results on the spectrum analyzer.

2. Identify the frequency of the components (harmonics) being displayed, their dB levels, and harmonic number. Do this by placing the slider directly over each harmonic. You should be seeing the 1st through 13th harmonics. The 1st harmonic is 1 kHz, the 3rd harmonic is 3 kHz, and so on.

frequency	____	____	____	____	____	____	____
harmonic	____	____	____	____	____	____	____
dB value	____	____	____	____	____	____	____

3. Which harmonic frequency has the greatest amplitude (dB value)?

_____

4. How many dB down is the 11th harmonic relative to the 1st harmonic?

_____

5. How many dB down is the 5th harmonic relative to the 1st harmonic?

_____

Part IV: Additional EWB Exercise on CD-ROM

This EWB laboratory has demonstrated that complex waveforms, such as a square wave or a triangle wave, generate multifrequency components called harmonics. It demonstrated how the spectrum analyzer can be used to observe and analyze the spectral content of a square wave. The simulation files for this exercise are provided on the textbook CD-ROM.[*]

FILE NAME
Lab2-sine_wave
Lab2-square_wave
Lab2-triangle_wave

[*]The files are also available at www.prenhall.com/beasley. Click on Companion Website, then Lab 2 files.

SIMULATION OF CLASS C AMPLIFIERS AND FREQUENCY MULTIPLIERS

OBJECTIVES:

1. To become acquainted with the use of Electronics Workbench Multisim in simulating class C amplifiers and frequency multiplier circuits.

2. To understand resonant circuits.

3. To analyze the class C amplifier and the frequency multiplier circuit.

VIRTUAL TEST EQUIPMENT:

Dual-trace oscilloscope   AC voltage source

Bode plotter   Virtual 2N2222A transistor

Virtual resistor   Virtual capacitor

Virtual inductor

PRELABORATORY:

Review Section 1-6, Information and Bandwidth, in your text.

PROCEDURE:

You will simulate a class C amplifier using EWB Multisim. You will use the EWB instruments to measure the operating characteristics of the amplifier.

1. Construct the class C amplifier provided in Fig. E3-1. This is the same circuit provided in Experiment 3, Fig. 3-4. Connect the EWB instruments to the amplifier as shown in Fig. E3-1.

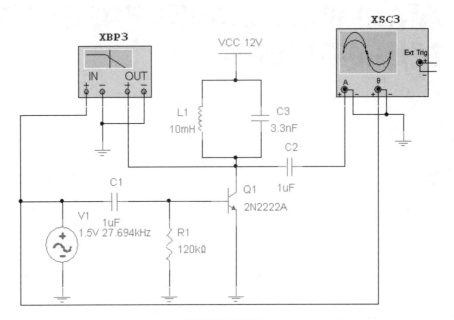

FIGURE E3-1

2. Calculate the resonant frequency of the tank circuit made up of $L1$ and $C3$. Use the equation $f_r = \dfrac{1}{2\pi\sqrt{LC}}$. Record your answer.

_____

3. Use the cursor on the Bode plotter to determine the resonant frequency of the class C amplifier. The resonant frequency, f_r, is the point on the plot where the amplitude is maximum. Record your answer and compare the test results obtained with the Bode plotter to the answer calculated in Part II. Your Bode plot should be similar to Fig. E3-2.

_____

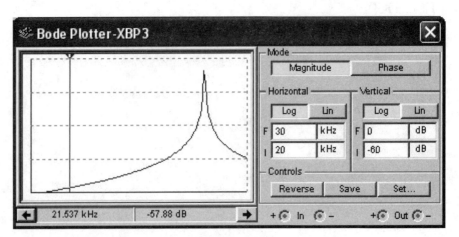

FIGURE E3-2

4. Use the Bode plotter to determine the bandwidth of the tank circuit. To determine the bandwidth, move the cursor so that it is sitting directly over the peak magnitude. Record the resonant frequency in the table provided. Next, move the marker to the left of the peak until the magnitude drops 3 dB of the peak value. Record the −3-dB frequency which will be indicated as $f(-)$. Repeat this procedure to determine the −3-dB frequency to the right of the peak magnitude and record your measurement. This point will be indicated as $f(+)$. Use the −3-dB measured frequencies to determine the bandwidth BW using the equation, $BW = f(+) - f(-)$.

f_r	$f(+)$	$f(-)$	BW

5. Use the measured values obtained from the Bode plotter to determine the Q of the amplifier. Record your answer.

$$Q = \frac{f_r}{BW}$$

_____

Part II: Frequency Multiplication

You will experiment with the use of the tank circuit to provide frequency multiplication. The tank circuit, or parallel resonant LC circuit, can be pulsed by a fast switch such as a transistor to turn the tank on and off.

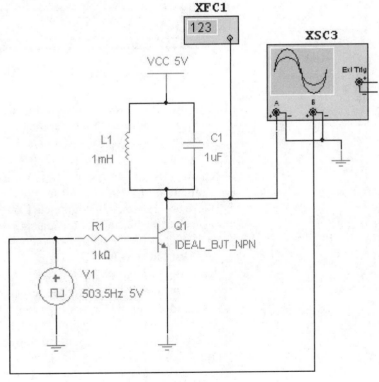

FIGURE E3-3

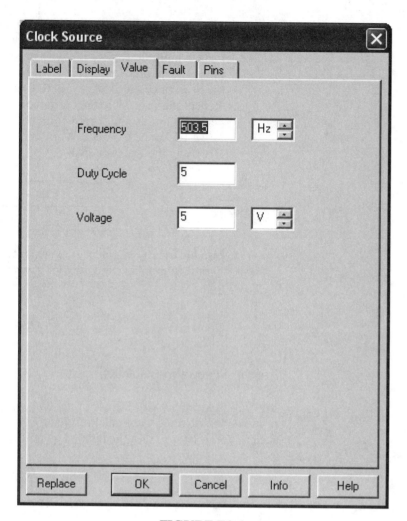

FIGURE E3-4

1. Construct the circuit shown in Fig. E3-3. Double-click on the square-wave generator and change the frequency to 503.5 Hz. Set the duty cycle to 5 and the voltage to 5 V as shown in Fig. E3-4.

2. The inductor $L1$ and the capacitor $C1$ used in the tank circuit have ideal characteristics. The use of these components will not provide an accurate simulation of a tank circuit in a real environment. Therefore, to provide a more accurate model, double-click on $L1$, click on the **Fault** tab, select **Leakage,** click on **1 2,** and set the leakage resistance to 500 Ω. Next, double-click on $C1$, click on the **Fault** tab, select **Leakage,** click on **1 2,** and set the leakage resistance to 750 kΩ.

3. Start the simulation and double-click on the oscilloscope. You should see a display similar to Fig. E3-5. Notice that the tank circuit pulsing is controlled by an ideal npn BJT transistor, which has been configured as a switch. The input to the transistor switch is a short duration pulse. This causes the transistor to saturate, momentarily connecting the collector to the emitter (ground) and causing the tank circuit to oscillate at a frequency defined by $L1$ and $C1$. Notice that the oscillation decays over time. This is due to the leakage of the inductor and capacitor.

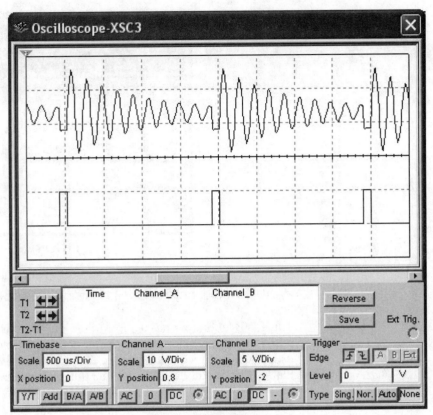

FIGURE E3-5

4. Use the values of $L1$ and $C1$ to calculate the resonant frequency of the tank.

$$f_r = \frac{1}{2\pi\sqrt{LC}}$$ _____

5. Use the cursors on the oscilloscope to measure the frequency of oscillation. This can be done by setting one cursor at the beginning of the cycle for a sine wave and setting the other cursor at the end of the cycle. The difference, in time (T), is displayed as $T2 - T1$ on the oscilloscope. Use the equation $f = \frac{1}{T}$ to calculate the frequency. Record your answer and compare the measured frequency to the calculated frequency. They should be close in value.

_____

6. Change the frequency of the square-wave pulse to 2.516 kHz, which is one-half the resonant frequency of the tank circuit. Set the duty cycle to 25 and the voltage to 3.0 V. You should observe that the output frequency is twice the input frequency. The tank circuit is being used to double the frequency.

7. With the frequency of the square wave set to 2.56 kHz, attach a frequency counter to the output of the multiplier. Double-click on the frequency counter. Change the sensitivity to 1 V and the trigger level to 1 V, and select Freq and AC coupling as shown in Fig. E3-6. Compare your measured value to the calculated value obtained in step 4. Use the frequency counter to determine the period of the output signal.

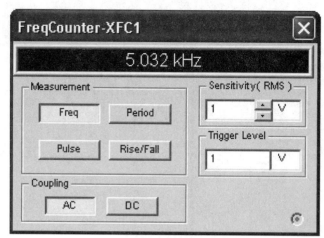

FIGURE E3-6

Part III: Additional EWB Exercise on CD-ROM

This EWB laboratory has demonstrated the analysis and simulation of a class C amplifier and the development of a frequency multiplier. The EWB instruments were used to confirm theoretical results. Your textbook EWB CD-ROM[*] contains the simulation files for this exercise and the full implementation of the frequency doubler.

FILE NAME
Lab3-classC_amplifier
Lab3-tank_circuit
Lab3-frequency_doubler

[*]The files are also available at www.prenhall.com/beasley. Click on Companion Website, then Lab 3 files.

SIMULATION OF A PHASE-SHIFT OSCILLATOR

OBJECTIVES:

1. To become acquainted with the use of Electronics Workbench Multisim in simulating and analyzing oscillator circuits.

2. To understand the phase-shift oscillator circuit.

3. To analyze the phase-shift oscillator and to better understand the feedback signal path in an oscillator circuit.

VIRTUAL TEST EQUIPMENT:

Dual-trace oscilloscope   AC voltage source

Virtual 741 op-amp   Virtual resistor

Virtual capacitor   DC voltage source

PRELABORATORY:

Review Section 1-8, Oscillators, in your text.

PROCEDURE:

You will construct and simulate a phase-shift oscillator circuit using EWB Multisim. To gain a better understanding of the operation of the circuit, you will use the EWB instruments to verify the operation and make measurements.

Part I: Analyzing and Testing the Components of the Phase-Shift Oscillator

Construct the phase-shift oscillator circuit provided in Fig. E5-1. This is the same circuit provided in Experiment 5, Fig. 5-1. Connect the EWB instruments to the amplifier as shown in Fig. E5-1. Notice that the circuit contains a virtual switch so that the feedback path can be opened. You can flip or rotate the components by selecting **Edit** from the Multisim menu and clicking on the desired change. Notice that the 741 op amps have been vertically shifted from their default orientation.

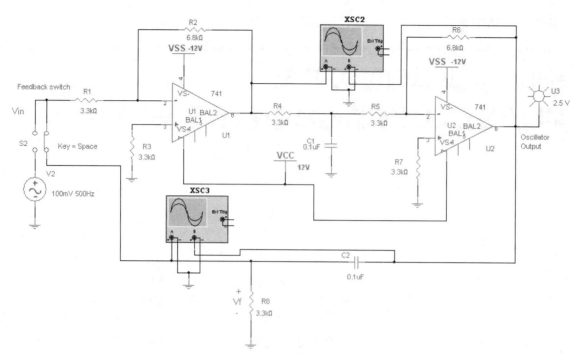

FIGURE E5-1

1. Use the EWB Multisim virtual multimeter to verify that the voltages to the operational amplifiers are set to ± 12 V dc. Double-click on the AC voltage source and set the voltage amplitude to 100 mV at a frequency of 500 Hz.

2. Set the double-pole single-throw (DPST) feedback switch so the feedback path from the oscillator is opened and the 500 Hz AC voltage source is connected to the input of operational amplifier $U1$ through resistor $R1$. The position of the virtual switch can be toggled by pressing the space bar on your keyboard. The circuit window in Multisim must be the active window for the space bar to change the switch. The two switch positions provide a way to test the operational amplifiers in an open-loop configuration. The switch can then be repositioned to close the feedback loop so that the oscillator can be tested.

3. Start the simulation and verify that $U1$ and $U2$ are working properly. Use the oscilloscope to measure and record the peak-to-peak voltage level of the signals at the output of $U1$ pin 6 ($U1$-6) and $U2$ pin 6 ($U2$-6).

   $U1$-6 _____ $U2$-6 _____

4. Next, use the oscilloscope to measure the phase difference of the input voltage V_{in} to the feedback voltage V_f. Refer to Fig. E5-1 for the location of V_{in} and V_f in the circuit. Connect V_{in} to channel A and V_f to channel B on the oscilloscope. Set the feedback switch so that the feedback loop is open and the AC voltage source is connected to the circuit. You will need to connect a 3.3-kΩ resistor ($R9$) in parallel with $R8$. The resistor $R9$ is being used to simulate the resistive load that the input operational amplifier places on the network if the oscillator is operating in the closed-loop mode.

Use the cursors on the oscilloscope to measure the time difference from the peak of the V_{in} input signal to the peak of the V_f feedback signal at each frequency. Use cursor $T1$ for V_{in} and cursor $T2$ for V_f. An example of the cursor placement for this measurement is provided in Fig. E5-2.

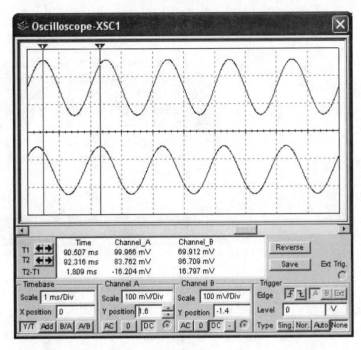

FIGURE E5-2

Use the following equation to calculate the amount of phase shift. Record the time and phase difference for V_{in} to V_f for the input frequencies of 500 Hz, 1000 Hz, 1500 Hz, and 2000 Hz in Table E5-1.

$$Phase\ shift = [(T2 - T1) / period] \times 360°$$

A difference of 1.8 ms is shown in Fig. E5-2, which equates to a phase shift of

$$[1.8\ ms / 2\ ms] \times 360° = 324°$$

TABLE E5-1   Test Results on the Phase Shift Circuit

FREQUENCY (Hz)	$T1(V_{in})$	$T2(V_f)$	$T2 - T1$	PHASE SHIFT
500				
1000				
1500				
2000				

5. Review the phase shift test results in Table E5-1 and explain what happens to the phase shift at 500, 1000, 1500, and 2000 Hz. At what frequency is the Barkhausen criteria for oscillation met in regards to the phase shift.

6. Verify your measurements and the results obtained in steps 4 and 5 with the EWB Bode plotter. Connect the input of the Bode plotter to the AC voltage source and the output of the Bode plotter to the node connecting the parallel resistors $R8$ and $R9$. Remember, $R9$ must be added in the open loop mode for testing the oscillator feedback. Sweep the open loop circuit from 1 Hz to 10 kHz. The resonant frequency of the circuit is indicated by the peak frequency. The expected result from the Bode plot test is provided in Fig. E5-3.

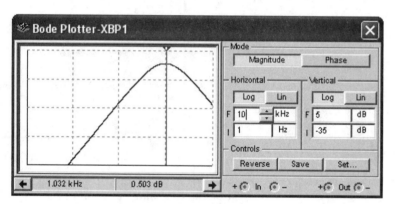

FIGURE E5-3

Record the frequency of the peak value in step 6.   _____

Record the phase of the peak frequency.   _____

Is the phase result consistent with the Barkhausen criteria? Why?

Part II: EWB Simulation of the Phase-Shift Oscillator

1. Change the feedback switch so that the loop for the oscillator is closed. $R8$ should now connect to the 3.3-kΩ resistor on the input of $U1$. This closes the feedback loop. Restart the simulation and observe the output of the oscillator. It may take up to .04 s of simulation time for the oscillation to start. Determine the oscillating frequency. Is this the frequency you expected? Why?

2. Connect a 3.3-kΩ resistor in parallel with $R8$. The addition of this resistor alters the feedback signal, which in turn prevents the circuit from maintaining oscillation. Double-click on the oscilloscope and restart the simulation. Press the space bar twice to toggle the feedback switch. This will momentarily start the oscillator. View the output of the oscillator to see the effect changing the feedback path has on the circuit. You should be seeing a damped sine wave similar to Fig. E5-4. Why does the circuit no longer maintain oscillation?

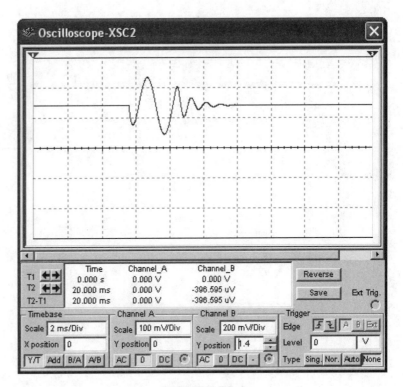

FIGURE E5-4

Part III: EWB Exercises on CD-ROM

This EWB laboratory has demonstrated the operation of the phase shift oscillator. The textbook EWB CD-ROM* contains the simulation files for this exercise.

FILE NAME
Lab5-phase_shift_oscillator
Lab5-phase_shift_test
Lab5-damped_osc
Lab5-phase_shift_Bode

*The files are also available at www.prenhall.com/beasley. Click on Companion Website, then Lab 5 files.

SIMULATION OF AN *LC* FEEDBACK OSCILLATOR

OBJECTIVES:

1. To become acquainted with the use of Electronics Workbench Multisim in simulating and analyzing oscillator circuits.

2. To understand the *LC* feedback oscillator circuit.

3. To analyze the *LC* feedback oscillator and develop a better understanding of the requirements for testing the feedback signal path in an oscillator circuit.

VIRTUAL TEST EQUIPMENT:

Dual-trace oscilloscope	AC voltage source
Virtual 741 op-amp	Virtual resistor
Virtual capacitor	DC voltage source
Virtual inductor	Bode plotter
Virtual switch	

PRELABORATORY:

Review Section 1-8, Oscillators, in your text.

PROCEDURE:

You will construct and simulate an *LC* oscillator circuit using EWB Multisim. You will use the EWB instruments to verify the operation of the oscillator, and you will conduct tests and make measurements to gain a better understanding of the operation of the circuit.

Part I: Analyzing and Testing the Components of the LC Oscillator

Construct the *LC* oscillator circuit provided in Fig. E6-1. This circuit is similar to the circuit provided in Experiment 6, Fig. 6-1. Connect the EWB instruments to the amplifier as shown in Fig. E6-1. The instruments include the Bode plotter and the oscilloscope. The circuit contains a virtual switch so that the feedback path can be opened and closed as needed for testing.

Note: In some circumstances you may need to change the orientation of a component. You can flip or rotate the components by selecting **Edit** from the Multisim menu and clicking on the desired change. Notice that the 741 op amps have been vertically flipped from their default orientation.

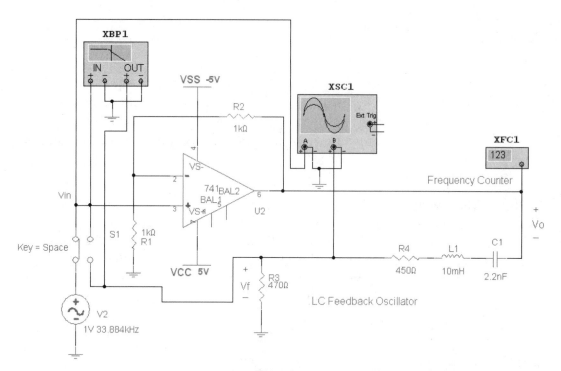

FIGURE E6-1

1. Construct the *LC* oscillator circuit shown in Fig. E6-1 in EWB Multisim. Connect the switch and the instruments to the circuit as shown.

2. Calculate the resonant frequency of the *LC* oscillator circuit shown in Fig. E6-1 using the equation below and record the value.

$$f = \frac{1}{2\pi\sqrt{L_1 C_1}}$$ _____

3. Use the EWB Multisim virtual multimeter to verify that the voltages to the operational amplifiers are set to ± 5-V dc. Double-click on the AC voltage source and set the voltage amplitude to 1.0 V at the frequency calculated in step 2. You will need to start the simulation to make the voltage measurements.

4. Verify your resonant frequency calculation from step 2 with the EWB Bode plotter. Connect the input of the Bode plotter to the AC voltage source and the output to the oscillator output, V_o. Make sure to conduct the test with the feedback switch set to the AC voltage source (feedback open). Sweep the open loop circuit from 1 Hz to 10 MHz. The resonant frequency of the circuit is indicated by the peak frequency on the Bode plot. The expected result from the Bode plot test is provided in Fig. E6-2.

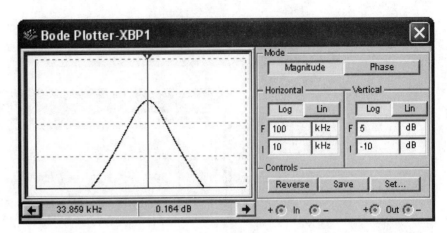

FIGURE E6-2

Record the peak frequency value measured with the Bode plotter and compare your result with the frequency calculated in step 2.

_____

5. Next, use the oscilloscope to measure the phase difference of the input voltage to the op amp V_{in} to the feedback voltage V_f. Refer to Fig. E6-1 for the location of V_{in} and V_f in the circuit. Connect the input voltage to channel A and V_f to channel B on the oscilloscope. Set the feedback switch so that the feedback loop is open and the AC voltage source is connected to the circuit.

Use the cursors on the oscilloscope to measure the time difference from the peak of the V_{in} input signal to the peak of the V_f feedback signal. Use cursor $T1$ for V_{in} and cursor $T2$ for V_f. An example of the cursor placement for this measurement is provided in Fig. E6-3.

Use the following equation to calculate the amount of phase shift. Record the time and phase difference for V_{in} to V_f for the input frequencies of 10 kHz, 25 kHz, 33.9 kHz, 65 kHz, and 100 kHz in Table E6-1.

$$Phase\ shift = [(T2 - T1) / period] \times 360°$$

where the period of a 10 kHz sine wave is 1/10 kHz = 100 μs.

A difference of 77.9 μs is shown in Fig. E6-3, which equates to a phase shift of

$$[77.9\ μs / 100\ μs] \times 360° = 280°$$

TABLE E6-1 Test Results on the Phase Shift Circuit

FREQUENCY (kHz)	$T1(V_{in})$	$T2(V_f)$	$T2 - T1$	PHASE SHIFT
10				
25				
33.9				
65				
100				

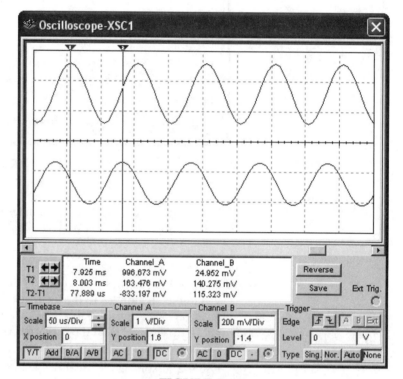

FIGURE E6-3

Part II: EWB Simulation of the LC Oscillator

1. Change the feedback switch so that the loop for the oscillator is closed. Restart the simulation and observe the output of the oscillator. It may take up to .02 s of simulation time for the oscillation to start. Determine the oscillating frequency. Is this the frequency you expected? Why?

2. Change $R4$ to 300 Ω and restart the simulation. What change do you observe with the oscillator? Why do you think this happened?

3. Change $R4$ to 480 Ω and restart the simulation. You will need to help start the oscillator by momentarily switching the input which opens and closes the feedback path. Change the time base on the oscilloscope to 100 μs/div to better see the output signal. What change do you observe with the oscillator? Why do you think this happened?

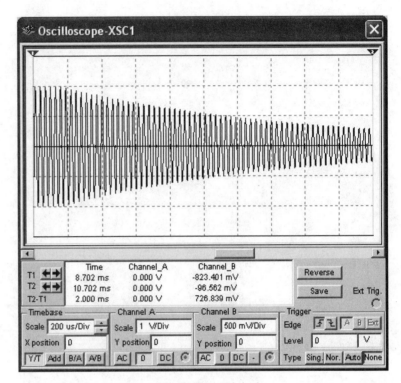

FIGURE E6-4

Part III: EWB Exercises on CD-ROM

This EWB laboratory has demonstrated the operation of the *LC* oscillator. The textbook EWB CD-ROM[*] contains the simulation files for this exercise.

FILE NAME
Lab6-LC_oscillator
Lab6-LC_R4_300
Lab6-LC_R4_480

[*]The files are also available at www.prenhall.com/beasley. Click on Companion Website, then Lab 6 files.

PERCENTAGE OF MODULATION MEASUREMENT OF AN AMPLITUDE MODULATED WAVEFORM

OBJECTIVES:

1. To become acquainted with the use of Electronics Workbench Multisim in simulating and measuring amplitude modulated waveforms.

2. To understand the EWB oscilloscope.

VIRTUAL TEST EQUIPMENT:

Dual-trace oscilloscope	AC voltage source
Virtual resistor	AM source

PRELABORATORY:

1. Review the amplitude modulation material in Chapters 2 and 3 of your text.

2. Review Experiment 11.

PROCEDURE:

You will use EWB Multisim to conduct tests on the simulation of an AM signal. You will use the EWB instruments to measure modulation level and frequency.

Part I: Trapezoidal Measurements of the AM Waveform

Construct the test circuit provided in Fig. E11-1. This circuit consists of an AM source, an AC voltage source, a resistive load, and an oscilloscope. The modulation (intelligence) frequencies for the AM source and the AC voltage must be equal.

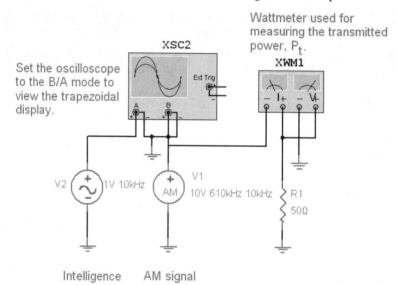

FIGURE E11-1

1. Double-click on the AM source and set the carrier frequency to 610 kHz, the modulation frequency (intelligence) to 10 kHz, the modulation index to 0.5, and the carrier amplitude to 1 V. The settings are shown in Fig. E11-2. Set the modulation frequency (intelligence) of the AC voltage source to 10 kHz and the amplitude to 1 V.

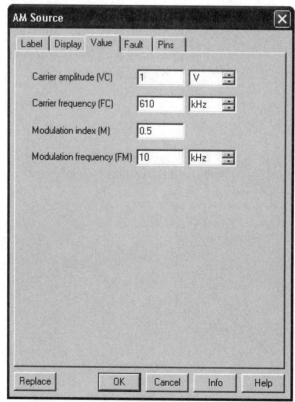

FIGURE E11-2

2. Double-click on the oscilloscope and set the display mode to B/A. This setting can be found in the lower left corner of the oscilloscope front panel. Set the scale for both channels A and B to 1 V/div. Start the simulation and observe the display on the oscilloscope. You should see a trapezoidal display similar to the one shown in Fig. E11-3. The A and B text will not be displayed on the oscilloscope.

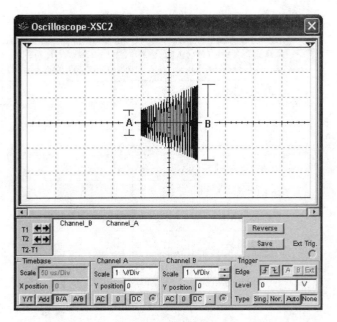

FIGURE E11-3

3. Use the techniques presented in the text to measure the percentage of modulation of the AM waveform. The maximum and minimum levels for the AM waveform are indicated as B and A in Fig. E11-3. These two values can be used to calculate the percentage of modulation (m) for the AM waveform. The equation for calculating the percentage of modulation is

$$\%m = \frac{B - A}{B + A} \times 100\%$$

For this example, $B = 3.0$V and $A = 1.0$V. Using these values to calculate the percentage of modulation gives

$$\%m = \frac{3 - 1}{3 + 1} \times 100\% = 50\%$$

4. In this step, you will open a Multisim simulation file on your textbook EWB CD-ROM.[*] You are to measure the percentage of modulation of the AM source for each simulation. Record your measurements in Table E11-1. You are also to calculate and measure the output power for each Multisim file. Refer back to Section 2-4 of the text for help with calculating P_T.

TABLE E11-1

FILE NAME (FROM EWB LAB CD-ROM)	PERCENTAGE OF MODULATION (%m)	OUTPUT POWER (P_T) CALCULATED	OUTPUT POWER (P_T) MEASURED
Lab11-AM-1			
Lab11-AM-2			
Lab11-AM-3			
Lab11-AM-4			
Lab11-AM-5			
Lab11-AM-6			
Lab11-AM-7			

Lab11-AM:7 is showing an example of what?
Note: The wattmeter measured value will vary as the carrier is being modulated by the intelligence. You may have to approximate the measured power.

[*]The files are also available at www.prenhall.com/beasley. Click on Companion Website, then Lab 11 files.

APPENDIX: MANUFACTURER DATA SHEETS

IRON POWDER TOROIDAL CORES

Iron Powder toroidal cores are available in numerous sizes ranging from .05 inches to more than 5 inches in outer diameter. There are two basic material groups : The Carbonyl Irons and the Hydrogen Reduced Irons.

The CARBONYL IRONS are especially noted for their stability over a wide range of temperatures and flux levels. Their permeability range is from less than 3 mu to 35 mu and can offer excellent 'Q' factors for the 50 KHz to 200 MHz frequency range. They are ideally suited for a variety of RF circuit applications where good stability and high 'Q' are essential.

The HYDROGEN REDUCED IRONS have permeabilities ranging from 35 mu to 90 mu. Somewhat lower 'Q' values can be expected from this group of cores and they are mainly used for EMI filters and low frequency chokes. In recent years they have been very much in demand for use in both input and output filters for switched-mode power supplies.

Toroidal cores, in general, are the most efficient of any core configuration. They are highly self-shielding since most of the lines of flux are contained within the toroidal form. The flux lines are essentially uniform over the entire magnetic path length and consequently stray magnetic fields will have very little effect on a toroidal inductor. It is seldom necessary to shield or isolate a toroidal inductor to prevent feedback or cross-talk. Toroidal inductors simply do not like to talk to each other.

The A_L values of Iron Powder toroidal cores will be found on the next few pages. Use these A_L values and the formula below to calculate the required number of turns for a given inductance value. The wire chart should then be consulted to determine if the required number of turns will fit on to the chosen core size.

Turns Formula

$$\text{Turns} = 100 \sqrt{\frac{\text{desired L (uh)}}{A_L \text{ (uh}/100\,t)}}$$

Key to part number

$$\underset{\text{toroid}}{T} - \underset{\text{outer-diameter}}{50} - \underset{\text{material}}{6}$$

Substantial quantities of most catalog items are maintained in stock for immediate delivery.

IRON POWDER TOROIDAL CORES
For Resonant Circuits

MATERIAL 2   Perm. 10   Freq. Range 1 - 30 MHz   Color code - Red

Core number \/	O.D. (inches)	I.D. (inches)	Hgt. (inches)	I_e (cm)	A_e (cm)^2	V_e (cm)^3	A_L Value uh/100 turns
T-12-2	.125	.062	.050	0.74	.010	.007	20
T-16-2	.160	.078	.060	0.95	.016	.015	22
T-20-2	.200	.088	.070	1.15	.025	.029	27
T-25-2	.255	.120	.096	1.50	.042	.063	34
T-30-2	.307	.151	.128	1.83	.065	.119	43
T-37-2	.375	.205	.128	2.32	.070	.162	40
T-44-2	.440	.229	.159	2.67	.107	.286	52
T-50-2	.500	.303	.190	3.03	.121	.367	49
T-68-2	.690	.370	.190	4.24	.196	.831	57
T-80-2	.795	.495	.250	5.15	.242	1.246	55
T-94-2	.942	.560	.312	6.00	.385	2.310	84
T-106-2	1.060	.570	.437	6.50	.690	4.485	135
T-130-2	1.300	.780	.437	8.29	.730	6.052	110
T-157-2	1.570	.950	.570	10.05	1.140	11.457	140
T-184-2	1.840	.950	.710	11.12	2.040	22.685	240
T-200-2	2.000	1.250	.550	12.97	1.330	17.250	120
T-200A-2	2.000	1.250	1.000	12.97	2.240	29.050	218
T-225 -2	2.250	1.405	.550	14.56	1.508	21.956	120
T-225A-2	2.250	1.485	1.000	14.56	2.730	39.749	215
T-300 -2	3.058	1.925	.500	19.83	1.810	35.892	800
T-300A-2	3.048	1.925	1.000	19.83	3.580	70.991	228
T-400 -2	4.000	2.250	.650	24.93	3.660	91.244	180
T-400A-2	4.000	2.250	1.300	24.93	7.432	185.280	360
T-520 -2	5.200	3.080	.800	33.16	5.460	181.000	207

MATERIAL 3   Perm 35   Freq. Range .05 -.5 MHz   Color code - Gray

Core number \/	O.D. (inches)	I.D. (inches)	Hgt. (inches)	I_e (cm)	A_e (cm)^2	V_e (cm)^3	A_L Value uh/100 turns
T-12-3	.125	.062	.050	0.74	.010	.007	60
T-16-3	.160	.078	.060	0.95	.016	.015	61
T-20-3	.200	.088	.070	1.15	.025	.029	90
T-25-3	.255	.120	.096	1.50	.042	.063	100
T-30-3	.307	.151	.128	1.83	.065	.119	140
T-37-3	.375	.205	.128	2.32	.070	.162	120
T-44-3	.440	.229	.159	2.67	.107	.286	180
T-50-3	.500	.303	.190	3.03	.121	.367	175
T-68-3	.690	.370	.190	4.24	.196	.831	195
T-80-3	.795	.495	.250	5.15	.242	1.246	180
T-94-3	.942	.560	.312	6.00	.385	2.310	248
T-106-3	1.060	.570	.437	6.50	.690	4.485	450
T-130-3	1.300	.780	.437	8.29	.730	6.052	350
T-157-3	1.570	.950	.570	10.05	1.140	11.457	420
T-184-3	1.840	.950	.710	11.12	2.040	22.685	720
T-200-3	2.000	1.250	.550	12.97	1.330	17.250	425
T-200A-3	2.000	1.250	1.000	12.97	2.240	29.050	460
T-225 -3	2.250	1.405	.550	14.56	1.508	21.956	425

AMIDON Associates · 12033 OTSEGO STREET · NORTH HOLLYWOOD, CALIF. 91607

INTRODUCTION TO CERAMIC FILTERS

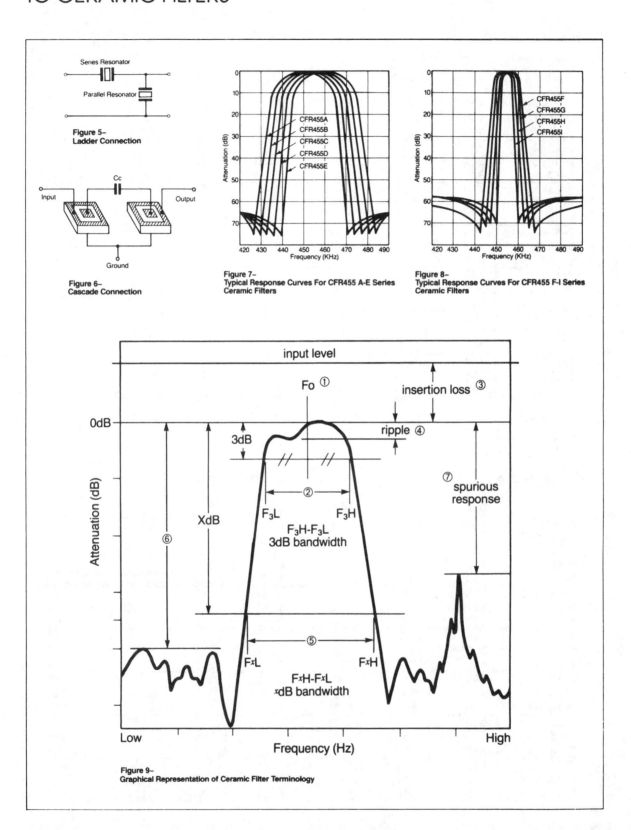

Figure 5–
Ladder Connection

Figure 6–
Cascade Connection

Figure 7–
Typical Response Curves For CFR455 A-E Series Ceramic Filters

Figure 8–
Typical Response Curves For CFR455 F-I Series Ceramic Filters

Figure 9–
Graphical Representation of Ceramic Filter Terminology

INTRODUCTION
TO CERAMIC FILTERS

CERAMIC FILTER TERMINOLOGY

Although the previous section has presented a concise discussion of piezoelectric theory as applied to ceramic filter technology, it is necessary that the respective terminology used in conjunction with ceramic filters be discussed before any further examination of ceramic filter technology is made.

Using Figure 9 as a typical model of a response curve for a ceramic filter, it can be seen that there are a number of relevant factors to be considered in specifying ceramic filters. These include: center frequency, pass-bandwidth, insertion loss, ripple, attenuation bandwidth, stopband attenuation, spurious response and selectivity. Although not all of these factors will apply to each filter design, these are the key specifications to consider with most filters. From the symbol key shown in Table 1 below, a thorough understanding of this basic terminology should be possible.

IMPEDANCE MATCHING

As it is imperative to properly match the impedances whenever any circuit is connected to another circuit, any component to another component, or any circuit to another component, it is also important that this be taken into account in using ceramic filters. Without proper impedance matching, the operational characteristics of the ceramic filters cannot be met.

Figure 12 illustrates a typical example of this requirement.

This example shows the changes produced in the frequency characteristics of the SFZ455A ceramic filter when the resistance values are altered. For instance, if the input/output impedances R_1 and R_2 are connected to lower values than those specified, the insertion loss increases, the center frequency shifts toward the low side and the ripple increases.

TABLE 1 - CERAMIC FILTER TERMINOLOGY CHART

Numbers In Fig. 9	Terminology	Symbol	Unit	Explanation of Term
1	Center Frequency	f_o	Hz	The frequency in the center of the pass-bandwidth. However, the center frequency for some products is expressed as the point where the loss is at its lowest point.
2	Pass-bandwidth (3dB bandwidth)	(3dB) B.W.	Hz	Signifies a difference between the two frequencies where the attenuation becomes 3dB from the level of the minimum loss point.
3	Insertion Loss	I.L.	dB	Expressed as the input/output ratio at the point of minimum loss. (The insertion loss for some products is expressed as the input/output ratio at the center frequency.) Insertion loss = 20 LOG (V_2/V_1) in dB.
4	Ripple	—	dB	If there are peaks and valleys in the pass-bandwidth, the ripple expresses the difference between the maximum peak and the minimum valley.
5	Attenuation Bandwidth (dB Bandwidth)	(20dB) B.W.	Hz	The bandwidth at a specified level of attenuation. Attenuation may be expressed as the ratio of the input signal strength to the output signal strength in decibels.
6	Stopband Attenuation	—	dB	The level of signal strength at a specified frequency outside of the passband.
7	Spurious Response	SR	dB	The difference in decibels between the insertion loss and the spurious signal in the stopband.
	Input/Output Impedance	—	Ohm	Internal impedance value of the input and output of the ceramic filter.
	Selectivity	—	dB	The ability of a filter to pass signals of one frequency and reject all others. A highly selective filter has an abrupt transition between a passband region and the stopband region. This is expressed as the shape factor—the attenuation bandwidth divided by the pass-bandwidth. The filter becomes more selective as the resultant value approaches one.

INTRODUCTION TO CERAMIC FILTERS

On the other hand, if R_1 and R_2 are connected to higher values other than those specified, the insertion loss will increase, the center frequency will shift toward the high side and the ripple will increase.

DEALING WITH SPURIOUS RESPONSE

Frequently in using 455 KHz filters, spurious will cause problems due to the fact that the resonance occurs under an alien vibrating mode or overtone deviating from the basic vibration characteristics. Among available solutions for dealing with spurious response are:

1. The use of a supplementary IFT together with the ceramic filter for suppression of the spurious.
2. The arrangement of two or more ceramic filters in parallel for the mutual cancellation of spurious.
3. The addition of a low-pass or high-pass LC filter for suppression of spurious.

Perhaps the most commonly used method of dealing with spurious is the use of a supplementary IFT in conjunction with the ceramic filter. The before and after effects of the use of an IFT are shown in Figures 10 and 11. In Figure 10, only a single SFZ455A ceramic filter is employed and spurious is a significant problem. With the addition of an IFT, the spurious problem is reduced as is shown in Figure 11.

Although spurious is a significant problem to contend with when using 455KHz ceramic filters, it is not a problem in 4.5MHz and 10.7MHz ceramic filters, as their vibration modes are significantly different.

CONSIDERATIONS FOR GAIN DISTRIBUTION

Since the impedance of both the input and output values of the ceramic filters are symmetric and small, it is necessary that the overall gain distribution within the circuit itself be taken into consideration. For instance, in the discussion concerning proper impedance matching, it was illustrated that a certain DC loss occurs if the recommended resistance values are not used. This can cause an overall reduction in the gain which could present a problem if no allowances have been made for the corresponding loss. To compensate for this problem, it is recommended that the following be done:

1. The amplifier stage should be designed to compensate for this loss.
2. The ceramic filter should be used in combination with the IFT for minimizing both matching and DC losses. The IFT should be used strictly as a matching transformer and the ceramic filter only for selectivity.

As the use of IC's has become more prevalent with ceramic filters, these considerations have been taken into account. It should be noted that few of the problems discussed above have been realized when more than three (3) IF stages have been employed.

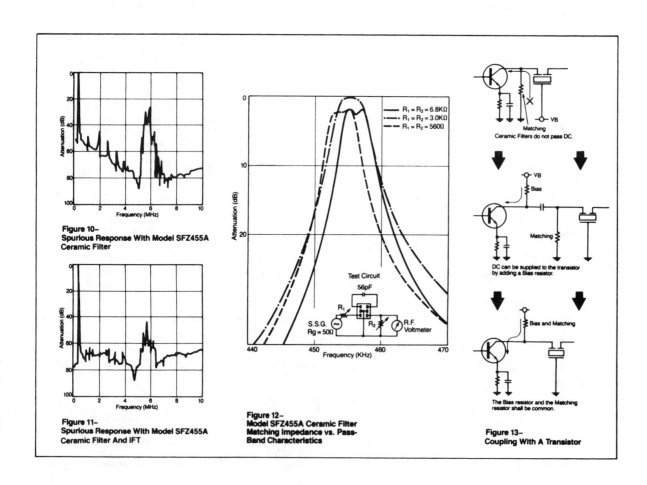

Figure 10—
Spurious Response With Model SFZ455A Ceramic Filter

Figure 11—
Spurious Response With Model SFZ455A Ceramic Filter And IFT

Figure 12—
Model SFZ455A Ceramic Filter Matching Impedance vs. Pass-Band Characteristics

Figure 13—
Coupling With A Transistor

CERAMIC FILTERS DO NOT PASS DC

It is important to note in designing circuits that ceramic filters are incapable of passing DC. As is illustrated in Figure 13, in a typical circuit where a transistor is used, a bias circuit will be required to drive the transistor. Since the ceramic filter requires matching resistance to operate properly, the matching resistor shown in the diagram can play a dual role as both a matching and bias resistor.

If the bias circuit is used, it is important that the parallel circuit of both the bias resistance and the transistor's internal resistance be taken into consideration in meeting the resistance values. This is necessary since the internal resistance of the transistor is changed by the bias resistance. However, when an IC is used, there is no need for an additional bias circuit since the IC has a bias circuit within itself.

Here it is recommended that an IFT be used for impedance matching with the ceramic filter when coupling with a mixer stage, as shown in Figure 14.

COUPLING CAPACITANCE

The SFZ455A is composed of two filter elements which must be connected by a coupling capacitor. Moreover, the frequency characteristic changes according to the coupling capacitance (Cc). As shown in Figure 15, the larger the coupling capacitance (Cc) becomes, the wider the bandwidth and more the ripple increases. Conversely, the smaller the coupling capacitance becomes, the narrower the bandwidth becomes and the more the insertion loss increases. Therefore, the specified value of the coupling capacitance in the catalog is desired in determining the specified passband characteristics.

GROUP DELAY TIME CHARACTERISTICS

Perhaps one of the most important characteristics of a transmitting element is to transmit a signal with the lowest possible distortion level. This distortion occurs when the phase shift of a signal which passes through a certain transmitting path is non-linear with respect to the frequency. For convenience, the group delay time (GDT) characteristic is used for the purpose of expressing non-linearity.

It is important to note the relationship between the amplitude and the GDT characteristics when using group delay time terminology. This relationship differs depending upon the filter characteristics. For example, in the Butterworth type, which has a relatively flat top, the passband is flat while the GDT characteristic is extremely curved, as shown in Figure 16. On the other hand, a Gaussian type, is curved in the passband, while the GDT characteristic is flat. With the flat GDT characteristics, the Gaussian type has excellent distortion characteristics.

Since the amplitude characteristics for the Butterworth type is flat in the passband the bandwidth does not change even at a low input level. With the amplitude characteristic for the Gaussian type being curved in the passband, the bandwidth becomes narrow at a low input level and the sensitivity is poor. Therefore, it should be noted that the Gaussian type has a desirable distortion factor while the Butterworth type has the desirable sensitivity.

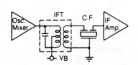

Figure 14–
Coupling From Mixer Stage

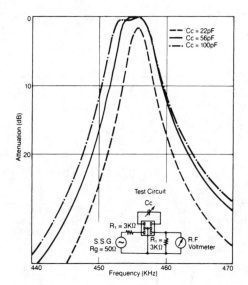

Figure 15–
Model SFZ455A Ceramic Filter
Coupling Capacitance vs. Passband
Characteristics

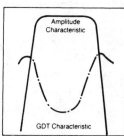

(A) Butterworth Characteristic

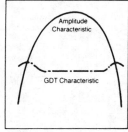

(B) Gaussian Characteristic

Figure 16–
Relationship Between Amplitude
And GDT Characteristics

FEATURES
■ High selectivity
■ High ultimate attenuation
■ A wide variety of pass-bandwidths
■ Small size
■ No peaking

CFM455

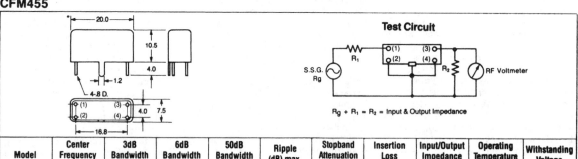

Test Circuit

$Rg + R_1 = R_2 =$ Input & Output Impedance

Model	Center Frequency (KHz)	3dB Bandwidth (KHz) min.	6dB Bandwidth (KHz) min.	50dB Bandwidth (KHz) max.	Ripple (dB) max.	Stopband Attenuation (dB) min.	Insertion Loss (dB) max.	Input/Output Impedance (ohms)	Operating Temperature Range	Withstanding Voltage
CFM455A	455±2	±13	±17.5	±30	3dB within 3dB B.W. and 6dB within 6dB B.W.	50	3	1000	−20°C to +80°C	50V DC
CFM455B	455±2	±10	±15	±25		50	3	1000		
CFM455C	455±2	±9	±13	±23		50	3	1000		
CFM455D	455±2	±7	±10	±20		50	3	1500		
CFM455E	455±2	±5.5	±8	±16		45	5	1500		
CFM455F	455±2	±4.2	±6	±12		45	5	2000		
CFM455G	455±2	—	±4	±10	*	45	5	2000		
CFM455H	455±2	—	±3	±7.5	*	45	6	2000		
CFM455I	455±2	—	±2	±5	*	45	7	2000		

*3dB Ripple in 6dB B.W.

CFR455

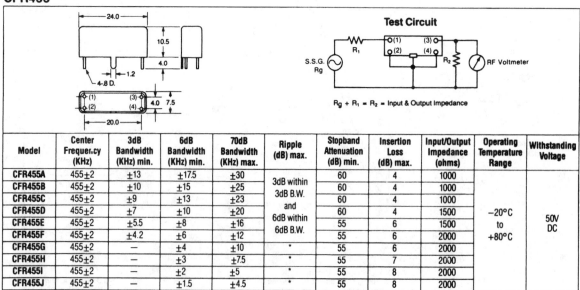

Test Circuit

$Rg + R_1 = R_2 =$ Input & Output Impedance

Model	Center Frequency (KHz)	3dB Bandwidth (KHz) min.	6dB Bandwidth (KHz) min.	70dB Bandwidth (KHz) max.	Ripple (dB) max.	Stopband Attenuation (dB) min.	Insertion Loss (dB) max.	Input/Output Impedance (ohms)	Operating Temperature Range	Withstanding Voltage
CFR455A	455±2	±13	±17.5	±30	3dB within 3dB B.W. and 6dB within 6dB B.W.	60	4	1000	−20°C to +80°C	50V DC
CFR455B	455±2	±10	±15	±25		60	4	1000		
CFR455C	455±2	±9	±13	±23		60	4	1000		
CFR455D	455±2	±7	±10	±20		60	4	1500		
CFR455E	455±2	±5.5	±8	±16		55	6	1500		
CFR455F	455±2	±4.2	±6	±12		55	6	2000		
CFR455G	455±2	—	±4	±10	*	55	6	2000		
CFR455H	455±2	—	±3	±7.5	*	55	7	2000		
CFR455I	455±2	—	±2	±5	*	55	8	2000		
CFR455J	455±2	—	±1.5	±4.5	*	55	8	2000		

*3dB Ripple in 6dB B.W.

National Semiconductor

LM741/LM741A/LM741C/LM741E Operational Amplifier

General Description

The LM741 series are general purpose operational amplifiers which feature improved performance over industry standards like the LM709. They are direct, plug-in replacements for the 709C, LM201, MC1439 and 748 in most applications.

The amplifiers offer many features which make their application nearly foolproof: overload pro-tection on the input and output, no latch-up when the common mode range is exceeded, as well as freedom from oscillations.

The LM741C/LM741E are identical to the LM741/LM741A except that the LM741C/LM741E have their performance guaranteed over a 0°C to +70°C temperature range, instead of −55°C to +125°C.

Schematic and Connection Diagrams (Top Views)

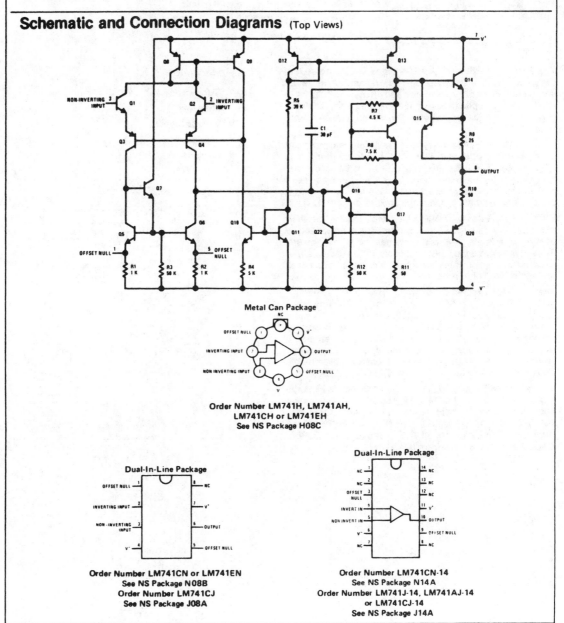

Metal Can Package

Order Number LM741H, LM741AH, LM741CH or LM741EH
See NS Package H08C

Dual-In-Line Package

Order Number LM741CN or LM741EN
See NS Package N08B
Order Number LM741CJ
See NS Package J08A

Dual-In-Line Package

Order Number LM741CN-14
See NS Package N14A
Order Number LM741J-14, LM741AJ-14 or LM741CJ-14
See NS Package J14A

**CA3080S
CA3080AS
8-LEAD TO-5
with Dual-In-Line
Formed Leads
"DIL-CAN"**

H-1787

**CA3080
CA3080A
8-LEAD TO-5**

H-1528

**CA3080E
CA3080AE
8-LEAD Dual-
In-Line Plastic
Package
"MINI-DIP"**

H-1817

Operational Transconductance Amplifiers (OTA's)

Gatable-Gain Blocks

Features:

- Slew rate (unity gain, compensated): 50 V/μs
- Adjustable power consumption: 10μW to 30 mW
- Flexible supply voltage range: $\pm$ 2 V to $\pm$ 15 V
- Fully adjustable gain: 0 to $g_m R_L$ limit
- Tight g_m spread: CA3080 (2:1), CA3080A (1.6:1)
- Extended g_m linearity: 3 decades

The RCA-CA3080 and CA3080A types are Gatable-Gain Blocks which utilize the unique operational-transconductance-amplifier (OTA) concept described in Application Note ICAN-6668, "Applications of the CA3080 and CA3080A High-Performance Operational Transconductance Amplifiers".

The CA3080 and CA3080A types have differential input and a single-ended, push-pull, class A output. In addition, these types have an amplifier bias input which may be used either for gating or for linear gain control. These types also have a high output impedance and their transconductance (g_m) is directly proportional to the amplifier bias current (I_{ABC}).

The CA3080 and CA3080A types are notable for their excellent slew rate (50 V/μs), which makes them especially useful for multiplex and fast unity-gain voltage followers. These types are especially applicable for multiplex applications because power is consumed only when the devices are in the "ON" channel state.

The CA3080A is rated for operation over the full military-temperature range (-55 to $+125°$C) and its characteristics are specifically controlled for applications such as sample-hold, gain-control, multiplex, etc. Operational transconductance amplifiers are also useful in programmable power-switch applications, e.g., as described in Application Note ICAN-6048, "Some Applications of a Programmable Power Switch/Amplifier" (CA3094, CA3094A, CA3094B).

These types are supplied in the 8-lead TO-5-style package (CA3080, CA3080A), and in the 8-lead TO-5-style package with dual-in-line formed leads ("DIL-CAN", CA3080S, CA3080AS). The CA3080 is also supplied in the 8-lead dual-in-line plastic ("MINI-DIP") package (CA3080E, CA3080AE), and in chip form (CA3080H).

Applications:

- Sample and hold
- Multiplex
- Voltage follower
- Multiplier
- Comparator

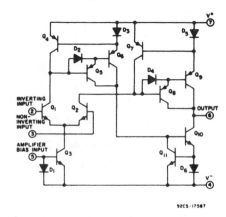

92CS-17587

*Fig.1 – Schematic diagram for CA3080
and CA3080A.*

Linear Integrated Circuits
CA3080, CA3080A Types

MAXIMUM RATINGS, *Absolute-Maximum Values:*

DC SUPPLY VOLTAGE (Between V$^+$ and V$^-$ terminals) . . . . 36 V
DIFFERENTIAL INPUT VOLTAGE . . . . ±5 V
DC INPUT VOLTAGE . . . . V$^+$ to V$^-$
INPUT SIGNAL CURRENT . . . . 1 mA
AMPLIFIER BIAS CURRENT . . . . 2 mA
OUTPUT SHORT-CIRCUIT DURATION* . . . . Indefinite
DEVICE DISSIPATION . . . . 125 mW
TEMPERATURE RANGE:
  Operating
  CA3080, CA3080E, CA3080S . . . . 0 to + 70 °C
  CA3080A, CA3080AE, CA3080AS . . . . −55 to + 125 °C
  Storage . . . . −65 to + 150 °C
LEAD TEMPERATURE (During Soldering):
  At distance 1/16 ± 1/32 in. (1.59 ± 0.79 mm)
  from case for 10 s max. . . . . + 265 °C

* Short circuit may be applied to ground or to either supply.

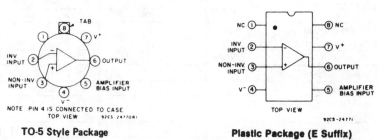

TO-5 Style Package　　　　　**Plastic Package (E Suffix)**

Fig. 2 — Functional diagrams.

TYPICAL CHARACTERISTICS CURVES AND TEST CIRCUITS FOR THE CA3080 AND CA3080A

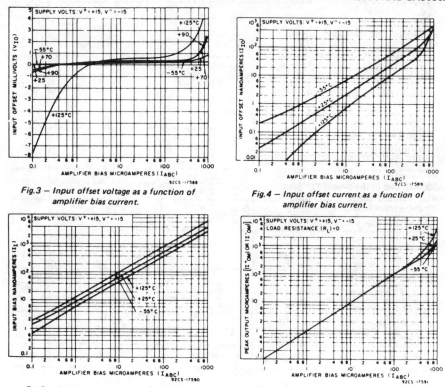

Fig. 3 – Input offset voltage as a function of
amplifier bias current.

Fig. 4 – Input offset current as a function of
amplifier bias current.

Fig. 5 – Input bias current as a function of
amplifier bias current.

Fig. 6 – Peak output current as a function of
amplifier bias current.

ELECTRICAL CHARACTERISTICS
For Equipment Design

CHARACTERISTIC		TEST CONDITIONS $V^+ = 15$ V, $V^- = -15$ V $I_{ABC} = 500\,\mu A$ $T_A = 25^oC$ (unless indicated otherwise)	CA3080 CA3080E CA3080S LIMITS Min.	Typ.	Max.	UNITS
Input Offset Voltage	V_{IO}		–	0.4	5	mV
		$T_A = 0$ to 70^oC	–	–	6	
Input Offset Current	I_{IO}		–	0.12	0.6	μA
Input Bias Current	I_I		–	2	5	μA
		$T_A = 0$ to 70^oC	–	–	7	
Forward Transconductance (large signal)	g_m		6700	9600	13000	μmho
		$T_A = 0$ to 70^oC	5400	–	–	
Peak Output Current	$|I_{OM}|$	$R_L = 0$	350	500	650	μA
		$R_L = 0$, $T_A = 0$ to 70^oC	300	–	–	
Peak Output Voltage: Positive	V^+_{OM}	$R_L = \infty$	12	13.5	–	V
Negative	V^-_{OM}		–12	–14.4	–	
Amplifier Supply Current	I_A		0.8	1	1.2	mA
Device Dissipation	P_D		24	30	36	mW
Input Offset Voltage Sensitivity: Positive	$\Delta V_{IO}/\Delta V^+$		–	–	150	$\mu V/V$
Negative	$\Delta V_{IO}/\Delta V^-$		–	–	150	
Common-Mode Rejection Ratio	CMRR		80	110	–	dB
Common-Mode Input-Voltage Range	V_{ICR}		12 to –12	13.6 to –14.6	–	V
Input Resistance	R_I		10	26	–	$k\Omega$

ELECTRICAL CHARACTERISTICS
Typical Values Intended Only for Design Guidance

CA3080
CA3080E
CA3080S

				UNITS
Input Offset Voltage	V_{IO}	$I_{ABC} = 5\,\mu A$	0.3	mV
Input Offset Voltage Change	$|\Delta V_{IO}|$	$I_{ABC} = 500\,\mu A$ to $I_{ABC} = 5\,\mu A$	0.2	mV
Peak Output Current	I_{OM}	$I_{ABC} = 5\,\mu A$	5	μA
Peak Output Voltage: Positive	V^+_{OM}	$I_{ABC} = 5\,\mu A$	13.8	V
Negative	V^-_{OM}		–14.5	
Magnitude of Leakage Current		$I_{ABC} = 0$, $V_{TP} = 0$	0.08	nA
		$I_{ABC} = 0$, $V_{TP} = 36$ V	0.3	
Differential Input Current		$I_{ABC} = 0$, $V_{DIFF} = 4$ V	0.008	nA
Amplifier Bias Voltage	V_{ABC}		0.71	V
Slew Rate: Maximum (uncompensated)	SR		75	V/μs
Unity Gain (compensated)			50	
Open-Loop Bandwidth	BW_{OL}		2	MHz
Input Capacitance	C_I	$f = 1$ MHz	3.6	pF
Output Capacitance	C_O	$f = 1$ MHz	5.6	pF
Output Resistance	R_O		15	$M\Omega$
Input-to-Output Capacitance	C_{I-O}	$f = 1$ MHz	0.024	pF
Propagation Delay	t_{PHL}, t_{PLH}	$I_{ABC} = 500\,\mu A$	45	ns

Linear Integrated Circuits
CA3080, CA3080A Types

ELECTRICAL CHARACTERISTICS
For Equipment Design

CHARACTERISTIC		TEST CONDITIONS $V^+ = 15$ V, $V^- = -15$ V $I_{ABC} = 500\,\mu A$ $T_A = 25^oC$ (unless indicated otherwise)	CA3080A CA3080AE CA3080AS LIMITS			UNITS
			Min.	Typ.	Max.	
Input Offset Voltage	V_{IO}	$I_{ABC} = 5\,\mu A$	–	0.3	2	mV
			–	0.4	2	
		$T_A = -55$ to $+125^oC$	–	–	5	
Input Offset Voltage Change	$|\Delta V_{IO}|$	$I_{ABC} = 500\,\mu A$ to $I_{ABC} = 5\,\mu A$	–	0.1	3	mV
Input Offset Current	I_{IO}		–	0.12	0.6	μA
Input Bias Current	I_I		–	2	5	μA
		$T_A = -55$ to $+125^oC$	–	–	8	
Forward Transconductance (large signal)	$9m$		7700	9600	12000	μmho
		$T_A = -55$ to $+125^oC$	4000	–	–	
Peak Output Current	$|I_{OM}|$	$I_{ABC} = 5\,\mu A$, $R_L = 0$	3	5	7	μA
		$R_L = 0$	350	500	650	
		$R_L = 0, T_A = -55$ to $+125^oC$	300	–	–	
Peak Output Voltage:						
Positive	V^+_{OM}	$I_{ABC} = 5\,\mu A$	12	13.8	–	
Negative	V^-_{OM}	$R_L = \infty$	–12	–14.5	–	V
Positive	V^+_{OM}	$R_L = \infty$	12	13.5	–	
Negative	V^-_{OM}		–12	–14.4	–	
Amplifier Supply Current	I_A		0.8	1	1.2	mA
Device Dissipation	P_D		24	30	36	mW
Input Offset Voltage Sensitivity:						
Positive	$\Delta V_{IO}/\Delta V^+$		–	–	150	$\mu V/V$
Negative	$\Delta V_{IO}/\Delta V^-$		–	–	150	
Magnitude of Leakage Current		$I_{ABC} = 0$, $V_{TP} = 0$	–	0.08	5	nA
		$I_{ABC} = 0$, $V_{TP} = 36$ V	–	0.3	5	
Differential Input Current		$I_{ABC} = 0$, $V_{DIFF} = 4$ V	–	0.008	5	nA
Common-Mode Rejection Ratio	CMRR		80	110	--	dB
Common-Mode Input-Voltage Range	V_{ICR}		12 to –12	13.6 to –14.6	–	V
Input Resistance	R_I		10	26	–	$k\Omega$

ELECTRICAL CHARACTERISTICS
Typical Values Intended Only for Design Guidance

CA3080A
CA3080AE
CA3080AS

Amplifier Bias Voltage	V_{ABC}		0.71	V
Slew Rate:				
Maximum (uncompensated)	SR		75	V/μs
Unity Gain (compensated)			50	
Open-Loop Bandwidth	BW_{OL}	–	2	MHz
Input Capacitance	C_I	$f = 1$ MHz	3.6	pF
Output Capacitance	C_O	$f = 1$ MHz	5.6	pF
Output Resistance	R_O		.15	$M\Omega$
Input-to-Output Capacitance	C_{I-O}	$f = 1$ MHz	0.024	pF
Input Offset Voltage Temperature Drift	$\Delta V_{IO}/\Delta T$	$I_{ABC} = 100\,\mu A$, $T_A = -55$ to $+125^oC$	3	$\mu V/^oC$
Propagation Delay	t_{PHL}, t_{PLH}	$I_{ABC} = 500\,\mu A$	45	ns

3N204

SILICON DUAL INSULATED-GATE FIELD-EFFECT TRANSISTOR
N-Channel Depletion Type With Integrated
Gate-Protection Circuits
For RF Amplifier Applications up to 400 MHz

RCA-40673 is an n-channel silicon, depletion type, dual insulated-gate field-effect transistor.

Special back-to-back diodes are diffused directly into the MOS* pellet and are electrically connected between each insulated gate and the FET's source. The diodes effectively bypass any voltage transients which exceed approximately ±10 volts. This protects the gates against damage in all normal handling and usage.

A feature of the back-to-back diode configuration is that it allows the 40673 to retain the wide input signal dynamic range inherent in the MOSFET. In addition, the low junction capacitance of these diodes adds little to the total capacitance shunting the signal gate.

The excellent overall performance characteristics of the RCA-40673 make it useful for a wide variety of rf-amplifier applications at frequencies up to 400 MHz. The two serially-connected channels with independent control gates make possible a greater dynamic range and lower cross-modulation than is normally achieved using devices having only a single control element.

The two gate arrangement of the 40673 also makes possible a desirable reduction in feedback capacitance by operating in

the common-source configuration and ac-grounding Gate No. 2. The reduced capacitance allows operation at maximum gain *without neutralization;* and, of special importance in rf-amplifiers, it reduces local oscillator feedthrough to the antenna.

The 40673 is hermetically sealed in the metal JEDEC TO-72 package.

*Metal-Oxide-Semiconductor.

Maximum Ratings, Absolute-Maximum Values, at $T_A = 25°C$

DRAIN-TO-SOURCE VOLTAGE, V_{DS}	-0.2 to +20	V
GATE No.1-TO-SOURCE VOLTAGE, V_{G1S}:		
Continuous (dc)	-6 to +1	V
Peak ac	-6 to +6	V
GATE No.2-TO-SOURCE VOLTAGE, V_{G2S}:		
Continuous (dc).................	-6 to 30% of V_{DS}	V
Peak ac	-6 to +6	V
DRAIN-TO-GATE VOLTAGE,		
V_{DG1} OR V_{DG2}	+20	V
DRAIN CURRENT, I_D	50	mA
TRANSISTOR DISSIPATION, P_T:		
At ambient up to 25°C	330	mW
temperatures above 25°C	derate linearly at 2.2 mW/°C	
AMBIENT TEMPERATURE RANGE:		
Storage and Operating	-65 to +175	°C
LEAD TEMPERATURE (During soldering):		
At distances ≥1/32 inch from		
seating surface for 10 seconds max.	265	°C

APPLICATIONS

- RF amplifier, mixer, and IF amplifier in military, industrial, and consumer communications equipment
- aircraft and marine vehicular receivers
- CATV and MATV equipment
- telemetry and multiplex equipment

PERFORMANCE FEATURES

- superior cross-modulation performance and greater dynamic range than bipolar or single-gate FETs
- wide dynamic range permits large-signal handling before overload
- dual-gate permits simplified agc circuitry
- virtually no agc power required
- greatly reduces spurious responses in fm receivers
- permits use of vacuum-tube biasing techniques
- excellent thermal stability

DEVICE FEATURES

- back-to-back diodes protect each gate against handling and in-circuit transients
- low gate leakage currents —— I_{G1SS} & I_{G2SS} = 20 nA(max.) at $T_A = 25°C$
- high forward transconductance —— g_{fs} = 12,000 µmho (typ.)
- high unneutralized RF power gain —— G_{ps} = 18 dB(typ.) at 200 MHz
- low VHF noise figure —— 3.5 dB(typ.) at 200 MHz

TERMINAL DIAGRAM

LEAD 1 - DRAIN
LEAD 2 - GATE No. 2
LEAD 3 - GATE No. 1
LEAD 4 - SOURCE, SUBSTRATE AND CASE

ELECTRICAL CHARACTERISTICS, at $T_A = 25°C$ unless otherwise specified

CHARACTERISTICS	SYMBOLS	TEST CONDITIONS	Min.	Typ.	Max.	UNITS
Gate-No.1-to-Source Cutoff Voltage	$V_{G1S(off)}$	V_{DS} = +15V, I_D = 200µA V_{G2S} = +4V	–	–2	–4	V
Gate-No.2-to-Source Cutoff Voltage	$V_{G2S(off)}$	V_{DS} = +15V, I_D = 200µA V_{G1S} = 0	–	–2	–4	V
Gate-No.1-Leakage Current	I_{G1SS}	V_{G1S} = +1 or–6 V V_{DS} = 0, V_{G2S} = 0	–	–	50	nA
Gate-No.2-Leakage Current	I_{G2SS}	V_{G2S} = ±6V V_{DS} = 0, V_{G1S} = 0	–	–	50	nA
Zero-Bias Drain Current	I_{DSS}	V_{DS} = +15V V_{G2S} = +4V V_{G1S} = 0	5	15	35	mA
Forward Transconductance (Gate-No.1-to-Drain)	g_{fs}	V_{DS} = +15V, I_D = 10mA V_{G2S} = +4V, f = 1kHz	–	12,000	–	µmho
Small-Signal, Short-Circuit Input Capacitance †	C_{iss}	V_{DS} = +15V, I_D = 10mA V_{G2S} = +4V, f=1MHz	–	6	–	pF
Small-Signal, Short-Circuit, Reverse Transfer Capacitance (Drain-to-Gate No.1) ◊	C_{rss}		0.005	0.02	0.03	pF
Small-Signal, Short-Circuit Output Capacitance	C_{oss}		–	2.0	–	pF
Power Gain (see Fig. 1)	G_{PS}		14	18	–	dB
Maximum Available Power Gain	MAG		–	20	–	dB
Maximum Usable Power Gain (unneutralized)	MUG		–	20*	–	dB
Noise Figure (see Fig. 1)	NF	V_{DS} = +15V, I_D = 10mA V_{G2S} = +4V, f = 200 MHz	–	3.5	6.0	dB
Magnitude of Forward Transadmittance	$|Y_{fs}|$		–	12,000	–	µmho
Phase Angle of Forward Trans-admittance	θ		–	–35	–	degrees
Input Resistance	r_{iss}		–	1.0	–	kΩ
Output Resistance	r_{oss}		–	2.8	–	kΩ
Protective Diode Knee Voltage	V_{knee}	$I_{DIODE(REVERSE)}$ = ±100µA	–	±10	–	V

*Limited only by practical design considerations.
†Capacitance between Gate No. 1 and all other terminals
◊Three-terminal measurement with Gate No. 2 and Source returned to guard terminal.

For characteristics curves, refer to type 3N187.

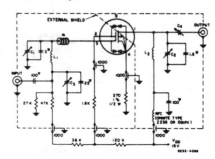

#Ferrite bead (4); Pyroferric Co. "Carbonyl J" 0.09 in. OD; 0.03 in. ID; 0.063 in. thickness.
All resistors in ohms
All capacitors in pF

Q = 40673
▼ Disc ceramic.
*Tubular ceramic.

C_1: 1.8 – 8.7 pF variable air capacitor: E.F. Johnson Type 160-104, or equivalent.

C_2: 1.5 – 5 pF variable air capacitor: E.F. Johnson Type 160-102, or equivalent.

C_3: 1 – 10 pF piston-type variable air capacitor: JFD Type VAM-010; Johanson Type 4335, or equivalent.

C_4: 0.8 – 4.5 pF piston type variable air capacitor: Erie 560-013 or equivalent.

L_1: 4 turns silver-plated 0.02-in. thick, 0.075-0.085-in. wide, copper ribbon. Internal diameter of winding = 0.25 in, winding length approx. 0.80 in.

L_2: 4½ turns silver-plated 0.02-in. thick, 0.085-0.095-in. wide, 5/16-in. ID. Coil ≈ .90 in. long.

Fig. 1. 200-MHz Power gain and noise-figure test circuit

![National Semiconductor]

LM386 Low Voltage Audio Power Amplifier

General Description

The LM386 is a power amplifier designed for use in low voltage consumer applications. The gain is internally set to 20 to keep external part count low, but the addition of an external resistor and capacitor between pins 1 and 8 will increase the gain to any value up to 200.

The inputs are ground referenced while the output is automatically biased to one half the supply voltage. The quiescent power drain is only 24 milliwatts when operating from a 6 volt supply, making the LM386 ideal for battery operation.

Features

- Battery operation
- Minimum external parts
- Wide supply voltage range   4V–12V or 5V–18V
- Low quiescent current drain   4 mA

- Voltage gains from 20 to 200
- Ground referenced input
- Self-centering output quiescent voltage
- Low distortion
- Eight pin dual-in-line package

Applications

- AM-FM radio amplifiers
- Portable tape player amplifiers
- Intercoms
- TV sound systems
- Line drivers
- Ultrasonic drivers
- Small servo drivers
- Power converters

Equivalent Schematic and Connection Diagrams

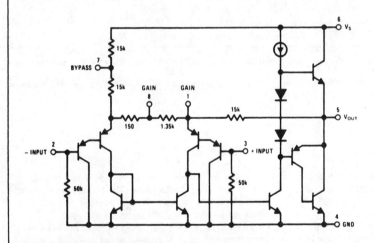

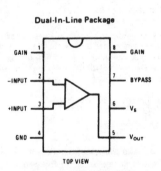

Dual-In-Line Package

TOP VIEW

Order Number LM386N-1, LM386N-3 or LM386N-4 See NS Package N08B

Typical Applications

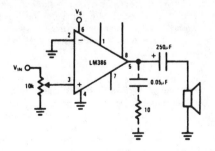

Amplifier with Gain = 20 Minimum Parts

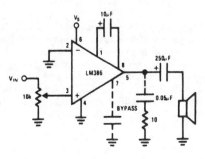

Amplifier with Gain = 200

Absolute Maximum Ratings

Supply Voltage (LM386N)	15V	Storage Temperature	−65°C to +150°C
Supply Voltage (LM386N-4)	22V	Operating Temperature	0°C to +70°C
Package Dissipation (Note 1) (LM386N-4)	1.25W	Junction Temperature	+150°C
Package Dissipation (Note 2) (LM386)	660 mW	Lead Temperature (Soldering, 10 seconds)	+300°C
Input Voltage	±0.4V		

Electrical Characteristics T_A = 25°C

PARAMETER	CONDITIONS	MIN	TYP	MAX	UNITS
Operating Supply Voltage (V_S)					
LM386		4		12	V
LM386N-4		5		18	V
Quiescent Current (I_Q)	V_S = 6V, V_{IN} = 0		4	8	mA
Output Power (P_{OUT})					
LM386N-1	V_S = 6V, R_L = 8Ω, THD = 10%	250	325		mW
LM386N-3	V_S = 9V, R_L = 8Ω, THD = 10%	500	700		mW
LM386N-4	V_S = 16V, R_L = 32Ω, THD = 10%	700	1000		mW
Voltage Gain (A_V)	V_S = 6V, f = 1 kHz		26		dB
	10μF from Pin 1 to 8		46		dB
Bandwidth (BW)	V_S = 6V, Pins 1 and 8 Open		300		kHz
Total Harmonic Distortion (THD)	V_S = 6V, R_L = 8Ω, P_{OUT} = 125 mW f = 1 kHz, Pins 1 and 8 Open		0.2		%
Power Supply Rejection Ratio (PSRR)	V_S = 6V, f = 1 kHz, C_{BYPASS} = 10μF Pins 1 and 8 Open, Referred to Output		50		dB
Input Resistance (R_{IN})			50		kΩ
Input Bias Current (I_{BIAS})	V_S = 6V, Pins 2 and 3 Open		250		nA

Note 1: For operation in ambient temperatures above 25°C, the device must be derated based on a 150°C maximum junction temperature and a thermal resistance of 100°C/W junction to ambient.

Note 2: For operation in ambient temperatures above 25°C, the device must be derated based on a 150°C maximum junction temperature and a thermal resistance of 187°C junction to ambient.

Application Hints

GAIN CONTROL

To make the LM386 a more versatile amplifier, two pins (1 and 8) are provided for gain control. With pins 1 and 8 open the 1.35 kΩ resistor sets the gain at 20 (26 dB). If a capacitor is put from pin 1 to 8, bypassing the 1.35 kΩ resistor, the gain will go up to 200 (46 dB). If a resistor is placed in series with the capacitor, the gain can be set to any value from 20 to 200. Gain control can also be done by capacitively coupling a resistor (or FET) from pin 1 to ground.

Additional external components can be placed in parallel with the internal feedback resistors to tailor the gain and frequency response for individual applications. For example, we can compensate poor speaker bass response by frequency shaping the feedback path. This is done with a series RC from pin 1 to 5 (paralleling the internal 15kΩ resistor). For 6 dB effective bass boost: R ≅ 15kΩ, the lowest value for good stable operation is R = 10 kΩ if pin 8 is open. If pins 1 and 8 are bypassed then R as low as 2 kΩ can be used. This restriction is because the amplifier is only compensated for closed-loop gains greater than 9.

INPUT BIASING

The schematic shows that both inputs are biased to ground with a 50 kΩ resistor. The base current of the input transistors is about 250 nA, so the inputs are at about 12.5 mV when left open. If the dc source resistance driving the LM386 is higher than 250 kΩ it will contribute very little additional offset (about 2.5 mV at the input, 50 mV at the output). If the dc source resistance is less than 10 kΩ, then shorting the unused input to ground will keep the offset low (about 2.5 mV at the input, 50 mV at the output). For dc source resistances between these values we can eliminate excess offset by putting a resistor from the unused input to ground, equal in value to the dc source resistance. Of course all offset problems are eliminated if the input is capacitively coupled.

When using the LM386 with higher gains (bypassing the 1.35 kΩ resistor between pins 1 and 8) it is necessary to bypass the unused input, preventing degradation of gain and possible instabilities. This is done with a 0.1μF capacitor or a short to ground depending on the dc source resistance on the driven input.

Reprinted with permission of National Semiconductor Corp.

Typical Performance Characteristics

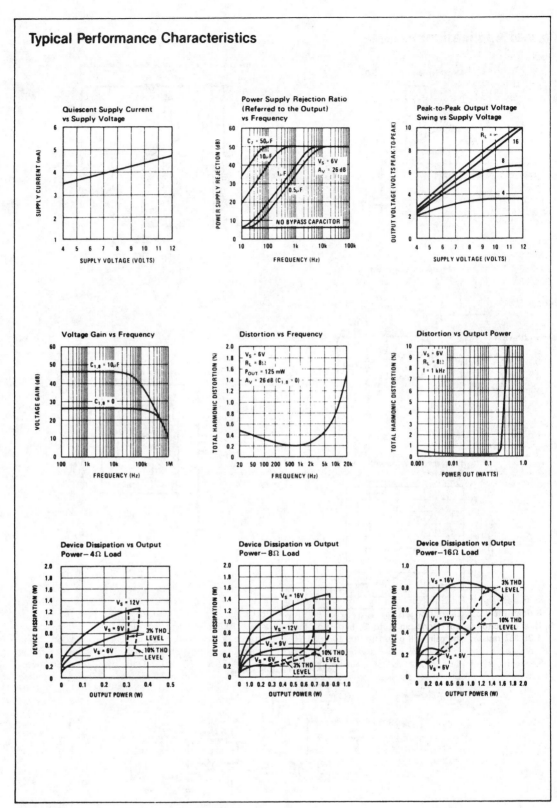

Typical Applications (Continued)

Amplifier with Gain = 50

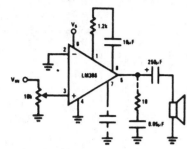

Low Distortion Power Wienbridge Oscillator

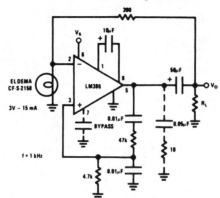

Amplifier with Bass Boost

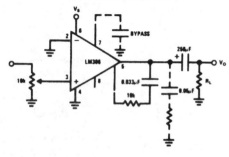

Frequency Response with Bass Boost

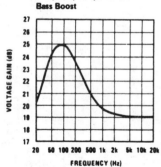

Square Wave Oscillator

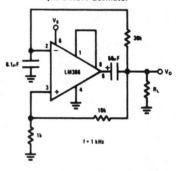

AM Radio Power Amplifier

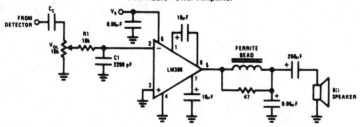

Note 1: Twist supply lead and supply ground very tightly.

Note 2: Twist speaker lead and ground very tightly.

Note 3: Ferrite bead is Ferroxcube K5-001-001/3B with 3 turns of wire.

Note 4: R1C1 band limits input signals.

Note 5: All components must be spaced very close to IC.

8-Lead TO-5-Style
"DIL-CAN" Package
H-1787

8-Lead
TO-5-Style
Package

8-Lead Dual-In-Line
Frit-Seal (Hermetic)
Package   H-1805

H-1528

DIFFERENTIAL/CASCODE AMPLIFIERS

For Communications and Industrial Equipment at Frequencies from DC to 120 MHz

FEATURES

- Controlled for Input Offset Voltage, Input Offset Current, and Input Bias Current (CA3028 Series only)
- Balanced Differential Amplifier Configuration with Controlled Constant-Current Source
- Single- and Dual-Ended Operation
- Operation from DC to 120 MHz
- Balanced-AGC Capability
- Wide Operating-Current Range

The CA3028A and CA3028B are differential/cascode amplifiers designed for use in communications and industrial equipment operating at frequencies from dc to 120 MHz.

The CA3028B is like the CA3028A but is capable of premium performance particularly in critical dc and differential amplifier applications requiring tight controls for input offset voltage, input offset current, and input bias current.

The CA3053 is similar to the CA3028A and CA3028B but is recommended for IF amplifier applications.

The CA3028A, CA3028B, and CA3053 are supplied in a hermetic 8-lead TO-5-style package. The "F" versions are supplied in a frit-seal TO-5 package, and the "S" versions in formed-lead (DIL-CAN) packages.

APPLICATIONS

- RF and IF Amplifiers (Differential or Cascode)
- DC, Audio, and Sense Amplifiers
- Converter in the Commercial FM Band
- Oscillator   • Mixer   • Limiter
- Companion Application Note, ICAN 5337 "Application of the RCA CA3028 Integrated Circuit Amplifier in the HF and VHF Ranges." This note covers characteristics of different operating modes, noise performance, mixer, limiter, and amplifier design considerations.

The CA3028A, CA3028B, and CA3053 are available in the packages shown below. When ordering these devices, it is important to add the appropriate suffix letter to the device.

Package 8-Lead TO-5	Suffix Letter	CA3028A	CA3028B	CA3053
TO-5	T	√	√	√
With Dual-In-Line Formed Leads (DIL-CAN)	S	√	√	√
Frit-Seal Ceramic	F	√	√	√
Beam-Lead	L	√		
Chip	H	√		

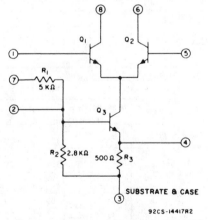

92CS-14417R2

Fig.1 - Schematic diagram for CA3028A, CA3028B and CA3053.

CA3028A, CA3028B, CA3053 Types

ABSOLUTE MAXIMUM RATINGS AT T_A = 25°C

DISSIPATION:

At T_A up to 55°C
(CA3028AF, CA3028BF,
CA3053F) . . . . 750 mW

At T_A > 55°C
(CA3028AF, CA3028BF,
CA3053F) . . . . Derate linearly 6.67 mW/°C

At T_A up to 85°C
(CA3028A, CA3028B, CA3053) . . . . 450 mW

At T_A > 85°C
(CA3028A, CA3028B, CA3053) Derate linearly 5 mW/°C

AMBIENT-TEMPERATURE RANGE:
Operating . . . . −55°C to +125°C
Storage . . . . −65°C to +150°C

LEAD TEMPERATURE (During Soldering):
At distance 1/16 ± 1/32" (1.59 ± 0.79 mm)
from case for 10 seconds max. . . . . +265°C

MAXIMUM VOLTAGE RATINGS at T_A = 25°C

MAXIMUM CURRENT RATINGS

TERM-INAL No.	1	2	3	4	5	6	7	8
1		0 to -15 ▲	0 to -15 ▲	+5 to -5	*	*	+20 ┼ to 0	
2			+5 to -11	+5 to 0	+15 ♦ to 0	*	+15 ♦ to 0	*
3 ┼				+10 to 0	+15 ♦ to 0	+30 ● to 0	+15 ♦ to 0	+30 ● to 0
4					+15 ♦ to 0	*	*	*
5						+20 ┼ to 0	*	*
6							*	*
7								*
8								

This chart gives the range of voltages which can be applied to the terminals listed horizontally with respect to the terminals listed vertically. For example, the voltage range of the horizontal terminal 4 with respect to terminal 2 is 1 to +5 volts

┼ Terminal #3 is connected to the substrate and case.
* Voltages are not normally applied between these terminals. Voltages appearing between these terminals will be safe, if the specified voltage limits between all other terminals are not exceeded.
▲ Limit is -12V for CA3053
┼ Limit is +15V for CA3053
♦ Limit is +12V for CA3053
● Limit is +24V for CA3028A and +18V for CA3053

TERM-INAL No.	I_{IN} mA	I_{OUT} mA
1	0.6	0.1
2	4	0.1
3	0.1	23
4	20	0.1
5	0.6	0.1
6	20	0.1
7	4	0.1
8	20	0.1

ELECTRICAL CHARACTERISTICS at T_A = 25°C

CHARACTERISTIC	SYMBOL	TEST CIR-CUIT Fig.	SPECIAL TEST CONDITIONS		LIMITS TYPE CA3028A			LIMITS TYPE CA3028B			LIMITS TYPE CA3053			UNITS	TYPICAL CHARAC-TERISTICS CURVES Fig.
			+V_{CC}	-V_{EE}	Min.	Typ.	Max.	Min.	Typ.	Max.	Min.	Typ.	Max.		
STATIC CHARACTERISTICS															
Input Offset Voltage	V_{IO}	2	6V 12V	6V 12V	- -	- -	- -	- -	0.98 0.89	5 5	- -	- -	- -	mV	4
Input Offset Current	I_{IO}	3a	6V 12V	6V 12V	- -	- -	- -	- -	0.56 1.06	5 6	- -	- -	- -	μA	4
Input Bias Current	I_I	3a	6V 12V	6V 12V	- -	16.6 36	70 106	- -	16.6 36	40 80	- -	- -	- -	μA	5a
		3b	9V 12V	- -	- -	- -	- -	- -	- -	- -	- -	29 36	85 125		5b
Quiescent U_r rating Current	I_6 or I_8	3a	6V 12V	6V 12V	0.8 2	1.25 3.3	2 5	1 2.5	1.25 3.3	1.5 4	- -	- -	- -	mA	6a 7
		3b	9V 12V	- -	- -	- -	- -	- -	- -	- -	1.2 2.0	2.2 3.3	3.5 5.0		6b
AGC Bias Current (Into Constant-Current Source Terminal No.7)	I_7	8a	12V 12V	V_{AGC}=+9 V_{AGC}=+12	- -	1.28 1.65	- -	- -	1.28 1.65	- -	- -	- -	- -	mA	8b
		-	9V 12V	- -	- -	- -	- -	- -	- -	- -	- -	1.15 1.55	- -		-
Input Current (Terminal No.7)	I_7	-	6V 12V	6V 12V	0.5 1	0.85 1.65	1 2.1	0.5 1	0.85 1.65	1 2.1	- -	- -	- -	mA	-
Device Dissipation	P_T	3a	6V 12V	6V 12V	24 120	36 175	54 260	24 120	36 175	42 220	- -	- -	- -	mW	9
		3b	9V 12V	- -	- -	- -	- -	- -	- -	- -	- -	50 100	80 150		-

ELECTRICAL CHARACTERISTICS at T_A = 25°C (cont'd)

CHARACTERISTIC	SYMBOL	TEST CIRCUIT Fig.	SPECIAL TEST CONDITIONS		LIMITS TYPE CA3028A Min.	Typ.	Max.	LIMITS TYPE CA3028B Min.	Typ.	Max.	LIMITS TYPE CA3053 Min.	Typ.	Max.	UNITS	TYPICAL CHARAC-TERISTICS CURVE Fig.
DYNAMIC CHARACTERISTICS															
Power Gain	G_P	10a	f = 100 MHz	Cascode	16	20	-	16	20	-	-	-	-	dB	10b
		11a,d	V_{CC} = +9V	Diff.-Ampl.	14	17	-	14	17	-	-	-	-		11b,e
		10a	f = 10.7 MHz	Cascode	35	39	-	35	39	-	35	39	-	dB	10b *
		11a	V_{CC} = +9V	Diff.-Ampl.	28	32	-	28	32	-	28	32	-		11b *
Noise Figure	NF	10a	f = 100 MHz	Cascode	-	7.2	9	-	7.2	9	-	-	-	dB	10c
		11a,d	V_{CC} = +9V	Diff.-Ampl.	-	6.7	9	-	6.7	9	-	-	-		11c,e
Input Admittance	Y_{11}	-		Cascode				-	0.6 + j 1.6	-				mmho	12
		-		Diff.-Ampl.				-	0.5 + j 0.5	-					13
Reverse Transfer Admittance	Y_{12}	-		Cascode				-	0.0003 - j0	-				mmho	14
		-	f = 10.7 MHz	Diff.-Ampl.				-	0.01 - j0.0002	-					15
Forward Transfer Admittance	Y_{21}	-	V_{CC} = +9V	Cascode				-	99 - j18	-				mmho	16
		-		Diff.-Ampl.				-	-37 + j0.5	-					17
Output Admittance	Y_{22}	-		Cascode				-	0. + j0.08	-				mmho	18
		-		Diff.-Ampl.				-	0.04 + j0.23	-					19
Power Output (Untuned)	P_o	20a	f = 10.7 MHz	Diff.-Ampl. 50 Ω Input-Output	-	5.7	-		5.7					µW	20b
AGC Range (Max. Power Gain to Full Cutoff)	AGC	21a	V_{CC} = +9V	Diff.-Ampl.	-	62	-		62	-	-			dB	21b
Voltage Gain — at f = 10.7 MHz	A	22a	f = 10.7 MHz	Cascode	-	40	-		40		-	40	-	dB	22b
		22c	V_{CC} = +9V R_L = 1 kΩ	Diff. Ampl.	-	30	-		30		-	30	-		22d
Voltage Gain — Differential at f = 1 kHz		23	V_{CC} = +6V, R_L = 2 kΩ, V_{EE} = -6V,		-		-	35	38	42	-			dB	-
			V_{CC} = +12V, R_L = 1.6 kΩ, V_{EE} = -12V,		-		-	40	42.5	45	-				-
Max. Peak-to-Peak Output Voltage at f = 1 kHz	$V_o(P-P)$	23	V_{CC} = +6V, R_L = 2 kΩ, V_{EE} = -6V,		-		-	7	11.5	-				V_{P-P}	-
			V_{CC} = +12V, R_L = 1.6 kΩ, V_{EE} = -12V		-		-	15	23	-					-
Bandwidth at -3 dB point	BW	23	V_{CC} = +6V, R_L = 2 kΩ, V_{EE} = -6V,		-		-		7.3	-	-		-	MHz	-
			V_{CC} = +12V, R_L = 1.6 kΩ, V_{EE} = -12V,		-		-		8	-	-				-
Common-Mode Input-Voltage Range	V_{CMR}	24	V_{CC} = +6V, V_{EE} = -6V V_{CC} = +12V, V_{EE} = -12V		-		-	-2.5 / -5	(-3.2 - 4.5) / (-7 - 9)	4 / 7	-			V	-
Common-Mode Rejection Ratio	CMR	24	V_{CC} = +6V, V_{EE} = -6V V_{CC} = +12V, V_{EE} = -12V		-		-	60 / 60	110 / 90	-	-			dB	-
Input Impedance at f = 1 kHz	Z_{IN}		V_{CC} = +6V, V_{EE} = -6V V_{CC} = +12V, V_{EE} = -12V		-		-		5.5 / 3	-	-			kΩ	-
Peak-to-Peak Output Current	I_{P-P}		V_{CC} = +9V	f = 10.7 MHz e_{in} = 400 mV Diff.-Ampl.	2	4	7	2.5	4	6	2	4	7	mA	
			V_{CC} = +12V		3.5	6	10	4.5	6	8	3.5	6	10		

* Does not apply to CA3053

CA3028A, CA3028B, CA3053 Types

DEFINITIONS OF TERMS

AGC Bias Current

The current drawn by the device from the AGC-voltage source, at maximum AGC voltage.

AGC Range

The total change in voltage gain (from maximum gain to complete cutoff) which may be achieved by application of the specified range of dc voltage to the AGC input terminal of the device.

Common-Mode Rejection Ratio

The ratio of the full differential voltage gain to the common-mode voltage gain.

Device Dissipation

The total power drain of the device with no signal applied and no external load current.

Input Bias Current

The average value (one-half the sum) of the currents at the two input terminals when the quiescent operating voltages at the two output terminals are equal.

Input Offset Current

The difference in the currents at the two input terminals when the quiescent operating voltages at the two output terminals are equal.

Input Offset Voltage

The difference in the dc voltages which must be applied to the input terminals to obtain equal quiescent operating voltages (zero output offset voltage) at the output terminals.

Noise Figure

The ratio of the total noise power of the device and a resistive signal source to the noise power of the signal source alone, the signal source representing a generator of zero impedance in series with the source resistance.

Power Gain

The ratio of the signal power developed at the output of the device to the signal power applied to the input, expressed in dB.

Quiescent Operating Current

The average (dc) value of the current in either output terminal.

Voltage Gain

The ratio of the change in output voltage at either output terminal with respect to ground, to a change in input voltage at either input terminal with respect to ground, with the other input terminal at ac ground.

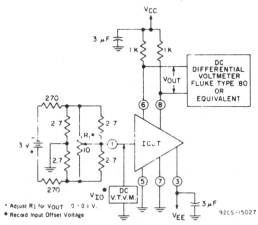

Fig.2 - Input offset voltage test circuit for CA3028B.

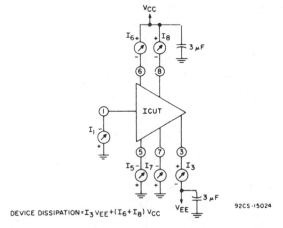

Fig.3a - Input offset current, input bias current, device dissipation, and quiescent operating current test circuit for CA3028A and CA3028B.

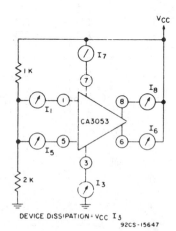

Fig.3b - Input bias current, device dissipation, and quiescent operating current test circuit for CA3053.

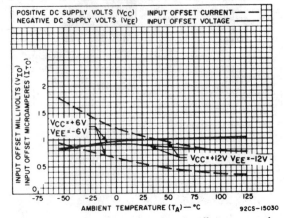

Fig.4 - Input offset voltage and input offset current for CA3028B.

ORDERING INFORMATION

Device	Temperature Range	Package
MC1496G	0°C to +70°C	Metal Can
MC1496L	0°C to +70°C	Ceramic DIP
MC1496P	0°C to +70°C	Plastic DIP
MC1596G	−55°C to +125°C	Metal Can
MC1596L	−55°C to +125°C	Ceramic DIP

MC1496
MC1596

BALANCED MODULATOR – DEMODULATOR

. . . designed for use where the output voltage is a product of an input voltage (signal) and a switching function (carrier). Typical applications include suppressed carrier and amplitude modulation, synchronous detection, FM detection, phase detection, and chopper applications. See Motorola Application Note AN-531 for additional design information.

* Excellent Carrier Suppression – 65 dB typ @ 0.5 MHz
                                 – 50 dB typ @ 10 MHz
* Adjustable Gain and Signal Handling
* Balanced Inputs and Outputs
* High Common Mode Rejection – 85 dB typ

BALANCED
MODULATOR – DEMODULATOR

SILICON MONOLITHIC
INTEGRATED CIRCUIT

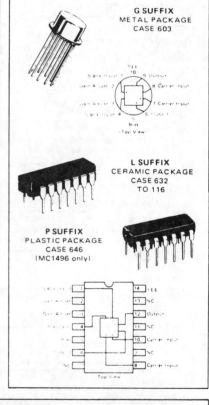

G SUFFIX
METAL PACKAGE
CASE 603

L SUFFIX
CERAMIC PACKAGE
CASE 632
TO-116

P SUFFIX
PLASTIC PACKAGE
CASE 646
(MC1496 only)

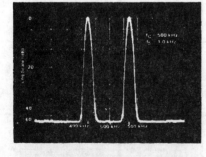

FIGURE 1
SUPPRESSED CARRIER
OUTPUT WAVEFORM

FIGURE 2
SUPPRESSED CARRIER
SPECTRUM

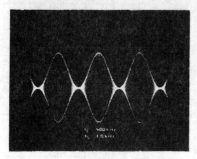

FIGURE 3 –
AMPLITUDE MODULATION
OUTPUT WAVEFORM

FIGURE 4 – AMPLITUDE-MODULATION SPECTRUM

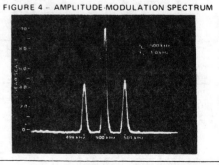

MAXIMUM RATINGS ($T_A = +25^{\circ}C$ unless otherwise noted)

Rating	Symbol	Value	Unit
Applied Voltage ($V_6 - V_7$, $V_8 - V_1$, $V_9 - V_7$, $V_9 - V_8$, $V_7 - V_4$, $V_7 - V_1$, $V_8 - V_4$, $V_6 - V_8$, $V_2 - V_5$, $V_3 - V_5$)	ΔV	30	Vdc
Differential Input Signal	$V_7 - V_8$ $V_4 - V_1$	$+5.0$ $\pm(5 + I_5 R_e)$	Vdc
Maximum Bias Current	I_5	10	mA

Power Dissipation (Package Limitation)
   Ceramic Dual In-Line Package
     Derate above $T_A = +25^{\circ}C$
   Metal Package
     Derate above $T_A = +25^{\circ}C$
	P_D	
		 575
		 3.85
		 680
		 4.6
			
			 mW
			 mW/$^{\circ}$C
			 mW
			 mW/$^{\circ}$C

Operating Temperature Range
                      MC1496
                      MC1596
	T_A	
		 0 to +70
		 -55 to +125
			$^{\circ}$C

Storage Temperature Range	T_{stg}	-65 to +150	$^{\circ}$C

ELECTRICAL CHARACTERISTICS ($V_{CC} = +12$ Vdc, $V_{EE} = -8.0$ Vdc, $I_5 = 1.0$ mAdc, $R_L = 3.9$ kΩ, $R_e = 1.0$ kΩ,
$T_A = +25^{\circ}$C unless otherwise noted) (All input and output characteristics are single-ended unless otherwise noted.)

Characteristic	Fig	Note	Symbol	MC1596 Min	MC1596 Typ	MC1596 Max	MC1496 Min	MC1496 Typ	MC1496 Max	Unit

Carrier Feedthrough
   $V_C = 60$ mV(rms) sine wave and $f_C = 1.0$ kHz
   offset adjusted to zero $f_C = 10$ MHz
   $V_C = 300$ mVp-p square wave:
   offset adjusted to zero $f_C = 1.0$ kHz
   offset not adjusted $f_C = 1.0$ kHz
	5	1	V_{CFT}	
				 -
				 -
				
				 -
				 -
					
					 40
					 140
					
					 0.04
					 20
						
						 -
						 -
						
						 0.2
						 100
							
							 -
							 -
							
							 -
							 -
								
								 40
								 140
								
								 0.04
								 20
									
									 -
									 -
									
									 0.4
									 200
										μV(rms)
										
										 mV(rms)

Carrier Suppression
   $f_S = 10$ kHz, 300 mV(rms)
   $f_C = 500$ kHz, 60 mV(rms) sine wave
   $f_C = 10$ MHz, 60 mV(rms) sine wave
	5	2	V_{CS}	
				
				 50
				 -
					
					 65
					 50
						
						 -
						 -
							
							 40
							 -
								
								 65
								 50
									
									 -
									 -
										dB
										
										 k

Transadmittance Bandwidth (Magnitude) ($R_L = 50$ ohms)
   Carrier Input Port, $V_C = 60$ mV(rms) sine wave
   $f_S = 1.0$ kHz, 300 mV(rms) sine wave
   Signal Input Port, $V_S = 300$ mV(rms) sine wave
   $|V_C| = 0.5$ Vdc
	8	8	BW_{3dB}	
				 -
				
				 -
					
					 300
					
					 80
						
						 -
						
						 -
							
							 -
							
							 -
								
								 300
								
								 80
									
									 -
									
									 -
										MHz

Signal Gain $V_S = 100$ mV(rms), $f = 1.0$ kHz, $|V_C| = 0.5$ Vdc	10	3	A_{VS}	2.5	3.5	-	2.5	3.5	-	V/V

Single-Ended Input Impedance, Signal Port, $f = 5.0$ MHz
   Parallel Input Resistance
   Parallel Input Capacitance
	6	-	
			 r_{ip}
			 c_{ip}
				
				 -
				 -
					
					 200
					 2.0
						
						 -
						 -
							
							 -
							 -
								
								 200
								 2.0
									
									 -
									 -
										
										 kΩ
										 pF

Single-Ended Output Impedance, $f = 10$ MHz
   Parallel Output Resistance
   Parallel Output Capacitance
	6	-	
			 r_{op}
			 c_{op}
				
				 -
				 -
					
					 40
					 5.0
						
						 -
						 -
							
							 -
							 -
								
								 40
								 5.0
									
									 -
									 -
										
										 kΩ
										 pF

Input Bias Current $I_{bS} = \dfrac{I_1 + I_4}{2}$; $I_{bC} = \dfrac{I_7 + I_8}{2}$	7	-	 I_{bS} I_{bC}	 - -	 12 12	 25 25	 - -	 12 12	 30 30	μA
Input Offset Current $I_{ioS} = I_1 - I_4$; $I_{ioC} = I_7 - I_8$	7	-	 $|I_{ioS}|$ $|I_{ioC}|$	 - -	 0.7 0.7	 5.0 5.0	 - -	 0.7 0.7	 7.0 7.0	μA
Average Temperature Coefficient of Input Offset Current ($T_A = -55^{\circ}$C to $+125^{\circ}$C)	7	-	$|TC_{Iio}|$	-	2.0	-	-	2.0	-	nA/$^{\circ}$C
Output Offset Current ($I_6 - I_9$)	7	-	$|I_{oo}|$	-	14	50	-	14	80	μA
Average Temperature Coefficient of Output Offset Current ($T_A = -55^{\circ}$C to $+125^{\circ}$C)	7	-	$|TC_{Ioo}|$	-	90	-	-	90	-	nA/$^{\circ}$C
Common-Mode Input Swing, Signal Port, $f_S = 1.0$ kHz	9	4	CMV	-	5.0	-	-	5.0	-	Vp-p
Common-Mode Gain, Signal Port, $f_S = 1.0$ kHz, $|V_C| = 0.5$ Vdc	9	-	ACM	-	-85	-	-	-85	-	dB
Common-Mode Quiescent Output Voltage (Pin 6 or Pin 9)	10	-	V_o	-	8.0	-	-	8.0	-	Vdc
Differential Output Voltage Swing Capability	10	-	V_{out}	-	8.0	-	-	8.0	-	Vp-p

Power Supply Current
   $I_6 + I_9$
 	
   I_{10}
	7	6	
			 I_{CC}
			
			 I_{EE}
				
				 -
				
				 -
					
					 2.0
					
					 3.0
						
						 3.0
						
						 4.0
							
							 -
							
							 -
								
								 2.0
								
								 3.0
									
									 4.0
									
									 5.0
										mAdc

DC Power Dissipation	7	5	P_D	-	33	-	-	33	-	mW

* Pin number references pertain to this device when packaged in a metal can. To ascertain the corresponding pin numbers for plastic or
ceramic packaged devices refer to the first page of this specification sheet.

MC1496, MC1596

GENERAL OPERATING INFORMATION *

Note 1 – Carrier Feedthrough

Carrier feedthrough is defined as the output voltage at carrier frequency with only the carrier applied (signal voltage = 0).

Carrier null is achieved by balancing the currents in the differential amplifier by means of a bias trim potentiometer (R_1 of Figure 5).

Note 2 – Carrier Suppression

Carrier suppression is defined as the ratio of each sideband output to carrier output for the carrier and signal voltage levels specified.

Carrier suppression is very dependent on carrier input level, as shown in Figure 22. A low value of the carrier does not fully switch the upper switching devices, and results in lower signal gain, hence lower carrier suppression. A higher than optimum carrier level results in unnecessary device and circuit carrier feedthrough, which again degenerates the suppression figure. The MC1596 has been characterized with a 60 mV(rms) sinewave carrier input signal. This level provides optimum carrier suppression at carrier frequencies in the vicinity of 500 kHz, and is generally recommended for balanced modulator applications.

Carrier feedthrough is independent of signal level, V_S. Thus carrier suppression can be maximized by operating with large signal levels. However, a linear operating mode must be maintained in the signal-input transistor pair – or harmonics of the modulating signal will be generated and appear in the device output as spurious sidebands of the suppressed carrier. This requirement places an upper limit on input-signal amplitude (see Note 3 and Figure 20). Note also that an optimum carrier level is recommended in Figure 22 for good carrier suppression and minimum spurious sideband generation.

At higher frequencies circuit layout is very important in order to minimize carrier feedthrough. Shielding may be necessary in order to prevent capacitive coupling between the carrier input leads and the output leads.

Note 3 – Signal Gain and Maximum Input Level

Signal gain (single-ended) at low frequencies is defined as the voltage gain,

$$A_{VS} = \frac{V_o}{V_S} = \frac{R_L}{R_e + 2r_e} \text{ where } r_e = \frac{26 \text{ mV}}{I_5 \text{ (mA)}}$$

A constant dc potential is applied to the carrier input terminals to fully switch two of the upper transistors "on" and two transistors "off" ($V_C = 0.5$ Vdc). This in effect forms a cascode differential amplifier.

Linear operation requires that the signal input be below a critical value determined by R_E and the bias current I_5.

$$V_S \leq I_5 R_E \text{ (Volts peak)}$$

Note that in the test circuit of Figure 10, V_S corresponds to a maximum value of 1 volt peak.

Note 4 – Common-Mode Swing

The common-mode swing is the voltage which may be applied to both bases of the signal differential amplifier, without saturating the current sources or without saturating the differential amplifier itself by swinging it into the upper switching devices. This swing is variable depending on the particular circuit and biasing conditions chosen (see Note 6).

Note 5 – Power Dissipation

Power dissipation, P_D, within the integrated circuit package should be calculated as the summation of the voltage-current products at each port, i.e. assuming $V_9 = V_6$, $I_5 = I_6 = I_9$ and ignoring

base current, $P_D = 2 I_5 (V_6 - V_{10}) + I_5 (V_5 - V_{10})$ where subscripts refer to pin numbers.

Note 6 – Design Equations

The following is a partial list of design equations needed to operate the circuit with other supply voltages and input conditions. See Note 3 for R_e equation.

A. Operating Current

The internal bias currents are set by the conditions at pin 5. Assume:

$$I_5 = I_6 = I_9$$

$$I_B \ll I_C \text{ for all transistors}$$

then:

$$R_5 = \frac{V^- - \phi}{I_5} - 500 \, \Omega \quad \text{where:} \quad \begin{array}{l} R_5 \text{ is the resistor between pin} \\ \text{5 and ground} \\ \phi = 0.75 \text{ V at } T_A = +25°C \end{array}$$

The MC1596 has been characterized for the condition $I_5 = 1.0$ mA and is the generally recommended value.

B. Common-Mode Quiescent Output Voltage

$$V_6 = V_9 = V^+ - I_5 R_L$$

Note 7 – Biasing

The MC1596 requires three dc bias voltage levels which must be set externally. Guidelines for setting up these three levels include maintaining at least 2 volts collector-base bias on all transistors while not exceeding the voltages given in the absolute maximum rating table;

$$30 \text{ Vdc} \geq [(V_6, V_9) - (V_7, V_8)] \geq 2 \text{ Vdc}$$

$$30 \text{ Vdc} \geq [(V_7, V_8) - (V_1, V_4)] \geq 2.7 \text{ Vdc}$$

$$30 \text{ Vdc} \geq [(V_1, V_4) - (V_5)] \geq 2.7 \text{ Vdc}$$

The foregoing conditions are based on the following approximations:

$$V_6 = V_9, \quad V_7 = V_8, \quad V_1 = V_4$$

Bias currents flowing into pins 1, 4, 7, and 8 are transistor base currents and can normally be neglected if external bias dividers are designed to carry 1.0 mA or more.

Note 8 – Transadmittance Bandwidth

Carrier transadmittance bandwidth is the 3-dB bandwidth of the device forward transadmittance as defined by:

$$Y_{21C} = \frac{i_o \text{ (each sideband)}}{v_s \text{ (signal)}} \Big|_{V_o = 0}$$

Signal transadmittance bandwidth is the 3-dB bandwidth of the device forward transadmittance as defined by:

$$Y_{21S} = \frac{i_o \text{ (signal)}}{v_s \text{ (signal)}} \Big|_{V_C = 0.5 \text{ Vdc}, V_o = 0}$$

*Pin number references pertain to this device when packaged in a metal can. To ascertain the corresponding pin numbers for plastic or ceramic packaged devices refer to the first page of this specification sheet.

MC1496, MC1596

Note 9 — Coupling and Bypass Capacitors C_1 and C_2

Capacitors C_1 and C_2 (Figure 5) should be selected for a reactance of less than 5.0 ohms at the carrier frequency.

Note 10 — Output Signal, V_o

The output signal is taken from pins 6 and 9, either balanced or single-ended. Figure 12 shows the output levels of each of the two output sidebands resulting from variations in both the carrier and modulating signal inputs with a single-ended output connection.

Note 11 — Signal Port Stability

Under certain values of driving source impedance, oscillation may occur. In this event, an RC suppression network should be connected directly to each input using short leads. This will reduce the Q of the source-tuned circuits that cause the oscillation.

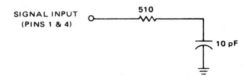

An alternate method for low-frequency applications is to insert a 1 k-ohm resistor in series with the inputs, pins 1 and 4. In this case input current drift may cause serious degradation of carrier suppression.

TEST CIRCUITS

FIGURE 5 — CARRIER REJECTION AND SUPPRESSION

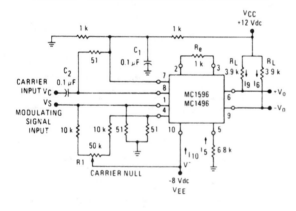

FIGURE 6 — INPUT-OUTPUT IMPEDANCE

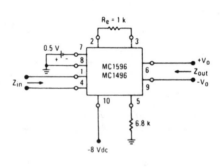

FIGURE 7 — BIAS AND OFFSET CURRENTS

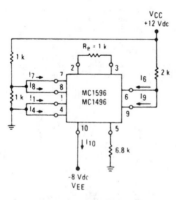

FIGURE 8 — TRANSCONDUCTANCE BANDWIDTH

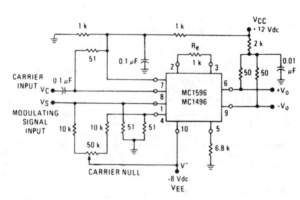

Pin number references pertain to this device when packaged in a metal can. To ascertain the corresponding pin numbers for plastic or ceramic packaged devices refer to the first page of this specification sheet.

MC1496, MC1596

TEST CIRCUITS (continued)

FIGURE 9 – COMMON-MODE GAIN

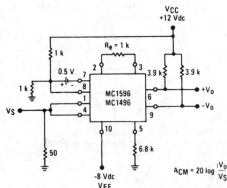

$$A_{CM} = 20 \log \frac{|V_0|}{V_S}$$

FIGURE 10 – SIGNAL GAIN AND OUTPUT SWING

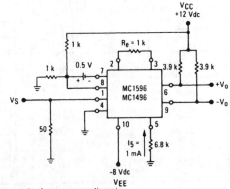

Pin number references pertain to this device when packaged in a metal can. To ascertain the corresponding pin numbers for plastic or ceramic packaged devices refer to the first page of this specification sheet.

TYPICAL CHARACTERISTICS (continued)

Typical characteristics were obtained with circuit shown in Figure 5, f_C = 500 kHz (sine wave),
V_C = 60 mV(rms), f_S = 1 kHz, V_S = 300 mV(rms), T_A = +25°C unless otherwise noted.

FIGURE 11 – SIDEBAND OUTPUT versus CARRIER LEVELS

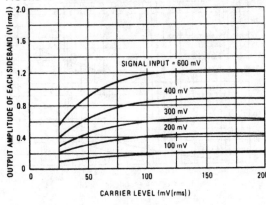

FIGURE 12 – SIGNAL-PORT PARALLEL-EQUIVALENT INPUT RESISTANCE versus FREQUENCY

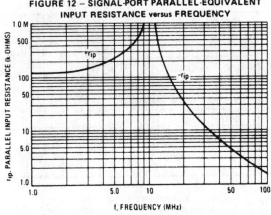

FIGURE 13 – SIGNAL-PORT PARALLEL-EQUIVALENT INPUT CAPACITANCE versus FREQUENCY

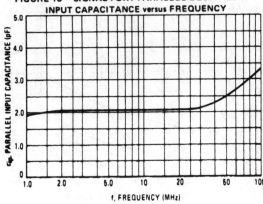

FIGURE 14 – SINGLE-ENDED OUTPUT IMPEDANCE versus FREQUENCY

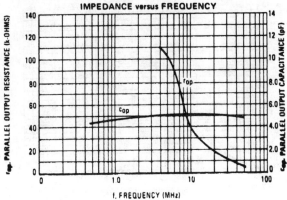

TYPICAL CHARACTERISTICS (continued)

Typical characteristics were obtained with circuit shown in Figure 5, f_C = 500 kHz (sine wave),
V_C = 60 mV(rms), f_S = 1 kHz, V_S = 300 mV(rms), T_A = +25°C unless otherwise noted.

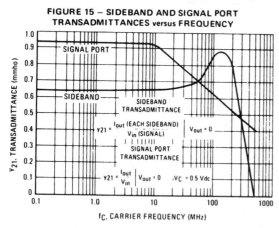

FIGURE 15 – SIDEBAND AND SIGNAL PORT
TRANSADMITTANCES versus FREQUENCY

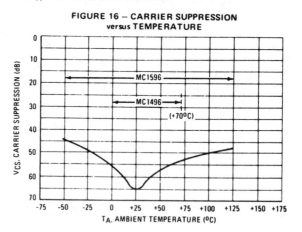

FIGURE 16 – CARRIER SUPPRESSION
versus TEMPERATURE

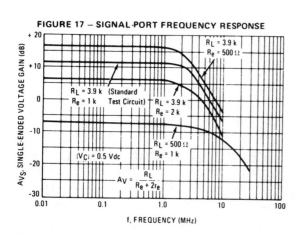

FIGURE 17 – SIGNAL-PORT FREQUENCY RESPONSE

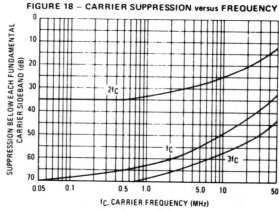

FIGURE 18 – CARRIER SUPPRESSION versus FREQUENCY

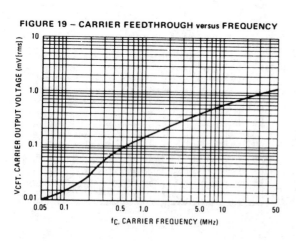

FIGURE 19 – CARRIER FEEDTHROUGH versus FREQUENCY

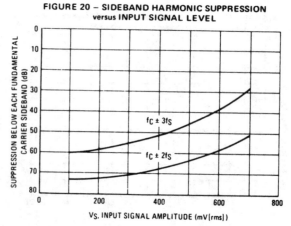

FIGURE 20 – SIDEBAND HARMONIC SUPPRESSION
versus INPUT SIGNAL LEVEL

MC1496, MC1596

TYPICAL CHARACTERISTICS (continued)

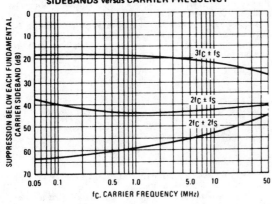

FIGURE 21 – SUPPRESSION OF CARRIER HARMONIC
SIDEBANDS versus CARRIER FREQUENCY

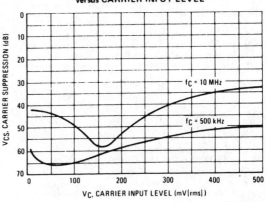

FIGURE 22 – CARRIER SUPPRESSION
versus CARRIER INPUT LEVEL

OPERATIONS INFORMATION

The MC1596/MC1496, a monolithic balanced modulator circuit, is shown in Figure 23.

This circuit consists of an upper quad differential amplifier driven by a standard differential amplifier with dual current sources. The output collectors are cross-coupled so that full-wave balanced multiplication of the two input voltages occurs. That is, the output signal is a constant times the product of the two input signals.

Mathematical analysis of linear ac signal multiplication indicates that the output spectrum will consist of only the sum and difference of the two input frequencies. Thus, the device may be used as a balanced modulator, doubly balanced mixer, product detector, frequency doubler, and other applications requiring these particular output signal characteristics.

The lower differential amplifier has its emitters connected to the package pins so that an external emitter resistance may be used. Also, external load resistors are employed at the device output.

Signal Levels

The upper quad differential amplifier may be operated either in a linear or a saturated mode. The lower differential amplifier is operated in a linear mode for most applications.

For low-level operation at both input ports, the output signal will contain sum and difference frequency components and have an amplitude which is a function of the product of the input signal amplitudes.

For high-level operation at the carrier input port and linear operation at the modulating signal port, the output signal will contain sum and difference frequency components of the modulating signal frequency and the fundamental and odd harmonics of the carrier frequency. The output amplitude will be a constant times the modulating signal amplitude. Any amplitude variations in the carrier signal will not appear in the output.

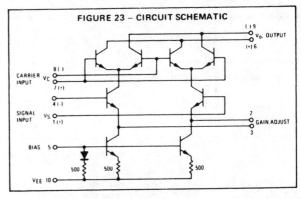

FIGURE 23 – CIRCUIT SCHEMATIC

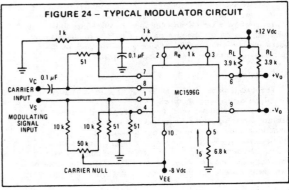

FIGURE 24 – TYPICAL MODULATOR CIRCUIT

Pin number references pertain to this device when packaged in a metal can. To ascertain the corresponding pin numbers for plastic or ceramic packaged devices refer to the first page of this specification sheet.

OPERATIONS INFORMATION (continued)

The linear signal handling capabilities of a differential amplifier are well defined. With no emitter degeneration, the maximum input voltage for linear operation is approximately 25 mV peak. Since the upper differential amplifier has its emitters internally connected, this voltage applies to the carrier input port for all conditions.

Since the lower differential amplifier has provisions for an external emitter resistance, its linear signal handling range may be adjusted by the user. The maximum input voltage for linear operation may be approximated from the following expression.

$$V = (I_5)(R_E) \text{ volts peak.}$$

This expression may be used to compute the minimum value of R_E for a given input voltage amplitude.

The gain from the modulating signal input port to the output is the MC1596/MC1496 gain parameter which is most often of interest to the designer. This gain has significance only when the lower differential amplifier is operated in a linear mode, but this includes most applications of the device.

As previously mentioned, the upper quad differential amplifier may be operated either in a linear or a saturated mode. Approximate gain expressions have been developed for the MC1596/MC1496 for a low-level modulating signal input and the following carrier input conditions:

1) Low-level dc
2) High-level dc
3) Low-level ac
4) High-level ac

These gains are summarized in Table 1, along with the frequency components contained in the output signal.

FIGURE 25 – TABLE 1
VOLTAGE GAIN AND OUTPUT FREQUENCIES

Carrier Input Signal (V_C)	Approximate Voltage Gain	Output Signal Frequency(s)
Low-level dc	$\dfrac{R_L V_C}{2(R_E + 2r_e)\left(\frac{KT}{q}\right)}$	f_M
High-level dc	$\dfrac{R_L}{R_E + 2r_e}$	f_M
Low-level ac	$\dfrac{R_L V_C(\text{rms})}{2\sqrt{2}\left(\frac{KT}{q}\right)(R_E + 2r_e)}$	$f_C \pm f_M$
High-level ac	$\dfrac{0.637 R_L}{R_E + 2r_e}$	$f_C \pm f_M, 3f_C \pm f_M, 5f_C \pm f_M, \ldots$

NOTES:
1. Low-level Modulating Signal, V_M, assumed in all cases. V_C is Carrier Input Voltage.
2. When the output signal contains multiple frequencies, the gain expression given is for the output amplitude of each of the two desired outputs, $f_C + f_M$ and $f_C - f_M$.
3. All gain expressions are for a single-ended output. For a differential output connection, multiply each expression by two.
4. R_L = Load resistance.
5. R_E = Emitter resistance between pins 2 and 3.
6. r_e = Transistor dynamic emitter resistance, at +25°C;

$$r_e \approx \frac{26 \text{ mV}}{I_5 \text{ (mA)}}$$

7. K = Boltzmann's Constant, T = temperature in degrees Kelvin, q = the charge on an electron.

$$\frac{KT}{q} \approx 26 \text{ mV at room temperature}$$

APPLICATIONS INFORMATION

Double sideband suppressed carrier modulation is the basic application of the MC1596/MC1496. The suggested circuit for this application is shown on the front page of this data sheet.

In some applications, it may be necessary to operate the MC1596/MC1496 with a single dc supply voltage instead of dual supplies. Figure 26 shows a balanced modulator designed for operation with a single +12 Vdc supply. Performance of this circuit is similar to that of the dual supply modulator.

AM Modulator

The circuit shown in Figure 27 may be used as an amplitude modulator with a minor modification.

All that is required to shift from suppressed carrier to AM operation is to adjust the carrier null potentiometer for the proper amount of carrier insertion in the output signal.

However, the suppressed carrier null circuitry as shown in Figure 27 does not have sufficient adjustment range. Therefore, the modulator may be modified for AM operation by changing two resistor values in the null circuit as shown in Figure 28.

Product Detector

The MC1596/MC1496 makes an excellent SSB product detector (see Figure 29).

This product detector has a sensitivity of 3.0 microvolts and a dynamic range of 90 dB when operating at an intermediate frequency of 9 MHz.

The detector is broadband for the entire high frequency range. For operation at very low intermediate frequencies down to 50 kHz the 0.1 μF capacitors on pins 7 and 8 should be increased to 1.0 μF. Also, the output filter at pin 9 can be tailored to a specific intermediate frequency and audio amplifier input impedance.

As in all applications of the MC1596/MC1496, the emitter resistance between pins 2 and 3 may be increased or decreased to adjust circuit gain, sensitivity, and dynamic range.

This circuit may also be used as an AM detector by introducing carrier signal at the carrier input and an AM signal at the SSB input.

The carrier signal may be derived from the intermediate frequency signal or generated locally. The carrier signal may be introduced with or without modulation, provided its level is sufficiently high to saturate the upper quad differential amplifier. If the carrier signal is modulated, a 300 mV(rms) input level is recommended.

MC1496, MC1596

APPLICATIONS INFORMATION (continued)

Doubly Balanced Mixer

The MC1596/MC1496 may be used as a doubly balanced mixer with either broadband or tuned narrow band input and output networks.

The local oscillator signal is introduced at the carrier input port with a recommended amplitude of 100 mV(rms).

Figure 30 shows a mixer with a broadband input and a tuned output.

Frequency Doubler

The MC1596/MC1496 will operate as a frequency doubler by introducing the same frequency at both input ports.

Figures 31 and 32 show a broadband frequency doubler and a tuned output very high frequency (VHF) doubler, respectively.

Phase Detection and FM Detection

The MC1596/MC1496 will function as a phase detector. High-level input signals are introduced at both inputs. When both inputs are at the same frequency the MC1596/MC1496 will deliver an output which is a function of the phase difference between the two input signals.

An FM detector may be constructed by using the phase detector principle. A tuned circuit is added at one of the inputs to cause the two input signals to vary in phase as a function of frequency. The MC1596/MC1496 will then provide an output which is a function of the input signal frequency.

Pin number references pertain to this device when packaged in a metal can. To ascertain the corresponding pin numbers for plastic or ceramic packaged devices refer to the first page of this specification sheet.

TYPICAL APPLICATIONS

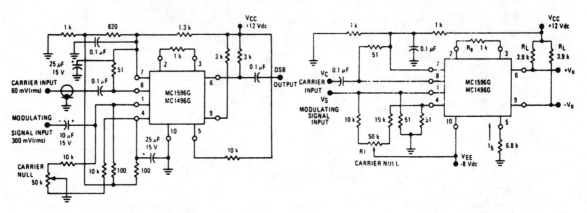

FIGURE 26 – BALANCED MODULATOR
(+12 Vdc SINGLE SUPPLY)

FIGURE 27 – BALANCED MODULATOR-DEMODULATOR

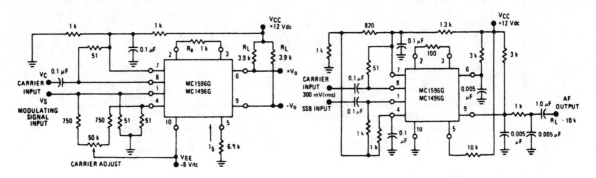

FIGURE 28 – AM MODULATOR CIRCUIT

FIGURE 29 – PRODUCT DETECTOR
(+12 Vdc SINGLE SUPPLY)

TYPICAL APPLICATIONS (continued)

**FIGURE 30 – DOUBLY BALANCED MIXER
(BROADBAND INPUTS, 9.0 MHz TUNED OUTPUT)**

FIGURE 31 – LOW-FREQUENCY DOUBLER

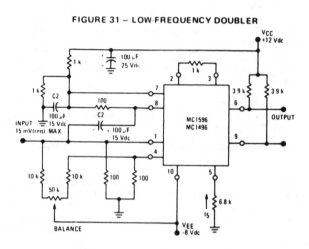

L1 44 TURNS AWG NO. 26 ENAMELED WIRE WOUND
ON MICROMETALS TYPE 44 6 TOROID CORE

FIGURE 32 – 150 to 300 MHz DOUBLER

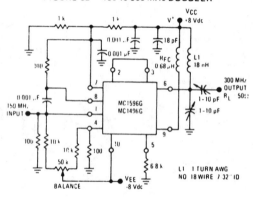

L1 1 TURN AWG
NO 18 WIRE / 32" ID

DEFINITIONS

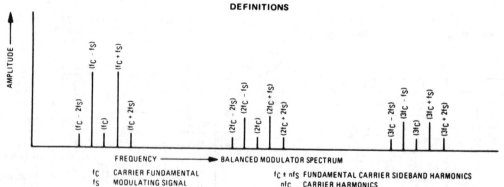

BALANCED MODULATOR SPECTRUM

fC	CARRIER FUNDAMENTAL
fS	MODULATING SIGNAL
fC ± fS	FUNDAMENTAL CARRIER SIDEBANDS

fC ± nfS	FUNDAMENTAL CARRIER SIDEBAND HARMONICS
nfC	CARRIER HARMONICS
nfC ± nfS	CARRIER HARMONIC SIDEBANDS

Pin number references pertain to this device when packaged in a metal can. To ascertain the corresponding pin numbers for plastic or ceramic packaged devices refer to the first page of this specification sheet.

LM565/LM565C phase locked loop

general description

The LM565 and LM565C are general purpose phase locked loops containing a stable, highly linear voltage controlled oscillator for low distortion FM demodulation, and a double balanced phase detector with good carrier suppression. The VCO frequency is set with an external resistor and capacitor, and a tuning range of 10:1 can be obtained with the same capacitor. The characteristics of the closed loop system—bandwidth, response speed, capture and pull in range—may be adjusted over a wide range with an external resistor and capacitor. The loop may be broken between the VCO and the phase detector for insertion of a digital frequency divider to obtain frequency multiplication.

The LM565H is specified for operation over the −55°C to +125°C military temperature range. The LM565CH and LM565CN are specified for operation over the 0°C to +70°C temperature range.

features

- 200 ppm/°C frequency stability of the VCO

- Power supply range of ±5 to ±12 volts with 100 ppm/% typical
- 0.2% linearity of demodulated output
- Linear triangle wave with in phase zero crossings available
- TTL and DTL compatible phase detector input and square wave output
- Adjustable hold in range from ±1% to > ±60%.

applications

- Data and tape synchronization
- Modems
- FSK demodulation
- FM demodulation
- Frequency synthesizer
- Tone decoding
- Frequency multiplication and division
- SCA demodulators
- Telemetry receivers
- Signal regeneration
- Coherent demodulators.

schematic and connection diagrams

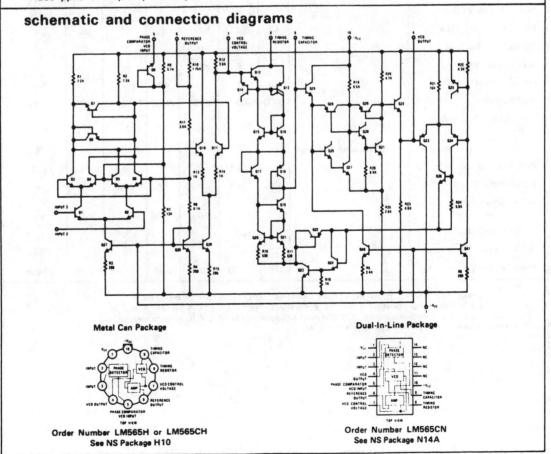

Metal Can Package

Order Number LM565H or LM565CH
See NS Package H10

Dual-In-Line Package

Order Number LM565CN
See NS Package N14A

absolute maximum ratings

Supply Voltage	±12V
Power Dissipation (Note 1)	300 mW
Differential Input Voltage	±1V
Operating Temperature Range LM565H	−55°C to +125°C
LM565CH, LM565CN	0°C to 70°C
Storage Temperature Range	−65°C to +150°C
Lead Temperature (Soldering, 10 sec)	300°C

electrical characteristics (AC Test Circuit, $T_A = 25°C$, $V_C = ±6V$)

PARAMETER	CONDITIONS	LM565			LM565C			UNITS
		MIN	TYP	MAX	MIN	TYP	MAX	
Power Supply Current			8.0	12.5		8.0	12.5	mA
Input Impedance (Pins 2, 3)	$-4V < V_2, V_3 < 0V$	7	10			5		kΩ
VCO Maximum Operating Frequency	$C_o = 2.7$ pF	300	500		250	500		kHz
Operating Frequency Temperature Coefficient			−100	300		−200	500	ppm/°C
Frequency Drift with Supply Voltage			0.01	0.1		0.05	0.2	%/V
Triangle Wave Output Voltage		2	2.4	3	2	2.4	3	V_{p-p}
Triangle Wave Output Linearity			0.2	0.75		0.5	1	%
Square Wave Output Level		4.7	5.4		4.7	5.4		V_{p-p}
Output Impedance (Pin 4)			5			5		kΩ
Square Wave Duty Cycle		45	50	55	40	50	60	%
Square Wave Rise Time			20	100		20		ns
Square Wave Fall Time			50	200		50		ns
Output Current Sink (Pin 4)		0.6	1		0.6	1		mA
VCO Sensitivity	$f_o = 10$ kHz	6400	6600	6800	6000	6600	7200	Hz/V
Demodulated Output Voltage (Pin 7)	±10% Frequency Deviation	250	300	350	200	300	400	mV_{pp}
Total Harmonic Distortion	±10% Frequency Deviation		0.2	0.75		0.2	1.5	%
Output Impedance (Pin 7)			3.5			3.5		kΩ
DC Level (Pin 7)		4.25	4.5	4.75	4.0	4.5	5.0	V
Output Offset Voltage $|V_7 - V_6|$			30	100		50	200	mV
Temperature Drift of $|V_7 - V_6|$			500			500		μV/°C
AM Rejection		30	40			40		dB
Phase Detector Sensitivity K_D		0.6	.68	0.9	0.55	.68	0.95	V/radian

Note 1: The maximum junction temperature of the LM565 is 150°C, while that of the LM565C and LM565CN is 100°C. For operation at elevated temperatures, devices in the TO-5 package must be derated based on a thermal resistance of 150°C/W junction to ambient or 45°C/W junction to case. Thermal resistance of the dual-in-line package is 100°C/W.

typical performance characteristics

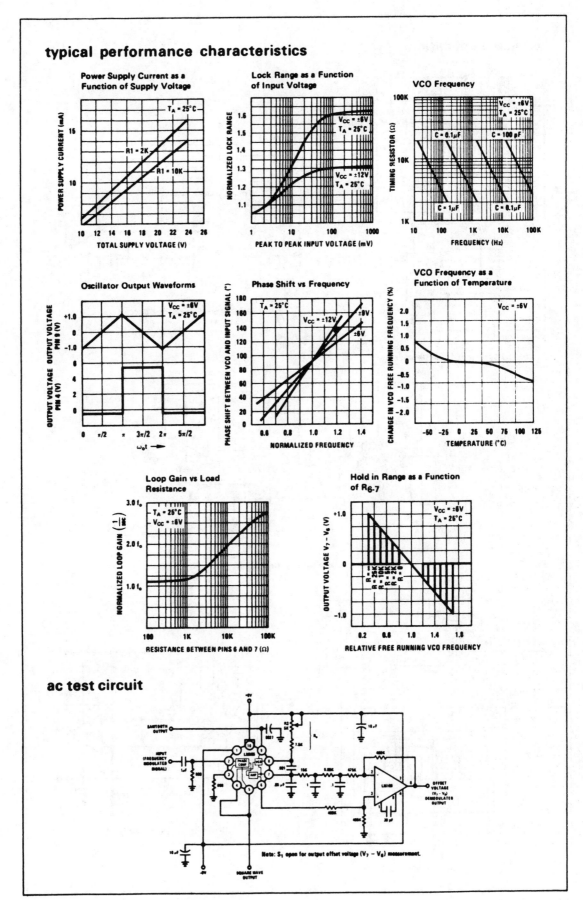

Power Supply Current as a Function of Supply Voltage

Lock Range as a Function of Input Voltage

VCO Frequency

Oscillator Output Waveforms

Phase Shift vs Frequency

VCO Frequency as a Function of Temperature

Loop Gain vs Load Resistance

Hold in Range as a Function of R$_{6-7}$

ac test circuit

Note: S$_1$ open for output offset voltage (V$_7$ − V$_6$) measurement.

Reprinted with permission of National Semiconductor Corp.

typical applications

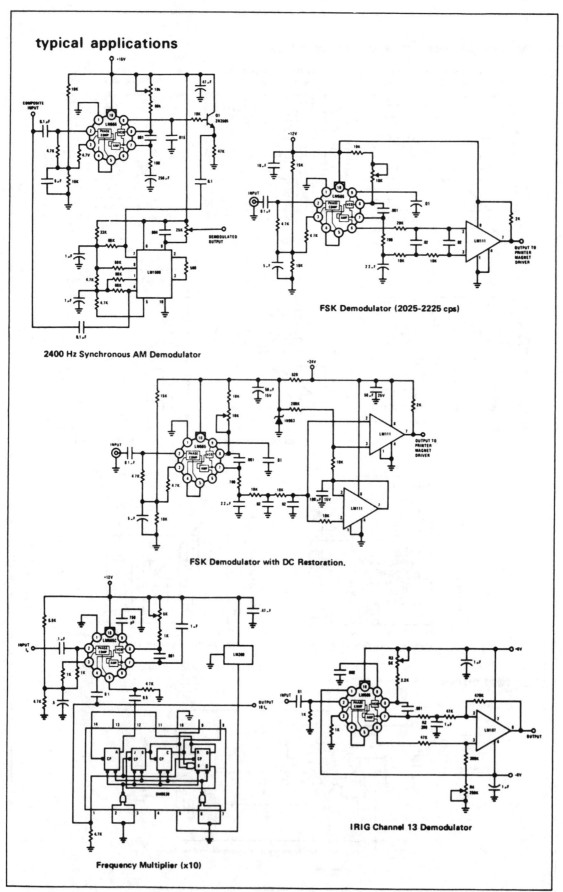

2400 Hz Synchronous AM Demodulator

FSK Demodulator (2025-2225 cps)

FSK Demodulator with DC Restoration.

Frequency Multiplier (x10)

IRIG Channel 13 Demodulator

applications information

In designing with phase locked loops such as the LM565, the important parameters of interest are:

FREE RUNNING FREQUENCY

$$f_o \cong \frac{1}{3.7 \, R_o C_o}$$

LOOP GAIN: relates the amount of phase change between the input signal and the VCO signal for a shift in input signal frequency (assuming the loop remains in lock). In servo theory, this is called the "velocity error coefficient".

Loop gain $= K_o K_D \left(\frac{1}{\text{sec}}\right)$

$K_o =$ oscillator sensitivity $\left(\frac{\text{radians/sec}}{\text{volt}}\right)$

$K_D =$ phase detector sensitivity $\left(\frac{\text{volts}}{\text{radian}}\right)$

The loop gain of the LM565 is dependent on supply voltage, and may be found from:

$$K_o K_D = \frac{33.6 \, f_o}{V_c}$$

$f_o =$ VCO frequency in Hz

$V_c =$ total supply voltage to circuit.

Loop gain may be reduced by connecting a resistor between pins 6 and 7; this reduces the load impedance on the output amplifier and hence the loop gain.

HOLD IN RANGE: the range of frequencies that the loop will remain in lock after initially being locked.

$$f_H = \pm \frac{8 \, f_o}{V_c}$$

$f_o =$ free running frequency of VCO

$V_c =$ total supply voltage to the circuit.

THE LOOP FILTER

In almost all applications, it will be desirable to filter the signal at the output of the phase detector (pin 7) this filter may take one of two forms:

Simple Lag Filter

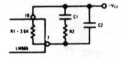
Lag-Lead Filter

A simple lag filter may be used for wide closed loop bandwidth applications such as modulation following where the frequency deviation of the carrier is fairly high (greater than 10%), or where wideband modulating signals must be followed.

The natural bandwidth of the closed loop response may be found from:

$$f_n = \frac{1}{2\pi} \sqrt{\frac{K_o K_D}{R_1 C_1}}$$

Associated with this is a damping factor:

$$\delta = \frac{1}{2} \sqrt{\frac{1}{R_1 C_1 K_o K_D}}$$

For narrow band applications where a narrow noise bandwidth is desired, such as applications involving tracking a slowly varying carrier, a lead lag filter should be used. In general, if $1/R_1 C_1 < K_o K_d$, the damping factor for the loop becomes quite small resulting in large overshoot and possible instability in the transient response of the loop. In this case, the natural frequency of the loop may be found from

$$f_n = \frac{1}{2\pi} \sqrt{\frac{K_o K_D}{\tau_1 + \tau_2}}$$

$$\tau_1 + \tau_2 = (R_1 + R_2) \, C_1$$

R_2 is selected to produce a desired damping factor δ, usually between 0.5 and 1.0. The damping factor is found from the approximation:

$$\delta \simeq \pi \tau_2 f_n$$

These two equations are plotted for convenience.

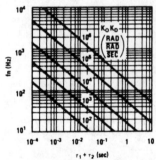
Filter Time Constant vs Natural Frequency

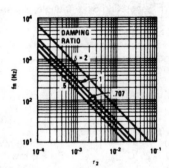
Damping Time Constant vs Natural Frequency

Capacitor C_2 should be much smaller than C_1 since its function is to provide filtering of carrier. In general $C_2 \leq 0.1 \, C_1$.

XR-8038

Precision Waveform Generator

GENERAL DESCRIPTION

The XR-8038 is a precision waveform generator IC capable of producing sine, square, triangular, sawtooth and pulse waveforms with a minimum number of external components and adjustments. Its operating frequency can be selected over nine decades of frequency, from 0.001 Hz to 1 MHz by the choice of external R-C components. The frequency of oscillation is highly stable over a wide range of temperature and supply voltage changes. The frequency control, sweep and modulation can be accomplished with an external control voltage, without affecting the quality of the output waveforms. Each of the three basic waveforms, i.e., sinewave, triangle and square wave outputs are available simultaneously, from independent output terminals.

The XR-8038 monolithic waveform generator uses advanced processing technology and Schottky-barrier diodes to enhance its frequency performance. It can be readily interfaced with a monolithic phase-detector circuit, such as the XR-2208, to form stable phase-locked loop circuits.

FEATURES

Direct Replacement for Intersil 8038
Low Frequency Drift—50 ppm/°C Max.
Simultaneous Sine, Triangle and Square-Wave Outputs
Low Distortion—THD ≈ 1%
High FM and Triangle Linearity
Wide Frequency Range—0.001 Hz to 1 MHz
Variable Duty-Cycle—2% to 98%

APPLICATIONS

Precision Waveform Generation Sine, Triangle, Square, Pulse
Sweep and FM Generation
Tone Generation
Instrumentation and Test Equipment Design
Precision PLL Design

ABSOLUTE MAXIMUM RATINGS

Power Supply	36V
Power Dissipation (package limitation)	
Ceramic package	750 mW
Derate above +25°C	6.0 mW/°C
Plastic package	625 mW
Derate above +25°C	5 mW/°C
Storage Temperature Range	−65°C to +150°C

FUNCTIONAL BLOCK DIAGRAM

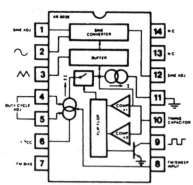

ORDERING INFORMATION

Part Number	Package	Operating Temperature
XR-8038M	Ceramic	−55°C to +125°C
XR-8038N	Ceramic	0°C to +70°C
XR-8038P	Plastic	0°C to +70°C
XR-8038CN	Ceramic	0°C to +70°C
XR-8038CP	Plastic	0°C to +70°C

SYSTEM DESCRIPTION

The XR-8038 precision waveform generator produces highly stable and sweepable square, triangle and sine waves across nine frequency decades. The device time base employs resistors and a capacitor for frequency and duty cycle determination. The generator contains dual comparators, a flip-flop driving a switch, current sources, a buffer amplifier and a sine wave converter. Three identical frequency waveforms are simultaneously available. Supply voltage can range from 10V to 30V, or ±5V with dual supplies.

Unadjusted sine wave distortion is typically less than 0.7%, with Pin 1 open and 8 kΩ from Pin 12 to Pin 11 ($-V_{EE}$ or ground). Sine wave distortion may be improved by including two 100 kΩ potentiometers between V_{CC} and V_{EE} (or ground), with one wiper connected to Pin 1 and the other connected to Pin 12.

Frequency sweeping or FM is accomplished by applying modulation to Pins 7 and 8 for small deviations, or only to Pin 8 for large shifts. Sweep range typically exceeds 1000:1.

The square wave output is an open collector transistor; output amplitude swing closely approaches the supply voltage. Triangle output amplitude is typically 1/3 of the supply, and sine wave output reaches 0.22 V_S.

XR-8038

ELECTRICAL CHARACTERISTICS

Test Conditions: $V_S = \pm 5V$ to $\pm 15V$, $T_A = 25°C$, $R_L = 1\ M\Omega$, $R_A = R_B = 10\ k\Omega$, $C_1 = 3300\ pF$, S_1 closed, unless otherwise specified. See Test Circuit of Figure 1.

PARAMETERS	XR-8038M/XR-8038 MIN	TYP	MAX	XR-8038C MIN	TYP	MAX	UNITS	CONDITIONS
GENERAL CHARACTERISTICS								
Supply Voltage, V_S								
Single Supply	10		30	10		30	V	
Dual Supplies	±5		±15	±5		±15	V	
Supply Current		12	15		12	20	mA	$V_S = \pm 10V$. See Note 1.
FREQUENCY CHARACTERISTICS (Measured at Pin 9)								
Range of Adjustment								
Max. Operating Frequency		1			1		MHz	$R_A = R_B = 500\Omega$, $C_1 = 0$, $R_L = 15\ k\Omega$
Lowest Practical Frequency		0.001			0.001		Hz	$R_A = R_B = 1\ M\Omega$, $C_1 = 500\ \mu F$
Max. FM Sweep Frequency		100			100		kHz	
FM Sweep Range		1000:1			1000:1			S_1 Open. See Notes 2 and 3.
FM Linearity		0.1			0.2		%	S_1 Open. See Note 3.
Range of Timing Resistors	0.5		1000	0.5		1000	kΩ	Values of R_A and R_B
Temperature Stability								
XR-8038M		20	50	—	—	—	ppm/°C	
XR-8038		50	100	—	—	—	ppm/°C	
XR-8038C	—	—	—		50		ppm/°C	
Power Supply Stability		0.05			0.05		%/V	See Note 4.
OUTPUT CHARACTERISTICS								
Square-Wave								Measured at Pin 9.
Amplitude	0.9	0.98		0.9	0.98		x V_S	$R_L = 100\ k\Omega$
Saturation Voltage		0.2	0.4		0.2	0.5	V	$I_{sink} = 2\ mA$
Rise Time		100			100		nsec	$R_L = 4.7\ k\Omega$
Fall Time		40			40		nsec	$R_L = 4.7\ k\Omega$
Duty Cycle Adj.	2		98	2		98	%	
Triangle/Sawtooth/Ramp								Measured at Pin 3.
Amplitude	0.3	0.33		0.3	0.33		x V_S	$R_L = 100\ k\Omega$
Linearity		0.05			0.1		%	
Output Impedance		200			200		Ω	$I_{out} = 5\ mA$
Sine-Wave Amplitude	0.2	0.22		0.2	0.22		x V_S	$R_L = 100\ k\Omega$
Distortion								
Unadjusted		0.7	1.5		0.8	3	%	$R_L = 1\ M\Omega$. See Note 5.
Adjusted		0.5			0.5		%	$R_L = 1\ M\Omega$

Note 1: Currents through R_A ad R_B not included.
Note 2: $V_S = 20V$, $f = 10\ kHz$, $R_A = R_B = 10k\Omega$.
Note 3: Apply sweep voltage at Pin 8.
    $(2/3\ V_S + 2V) \le V_{sweep} \le V_S$
Note 4: $10V \le V_S \le 30V$ or $\pm 5V \le V_S \le \pm 15V$.
Note 5: 81 kΩ resistor connected between Pins 11 and 12.

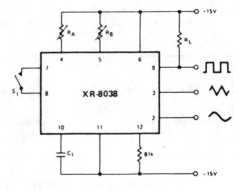

Figure 1. Generalized Test Circuit

CHARACTERISTIC CURVES

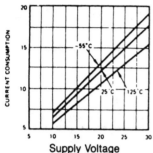

Power Dissipation vs. Supply Voltage

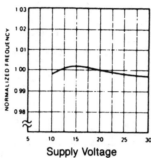

Frequency Drift vs. Power Supply

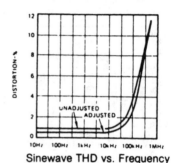

Sinewave THD vs. Frequency

WAVEFORM ADJUSTMENT

The *symmetry* of all waveforms can be adjusted with the external timing resistors. Two possible ways to accomplish this are shown in Figure 2. Best results are obtained by keeping the timing resistors R_A and R_B separate (a). R_A controls the rising portion of the triangle and sine-wave and the "Low" state of the square wave.

The magnitude of the triangle waveform is set at 1/3 V_{CC}; therefore, the duration of the rising portion of the triangle is:

$$t_1 = \frac{C \times V}{I} = \frac{C \times 1/3 \times V_{CC} \times R_A}{1/5 \times V_{CC}} = \frac{5}{3} R_A \times C$$

The duration of the falling portion of the triangle and the sinewave, and the "High" state of the square-wave is:

$$t_2 = \frac{C \times V}{I} = \frac{C \times 1/3\, V_{CC}}{\frac{2}{5} \times \frac{V_{CC}}{R_B} - \frac{1}{5} \times \frac{V_{CC}}{R_A}} = \frac{5}{3} \times \frac{R_A R_B C}{2R_A - R_B}$$

Thus a 50% duty cycle is achieved when $R_A = R_B$.

If the duty-cycle is to be varied over a small range about 50% only, the connection shown in Figure 2b is slightly more convenient. If no adjustment of the duty cycle is desired, terminals 4 and 5 can be shorted together, as shown in Figure 2c. This connection, however, carries an inherently larger variation of the duty-cycle.

With two separate timing resistors, the *frequency* is given by

$$f = \frac{1}{t_1 + t_2} = \frac{1}{\frac{5}{3} R_A C \left(1 + \frac{R_B}{2R_A - R_B}\right)}$$

or, if $R_A = R_B = R$

$$f = 0.3/RC \text{ (for Figure 2a)}$$

If a single timing resistor is used (Figures 2b and c), the frequency is

$$f = 0.15/RC$$

The frequency of oscillation is independent of supply voltage, even though none of the voltages are regulated inside the integrated circuit. This is due to the fact that both currents *and* thresholds are direct, linear function of the supply voltage and thus their effects cancel.

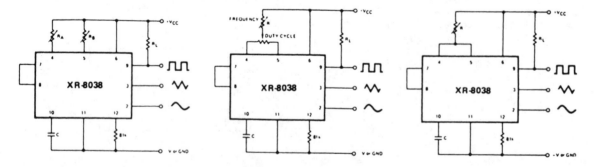

Figure 2. Possible Connections for the External Timing Resistors.

XR-8038

DISTORTION ADJUSTMENT

To minimize *sine-wave* distortion the 81 kΩ resistor between pins 11 and 12 is best made a variable one. With this arrangement distortion of less than 1% is achievable. To reduce this even further, two potentiometers can be connected as shown in Figure 3. This configuration allows a reduction of sine-wave distortion close to 0.5%

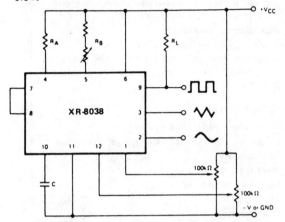

Figure 3. Connection to Achieve Minimum Sine-Wave Distortion.

SELECTING TIMING COMPONENTS

For any given output frequency, there is a wide range of RC combinations that will work. However certain constraints are placed upon the magnitude of the charging current for optimum performance. At the low end, currents of less than 0.1 μA are undesirable because circuit leakages will contribute significant errors at high temperatures. At higher currents (1 > 5 mA), transistor betas and saturation voltages will contribute increasingly larger errors. Optimum performance will be obtained for charging currents of 1 μ to 1 mA. If pins 7 and 8 are shorted together the magnitude of the charging current due to R_A can be calculated from:

$$1 = \frac{R_1 \times V_{CC}}{(R_1 + R_2)} \times \frac{1}{R_A} = \frac{V_{CC}}{5R_A}$$

A similar calculation holds for R_B.

SINGLE-SUPPLY AND SPLIT-SUPPLY OPERATION

The waveform generator can be operated either from a single power-supply (10 to 30 Volts) or a dual power-supply (±5 to ±15 Volts). With a single power-supply the average levels of the triangle and sine-wave are at exactly one-half of the supply voltage, while the square-wave alternates between +V_{CC} and ground. A split power supply has the advantage that all waveforms move symmetrically about ground.

The square-wave output is not committed. A load resistor can be connected to a different power-supply, as long as the applied voltage remains within the breakdown capability of the waveform generator (30V). In this way, the square-wave output will be TTL compatible (load resistor connected to +5 Volts) while the waveform generator itself is powered from a higher supply voltage.

FREQUENCY MODULATION AND SWEEP

The frequency of the waveform generator is a direct function of the DC voltage at terminal 8 (measured from +V_{CC}). By altering this voltage, frequency modulation is performed.

For small deviations (e.g., ±10%) the modulating signal can be applied directly to pin 8 by merely providing ac coupling with a capacitor, as shown in Figure 4a. An external resistor between pins 7 and 8 is not necessary, but it can be used to increase input impedance. Without it (i.e. terminals 7 and 8 connected together), the input impedance is 8kΩ; with it, this impedance increases to (R + 8kΩ).

For larger FM deviations or for frequency sweeping, the modulating signal is applied between the positive supply voltage and pin 8 (Figure 4b). In this way the entire bias for the current sources is created by the modulating signal and a very large (e.g., 1000:1) sweep range is obtained (f = 0 at V_{sweep} = 0). Care must be taken, however, to regulate the supply voltage; in this configuration the charge current is no longer a function of the supply voltage (yet the trigger thresholds still are) and thus the frequency becomes dependent on the supply voltage. The potential on Pin 8 may be swept from V_{CC} to 2/3 V_{CC} +2V.

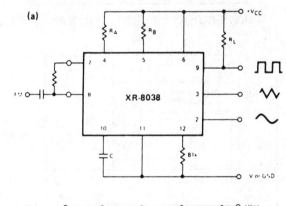

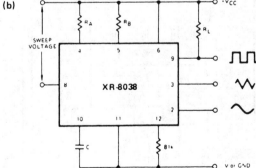

Figure 4. Connections for Frequency Modulation (a) and Sweep (b).

LM2907, LM2917 Frequency to Voltage Converter

General Description

The LM2907, LM2917 series are monolithic frequency to voltage converters with a high gain op amp/comparator designed to operate a relay, lamp, or other load when the input frequency reaches or exceeds a selected rate. The tachometer uses a charge pump technique and offers frequency doubling for low ripple, full input protection in two versions (LM2907-8, LM2917-8) and its output swings to ground for a zero frequency input.

Advantages

- Output swings to ground for zero frequency input
- Easy to use; $V_{OUT} = f_{IN} \times V_{CC} \times R1 \times C1$
- Only one RC network provides frequency doubling
- Zener regulator on chip allows accurate and stable frequency to voltage or current conversion. (LM2917)

Features

- Ground referenced tachometer input interfaces directly with variable reluctance magnetic pickups
- Op amp/comparator has floating transistor output
- 50 mA sink or source to operate relays, solenoids, meters, or LEDs

- Frequency doubling for low ripple
- Tachometer has built-in hysteresis with either differential input or ground referenced input
- Built-in zener on LM2917
- ±0.3% linearity typical
- Ground referenced tachometer is fully protected from damage due to swings above V_{CC} and below ground

Applications

- Over/under speed sensing
- Frequency to voltage conversion (tachometer)
- Speedometers
- Breaker point dwell meters
- Hand-held tachometer
- Speed governors
- Cruise control
- Automotive door lock control
- Clutch control
- Horn control
- Touch or sound switches

Block and Connection Diagrams Dual-In-Line Packages, Top Views

Order Number LM2907N-8
See NS Package N08B

Order Number LM2917N-8
See NS Package N08B

Order Number LM2907J
See NS Package J14A
Order Number LM2907N
See NS Package N14A

Order Number LM2917J
See NS Package J14A
Order Number LM2917N
See NS Package N14A

Voltage Comparators

LM311 Voltage Comparator

General Description

The LM311 is a voltage comparator that has input currents more than a hundred times lower than devices like the LM306 or LM710C. It is also designed to operate over a wider range of supply voltages: from standard ±15V op amp supplies down to the single 5V supply used for IC logic. Its output is compatible with RTL, DTL and TTL as well as MOS circuits. Further, it can drive lamps or relays, switching voltages up to 40V at currents as high as 50 mA.

Features

- Operates from single 5V supply
- Maximum input current: 250 nA
- Maximum offset current: 50 nA

- Differential input voltage range: ±30V
- Power consumption: 135 mW at ±15V

Both the input and the output of the LM311 can be isolated from system ground, and the output can drive loads referred to ground, the positive supply or the negative supply. Offset balancing and strobe capability are provided and outputs can be wire OR'ed. Although slower than the LM306 and LM710C (200 ns response time vs 40 ns) the device is also much less prone to spurious oscillations. The LM311 has the same pin configuration as the LM306 and LM710C. See the "application hints" of the LM311 for application help.

Auxiliary Circuits**

**Note: Pin connections shown on schematic diagram and typical applications are for TO-5 package.

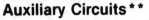

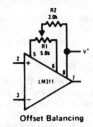

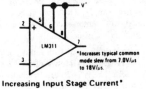

Offset Balancing   Strobing   Increasing Input Stage Current*

Typical Applications**

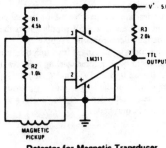

Detector for Magnetic Transducer

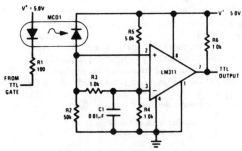

Digital Transmission Isolator

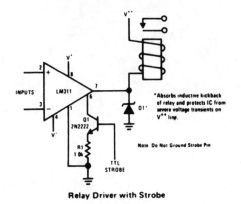

Relay Driver with Strobe   Strobing off Both Input* and Output Stages

Reprinted with permission of National Semiconductor Corp.   5-48

Low Cost Signal Conditioning 8-Bit ADC
AD670

FEATURES
Complete 8-Bit Signal Conditioning A/D Converter Including Instrumentation Amp and Reference
Microprocessor Bus Interface
10μs Conversion Speed
Flexible Input Stage: Instrumentation Amp Front End Provides Differential Inputs and High Common-Mode Rejection
No User Trims Required
No Missing Codes Over Temperature
Single +5V Supply Operation
Convenient Input Ranges
20-Pin DIP or Surface-Mount Package
Low Cost Monolithic Construction
MIL-STD-883B Compliant Versions Available

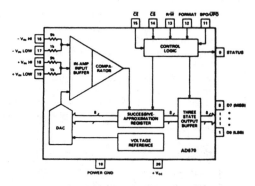

GENERAL DESCRIPTION
The AD670 is a complete 8-bit signal conditioning analog-to-digital converter. It consists of an instrumentation amplifier front end along with a DAC, comparator, successive approximation register (SAR), precision voltage reference, and a three-state output buffer on a single monolithic chip. No external components or user trims are required to interface, with full accuracy, an analog system to an 8-bit data bus. The AD670 will operate on the +5V system supply. The input stage provides differential inputs with excellent common-mode rejection and allows direct interface to a variety of transducers.

The device is configured with input scaling resistors to permit two input ranges: 0 to 255mV (1mV/LSB) and 0 to 2.55V (10mV/LSB). The AD670 can be configured for both unipolar and bipolar inputs over these ranges. The differential inputs and common-mode rejection of this front end are useful in applications such as conversion of transducer signals superimposed on common-mode voltages.

The AD670 incorporates advanced circuit design and proven processing technology. The successive approximation function is implemented with I^2L (integrated injection logic). Thin-film SiCr resistors provide the stability required to prevent missing codes over the entire operating temperature range while laser wafer trimming of the resistor ladder permits calibration of the device to within ±1LSB. Thus, no user trims for gain or offset are required. Conversion time of the device is 10μs.

The AD670 is available in four package types and five grades. The J and K grades are specified over 0 to +70°C and come in 20-pin plastic DIP packages or 20-terminal PLCC packages. The A and B grades (−40°C to +85°C) and the S grade (−55°C to +125°C) come in 20-pin ceramic DIP packages.

The S grade is also available with optional processing to MIL-STD-883 in 20-pin ceramic DIP or 20-terminal LCC packages. The Analog Devices Military Products Databook should be consulted for detailed specifications.

PRODUCT HIGHLIGHTS
1. The AD670 is a complete 8-bit A/D including three-state outputs and microprocessor control for direct connection to 8-bit data buses. No external components are required to perform a conversion.
2. The flexible input stage features a differential instrumentation amp input with excellent common-mode rejection. This allows direct interface to a variety of transducers without preamplification.
3. No user trims are required for 8-bit accurate performance.
4. Operation from a single +5V supply allows the AD670 to run off of the microprocessor's supply.
5. Four convenient input ranges (two unipolar and two bipolar) are available through internal scaling resistors: 0 to 255mV (1mV/LSB) and 0 to 2.55V (10mV/LSB).
6. Software control of the output mode is provided. The user can easily select unipolar or bipolar inputs and binary or 2's complement output codes.

AD670

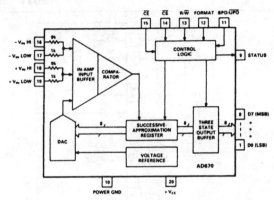

Figure 1. AD670 Block Diagram and Terminal Configuration
(All Packages)

ABSOLUTE MAXIMUM RATINGS*

V_{CC} to Ground . . . . 0V to +7.5V
Digital Inputs (Pins 11-15) . . . . -0.5V to V_{CC} +0.5V
Digital Outputs (Pins 1-9) . Momentary Short to V_{CC} or Ground
Analog Inputs (Pins 16-19) . . . . -30V to +30V
Power Dissipation 450mW
Storage Temperature Range . . . . $-65°$C to +150°C
Lead Temperature (Soldering, 10sec) . . . . +300°C

*Stresses above those listed under "Absolute Maximum Ratings" may cause permanent damage to the device. This is a stress rating only and functional operation of the device at these or any other conditions above those indicated in the operational sections of this specification is not implied. Exposure to absolute maximum rating conditions for extended periods may affect device reliability.

ORDERING GUIDE

Model[1]	Temperature Range	Relative Accuracy @ +25°C	Gain Accuracy @ +25°C	Package Option[2]
AD670JN	0 to +70°C	±1/2LSB	±1.5LSB	Plastic DIP (N-20)
AD670JP	0 to +70°C	±1/2LSB	±1.5LSB	PLCC (P-20A)
AD670KN	0 to +70°C	±1/4LSB	±0.75LSB	Plastic DIP (N-20)
AD670KP	0 to +70°C	±1/4LSB	±0.75LSB	PLCC (P-20A)
AD670AD	−40°C to +85°C	±1/2LSB	±1.5LSB	Ceramic DIP (D-20)
AD670BD	−40°C to +85°C	±1/4LSB	±0.75LSB	Ceramic DIP (D-20)
AD670SD	−55°C to +125°C	±1/2LSB	±1.5LSB	Ceramic DIP (D-20)

NOTES
[1]For details on grade and package offerings screened in accordance with MIL-STD-883, refer to the Analog Devices Military Products Databook.
[2]D = Ceramic DIP; N = Plastic DIP; P = Plastic Leaded Chip Carrier. For outline information see Package Information section.

CIRCUIT OPERATION/FUNCTIONAL DESCRIPTION

The AD670 is a functionally complete 8-bit signal conditioning A/D converter with microprocessor compatibility. The input section uses an instrumentation amplifier to accomplish the voltage to current conversion. This front end provides a high impedance, low bias current differential amplifier. The common-mode range allows the user to directly interface the device to a variety of transducers.

The A/D conversions are controlled by R/$\overline{W}$, $\overline{CS}$, and $\overline{CE}$. The R/$\overline{W}$ line directs the converter to read or start a conversion. A minimum write/start pulse of 300ns is required on either $\overline{CE}$ or $\overline{CS}$. The STATUS line goes high, indicating that a conversion is in process. The conversion thus begun, the internal 8-bit DAC is sequenced from MSB to LSB using a novel successive approximation technique. In conventional designs, the DAC is stepped through the bits by a clock. This can be thought of as a static design since the speed at which the DAC is sequenced is determined solely by the clock. No clock is used in the AD670. Instead, a "dynamic SAR" is created consisting of a string of inverters with taps along the delay line. Sections of the delay line between taps act as one shots. The pulses are used to set and reset the DAC's bits and strobe the comparator. When strobed, the comparator then determines whether the addition of each successively weighted bit current causes the DAC current

sum to be greater or less than the input current. If the sum is less, the bit is turned off. After all bits are tested, the SAR holds an 8-bit code representing the input signal to within 1/2LSB accuracy. Ease of implementation and reduced dependence on process related variables make this an attractive approach to a successive approximation design.

The SAR provides an end-of-conversion signal to the control logic which then brings the STATUS line low. Data outputs remain in a high impedance state until R/$\overline{W}$ is brought high with $\overline{CE}$ and $\overline{CS}$ low and allows the converter to be read. Bringing $\overline{CE}$ or $\overline{CS}$ high during the valid data period ends the read cycle. The output buffers cannot be enabled during a conversion. Any convert start commands will be ignored until the conversion cycle is completed; once a conversion cycle has been started it cannot be stopped or restarted.

The AD670 provides the user with a great deal of flexibility by offering two input spans and formats and a choice of output codes. Input format and input range can each be selected. The BPO/$\overline{UPO}$ pin controls a switch which injects a bipolar offset current of a value equal to the MSB less 1/2LSB into the summing node of the comparator to offset the DAC output. Two precision 10 to 1 attenuators are included on board to provide input range selection of 0 to 2.55V or 0 to 255mV. Additional ranges of

−1.28 to 1.27V and −128 to 127mV are possible if the BPO/$\overline{\text{UPO}}$ switch is high when the conversion is started. Finally, output coding can be chosen using the FORMAT pin when the conversion is started. In the bipolar mode and with a logic 1 on FORMAT, the output is in two's complement; with a logic 0, the output is offset binary.

CONNECTING THE AD670

The AD670 has been designed for ease of use. All active components required to perform a complete A/D conversion are on board and are connected internally. In addition, all calibration trims are performed at the factory, assuring specified accuracy without user trims. There are, however, a number of options and connections that should be considered to obtain maximum flexibility from the part.

INPUT CONNECTIONS

Standard connections are shown in the figures that follow. An input range of 0 to 2.55V may be configured as shown in Figure 2a. This will provide a one LSB change for each 10mV of input change. The input range of 0 to 255mV is configured as shown in Figure 2b. In this case, each LSB represents 1mV of input change. When unipolar input signals are used, Pin 11, BPO/$\overline{\text{UPO}}$, should be grounded. Pin 11 selects the input format for either unipolar or bipolar signals. Figures 3a and 3b show the input connections for bipolar signals. Pin 11 should be tied to +V$_{CC}$ for bipolar inputs.

Although the instrumentation amplifier has a differential input, there must be a return path to ground for the bias currents. If it is not provided, these currents will charge stray capacitances and cause internal circuit nodes to drift uncontrollably causing the digital output to change. Such a return path is provided in Figures 2a and 3a (larger input ranges) since the 1k resistor leg

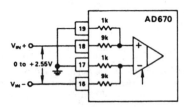

2a. 0 to 2.55V (10mV/LSB)

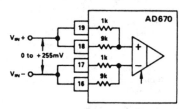

2b. 0 to 255mV (1mV/LSB)

NOTE: PIN 11, BPO/$\overline{\text{UPO}}$ SHOULD BE LOW WHEN CONVERSION IS STARTED.

Figure 2. Unipolar Input Connections

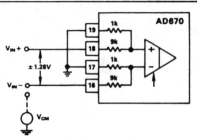

3a. ±1.28V Range

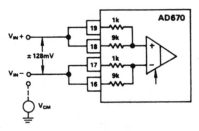

3b. ±128mV Range

NOTE: PIN 11, BPO/$\overline{\text{UPO}}$ SHOULD BE HIGH WHEN CONVERSION IS STARTED.

Figure 3. Bipolar Input Connections

is tied to ground. This is not the case for Figures 2b and 3b (the lower input ranges). When connecting the AD670 inputs to floating sources, such as transformers and ac-coupled sources, there must still be a dc path from each input to common. This can be accomplished by connecting a 10kΩ resistor from each input to ground.

Bipolar Operation

Through special design of the instrumentation amplifier, the AD670 accommodates input signal excursions below ground, even though it operates from a single 5V supply. To the user, this means that true bipolar input signals can be used without the need for any additional external components. Bipolar signals can be applied differentially across both inputs, or one of the inputs can be grounded and a bipolar signal applied to the other.

Common-Mode Performance

The AD670 is designed to reject dc and ac common-mode voltages. In some applications it is useful to apply a differential input signal V$_{IN}$ in the presence of a dc common-mode voltage V$_{CM}$. The user must observe the absolute input signal limits listed in the specifications, which represent the maximum voltage V$_{IN}$ + V$_{CM}$ that can be applied to either input without affecting proper operation. Exceeding these limits (within the range of absolute maximum ratings), however, will not cause permanent damage.

The excellent common-mode rejection of the AD670 is due to the instrumentation amplifier front end, which maintains the differential signal until it reaches the output of the comparator. In contrast to a standard operational amplifier, the instrumentation amplifier front end provides significantly improved CMRR over a wide frequency range (Figure 4a).

ANALOG DEVICES

DACPORT Low Cost, Complete μP-Compatible 8-Bit DAC

AD557

FEATURES
Complete 8-Bit DAC
Voltage Output – 0 to 2.56V
Internal Precision Band-Gap Reference
Single-Supply Operation: +5V (±10%)
Full Microprocessor Interface
Fast: 1μs Voltage Settling to ±1/2LSB
Low Power: 75mW
No User Trims Required
Guaranteed Monotonic Over Temperature
All Errors Specified T_{min} to T_{max}
Small 16-Pin DIP or 20-Pin PLCC Package
Low Cost

FUNCTIONAL BLOCK DIAGRAM

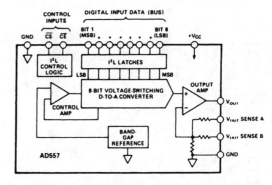

PRODUCT DESCRIPTION
The AD557 DACPORT™ is a complete voltage-output 8-bit digital-to-analog converter, including output amplifier, full microprocessor interface and precision voltage reference on a single monolithic chip. No external components or trims are required to interface, with full accuracy, an 8-bit data bus to an analog system.

The low cost and versatility of the AD557 DACPORT are the result of continued development in monolithic bipolar technologies.

The complete microprocessor interface and control logic is implemented with integrated injection logic (I^2L), an extremely dense and low-power logic structure that is process-compatible with linear bipolar fabrication. The internal precision voltage reference is the patented low-voltage band-gap circuit which permits full-accuracy performance on a single +5V power supply. Thin-film silicon-chromium resistors provide the stability required for guaranteed monotonic operation over the entire operating temperature range, while laser-wafer trimming of these thin-film resistors permits absolute calibration at the factory to within ±2.5LSB; thus, no user-trims for gain or offset are required. A new circuit design provides voltage settling to ±1/2LSB for a full-scale step in 800ns.

The AD557 is available in two package configurations. The AD557JN is packaged in a 16-pin plastic, 0.3"-wide DIP. For surface mount applications, the AD557JP is packaged in a 20-pin JEDEC standard PLCC. Both versions are specified over the operating temperature range of 0 to +70°C.

PRODUCT HIGHLIGHTS
1. The 8-bit I^2L input register and fully microprocessor-compatible control logic allow the AD557 to be directly connected to 8- or 16-bit data buses and operated with standard control signals. The latch may be disabled for direct DAC interfacing.

2. The laser-trimmed on-chip SiCr thin-film resistors are calibrated for absolute accuracy and linearity at the factory. Therefore, no user trims are necessary for full rated accuracy over the operating temperature range.

3. The inclusion of a precision low-voltage band-gap reference eliminates the need to specify and apply a separate reference source.

4. The AD557 is designed and specified to operate from a single +4.5V to +5.5V power supply.

5. Low digital input currents, 100μA max, minimize bus loading. Input thresholds are TTL/low voltage CMOS compatible.

6. The single-chip, low power I^2L design of the AD557 is inherently more reliable than hybrid multichip or conventional single-chip bipolar designs.

AD557 — SPECIFICATIONS (@ T_A = +25C, V_{CC} = +5V unless otherwise specified)

Model	AD557J Min	Typ	Max	Units
RESOLUTION			8	Bits
RELATIVE ACCURACY[1]				
0 to +70°C		±1/2	1	LSB
OUTPUT				
Ranges		0 to +2.56		V
Current Source	+5			mA
Sink		Internal Passive Pull-Down to Ground[2]		
OUTPUT SETTLING TIME[3]		0.8	1.5	µs
FULL SCALE ACCURACY[4]				
@25°C		±1.5	±2.5	LSB
T_{min} to T_{max}		±2.5	±4.0	LSB
ZERO ERROR				
@25°C			±1	LSB
T_{min} to T_{max}			±3	LSB
MONOTONICITY[5]				
T_{min} to T_{max}		Guaranteed		
DIGITAL INPUTS				
T_{min} to T_{max}				
Input Current			±100	µA
Data Inputs, Voltage				
Bit On – Logic "1"	2.0			V
Bit On – Logic "0"	0		0.8	V
Control Inputs, Voltage				
On – Logic "1"	2.0			V
On – Logic "0"	0		0.8	V
Input Capacitance		4		pF
TIMING[6]				
t_W Strobe Pulse Width	225			ns
T_{min} to T_{max}	**300**			ns
t_{DH} Data Hold Time	10			ns
T_{min} to T_{max}	**10**			ns
t_{DS} Data Setup Time	225			ns
T_{min} to T_{max}	**300**			ns
POWER SUPPLY				
Operating Voltage Range (V_{CC})				
2.56 Volt Range	+4.5		+5.5	V
Current (I_{CC})		15	25	mA
Rejection Ratio			0.03	%/%
POWER DISSIPATION, V_{CC} = 5V		75	125	mW
OPERATING TEMPERATURE RANGE	0		+70	°C

NOTES

[1]Relative Accuracy is defined as the deviation of the code transition points from the ideal transfer point on a straight line from the offset to the full scale of the device. See "Measuring Offset Error" on AD558 data sheet.
[2]Passive pull-down resistance is 2kΩ.
[3]Settling time is specified for a positive-going full-scale step to +1/2LSB. Negative-going steps to zero are slower, but can be improved with an external pull-down.
[4]The full-scale output voltage is 2.55V and is guaranteed with a +5V supply.
[5]A monotonic converter has a maximum differential linearity error of ±1LSB.
[6]See Figure 7.

Specifications shown in **boldface** are tested on all production units at final electrical test.

Specifications subject to change without notice.

PIN CONFIGURATIONS

DIP

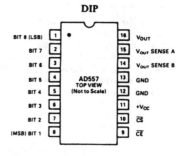

PLCC

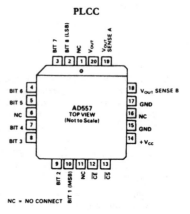

NC = NO CONNECT

ORDERING GUIDE

Model	Package Option*	Temperature
AD557JN	Plastic (N-16)	0 to +70°C
AD557JP	PLCC (P-20A)	0 to +70°C

*N = Plastic DIP; P = Plastic Leaded Chip Carrier. For outline information see Package Information section.

ABSOLUTE MAXIMUM RATINGS*

V_{CC} to Ground . . . . 0V to +18V
Digital Inputs (Pins 1-10) . . . . 0 to +7.0V
V_{OUT} . . . . Indefinite Short to Ground
    Momentary Short to V_{CC}
Power Dissipation 450mW
Storage Temperature Range
    N/P (Plastic) Packages . . . . −25°C to +100°C
Lead Temperature (soldering, 10 sec) . . . . 300°C

Thermal Resistance
    Junction to Ambient/Junction to Case
    N/P (Plastic) Packages . . . . 140/55°C/W

*Stresses above those listed under "Absolute Maximum Ratings" may cause permanent damage to the device. This is a stress rating only and functional operation of the device at these or any other conditions above those indicated in the operational sections of this specification is not implied. Exposure to absolute maximum rating conditions for extended periods may affect device reliability.

CIRCUIT DESCRIPTION

The AD557 consists of four major functional blocks fabricated on a single monolithic chip (see Figure 1). The main D/A converter section uses eight equally weighted laser-trimmed current sources switched into a silicon-chromium thin-film R/2R resistor ladder network to give a direct but unbuffered 0mV to 400mV output range. The transistors that form the DAC switches are PNPs; this allows direct positive-voltage logic interface and a zero-based output range.

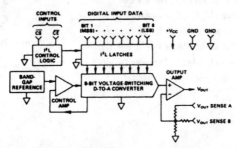

Figure 1. Functional Block Diagram

The high-speed output buffer amplifier is operated in the noninverting mode with gain determined by the user-connections at the output range select pin. The gain-setting application resistors are thin film laser trimmed to match and track the DAC resistors and to assure precise initial calibration of the output range, 0V to 2.56V. The amplifier output stage is an NPN transistor with passive pull-down for zero-based output capability with a single power supply.

The internal precision voltage reference is of the patented band-gap type. This design produces a reference voltage of 1.2V and thus, unlike 6.3V temperature-compensated zeners, may be operated from a single, low-voltage logic power supply. The microprocessor interface logic consists of an 8-bit data latch and control circuitry. Low power, small geometry and high speed are advantages of the I²L design as applied to this section. I²L is bipolar process compatible so that the performance of the analog sections need not be compromised to provide on-chip logic capabilities. The control logic allows the latches to be operated from a decoded microprocessor address and write signal. If the application does not involve a μP or data bus, wiring $\overline{CS}$ and $\overline{CE}$ to ground renders the latches "transparent" for direct DAC access.

Digital Input Code			Output
Binary	Hexadecimal	Decimal	Voltage
0000 0000	00	0	0
0000 0001	01	1	0.010V
0000 0010	02	2	0.020V
0000 1111	0F	15	0.150V
0001 0000	10	16	0.160V
0111 1111	7F	127	1.270V
1000 0000	80	128	1.280V
1100 0000	C0	192	1.920V
1111 1111	FF	255	2.55V

CONNECTING THE AD557

The AD557 has been configured for low cost and ease of application. All reference, output amplifier and logic connections are made internally. In addition, all calibration trims are performed at the factory assuring specified accuracy without user trims. The only connection decision to be made by the user is whether the output range desired is unipolar or bipolar. Clean circuit board layout is facilitated by isolating all digital bit inputs on one side of the package; analog outputs are on the opposite side.

UNIPOLAR 0 TO +2.56V OUTPUT RANGE

Figure 2 shows the configuration for the 0 to +2.56V full-scale output range. Because of its precise factory calibration, the AD557 is intended to be operated without user trims for gain and offset; therefore, no provisions have been made for such user trims. If a small increase in scale is required, however, it may be accomplished by slightly altering the effective gain of the output buffer. A resistor in series with V_OUT SENSE will increase the output range. Note that decreasing the scale by putting a resistor in series with GND will not work properly due to the code-dependent currents in GND. Adjusting offset by injecting dc at GND is not recommended for the same reason.

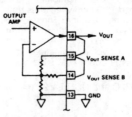

Figure 2. 0 to 2.56V Output Range

BIPOLAR −1.28V TO +1.28V OUTPUT RANGE

The AD557 was designed for operation from a single power supply and is thus capable of providing only a unipolar 0 to +2.56V output range. If a negative supply is available, bipolar output ranges may be achieved by suitable output offsetting and scaling. Figure 3 shows how a ±1.28V output range may be achieved when a −5V power supply is available. The offset is provided by the AD589 precision 1.2V reference which will operate from a +5V supply. The AD711 output amplifier can provide the necessary ±1.28V output swing from ±5V supplies. Coding is complementary offset binary.

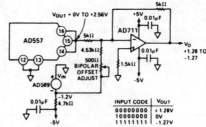

Figure 3. Bipolar Operation of AD557 from ±5V Supplies

/VI/IXI/VI

+5V Powered
RS-232 Drivers/Receivers

MAX230-241*

General Description

Maxim's family of line drivers/receivers are intended for all RS-232 and V.28/V.24 communications interfaces, and in particular, for those applications where ±12V is not available. The MAX230, MAX236, MAX240 and MAX241 are particularly useful in battery powered systems since their low power shutdown mode reduces power dissipation to less than 5µW. The MAX233 and MAX235 use no external components and are recommended for applications where printed circuit board space is critical.

All members of the family except the MAX231 and MAX239 need only a single +5V supply for operation. The RS-232 drivers/receivers have on-board charge pump voltage converters which convert the +5V input power to the ±10V needed to generate the RS-232 output levels. The MAX231 and MAX239, designed to operate from +5V and +12V, contain a +12V to -12V charge pump voltage converter.

Since nearly all RS-232 applications need both line drivers and receivers, the family includes both receivers and drivers in one package. The wide variety of RS-232 applications require differing numbers of drivers and receivers. Maxim offers a wide selection of RS-232 driver/receiver combinations in order to minimize the package count (see table below).

Both the receivers and the line drivers (transmitters) meet all EIA RS-232C and CCITT V.28 specifications.

Features

- ◆ **Operates from Single 5V Power Supply**
  (+5V and +12V — MAX231 and MAX239)
- ◆ **Meets All RS-232C and V.28 Specifications**
- ◆ **Multiple Drivers and Receivers**
- ◆ **Onboard DC-DC Converters**
- ◆ **±9V Output Swing with +5V Supply**
- ◆ **Low Power Shutdown — <1µA (typ)**
- ◆ **3-State TTL/CMOS Receiver Outputs**
- ◆ **±30V Receiver Input Levels**

Applications

Computers

Peripherals

Modems

Printers

Instruments

2

Selection Table

Part Number	Power Supply Voltage	No. of RS-232 Drivers	No. of RS-232 Receivers	External Components	Low Power Shutdown /TTL 3-State	No. of Pins
MAX230	+5V	5	0	4 capacitors	Yes/No	20
MAX231	+5V and +7.5V to 13.2V	2	2	2 capacitors	No/No	14
MAX232	+5V	2	2	4 capacitors	No/No	16
MAX233	+5V	2	2	None	No/No	20
MAX234	+5V	4	0	4 capacitors	No/No	16
MAX235	+5V	5	5	None	Yes/Yes	24
MAX236	+5V	4	3	4 capacitors	Yes/Yes	24
MAX237	+5V	5	3	4 capacitors	No/No	24
MAX238	+5V	4	4	4 capacitors	No/No	24
MAX239	+5V and +7.5V to 13.2V	3	5	2 capacitors	No/Yes	24
MAX240	+5V	5	5	4 capacitors	Yes/Yes	44 (Flatpak)
MAX241	+5V	4	5	4 capacitors	Yes/Yes	28 (Small Outline)

*Patent Pending

/VI/IXI/VI _____ **Maxim Integrated Products**

/VI/IXI/VI is a registered trademark of Maxim Integrated Products.

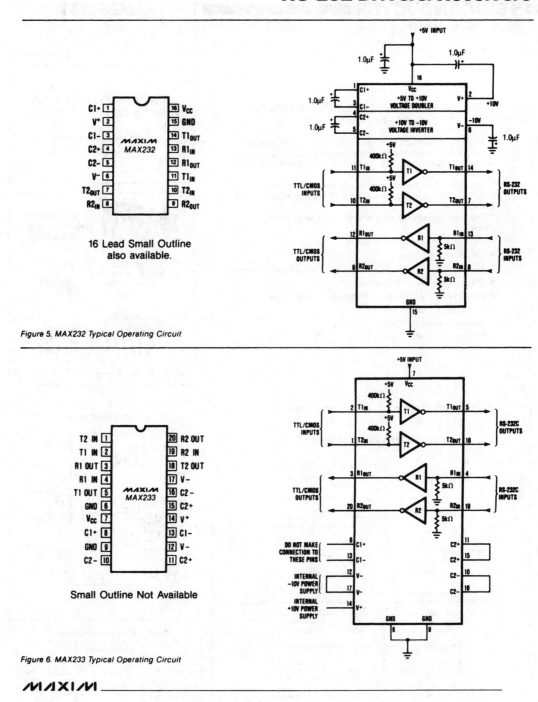

Figure 5. MAX232 Typical Operating Circuit

16 Lead Small Outline
also available.

Small Outline Not Available

Figure 6. MAX233 Typical Operating Circuit

/\/\/\X\/\/\

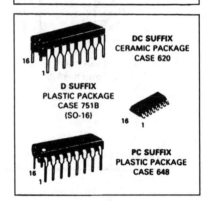

AM26LS31

QUAD EIA-422 LINE DRIVER WITH THREE-STATE OUTPUTS

**SILICON MONOLITHIC
INTEGRATED CIRCUIT**

QUAD LINE DRIVER WITH NAND ENABLED THREE-STATE OUTPUTS

The Motorola AM26LS31 is a quad differential line driver intended for digital data transmission over balanced lines. It meets all the requirements of EIA-422 Standard and Federal Standard 1020.

The AM26LS31 provides an enable/disable function common to all four drivers as opposed to the split enables on the MC3487 EIA-422 driver.

The high impedance output state is assured during power down.

- Full EIA-422 Standard Compliance
- Single +5.0 V Supply
- Meets Full V_O = 6.0 V, V_{CC} = 0 V, I_O < 100 μA Requirement
- Output Short Circuit Protection
- Complementary Outputs for Balanced Line Operation
- High Output Drive Capability
- Advanced LS Processing
- PNP Inputs for MOS Compatibility

**DC SUFFIX
CERAMIC PACKAGE
CASE 620**

**D SUFFIX
PLASTIC PACKAGE
CASE 751B
(SO-16)**

**PC SUFFIX
PLASTIC PACKAGE
CASE 648**

DRIVER BLOCK DIAGRAM

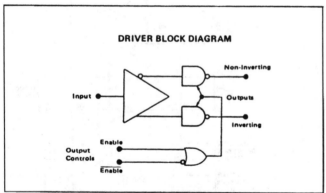

PIN CONNECTIONS

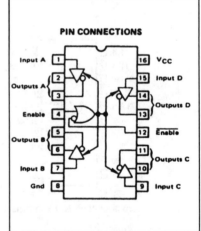

TRUTH TABLE

Input	Control Inputs ($E/\overline{E}$)	Non-Inverting Output	Inverting Output
H	H/L	H	L
L	H/L	L	H
X	L/H	Z	Z

L = Low Logic State   X = Irrelevant
H = High Logic State   Z = Third-State (High Impedance)

ORDERING INFORMATION

Device	Temperature Range	Package
AM26LS31DC		Ceramic DIP
AM26LS31PC	0 to 70°C	Plastic DIP
MC26LS31D*		SO-16

*Note that the surface mount MC26LS31D devices use the same die as in the ceramic and plastic DIP AM26LS31DC devices, but with an MC prefix to prevent confusion with the package suffixes.

MOTOROLA LINEAR/INTERFACE ICs DEVICE DATA

MOTOROLA
■ SEMICONDUCTOR ■
TECHNICAL DATA

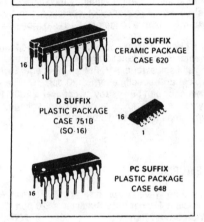

AM26LS32

QUAD EIA-422/423 LINE RECEIVER

Motorola's Quad EIA-422/3 Receiver features four independent receiver chains which comply with EIA Standards for the Electrical Characteristics of Balanced/Unbalanced Voltage Digital Interface Circuits. Receiver outputs are 74LS compatible, three-state structures which are forced to a high impedance state when Pin 4 is a Logic "0" and Pin 12 is a Logic "1." A PNP device buffers each output control pin to assure minimum loading for either Logic "1" or Logic "0" inputs. In addition, each receiver chain has internal hysteresis circuitry to improve noise margin and discourage output instability for slowly changing input waveforms. A summary of AM26LS32 features include:

- Four Independent Receiver Chains
- Three-State Outputs
- High Impedance Output Control Inputs (PIA Compatible)
- Internal Hysteresis — 30 mV (Typ) @ Zero Volts Common Mode
- Fast Propagation Times — 25 ns (Typ)
- TTL Compatible
- Single 5 V Supply Voltage
- Fail-Safe Input-Output Relationship. Output Always High When Inputs Are Open, Terminated or Shorted
- 6 k Minimum Input Impedance

QUAD EIA-422/3 LINE RECEIVER WITH THREE-STATE OUTPUTS

SILICON MONOLITHIC INTEGRATED CIRCUIT

DC SUFFIX
CERAMIC PACKAGE
CASE 620

D SUFFIX
PLASTIC PACKAGE
CASE 751B
(SO-16)

PC SUFFIX
PLASTIC PACKAGE
CASE 648

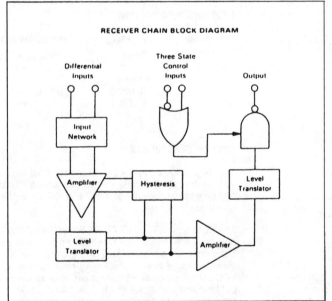

RECEIVER CHAIN BLOCK DIAGRAM

*Note that the surface mount MC26LS32D devices use the same die as in the ceramic and plastic DIP AM26LS32DC devices, but with an MC prefix to prevent confusion with the package suffixes.

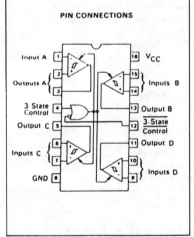

PIN CONNECTIONS

ORDERING INFORMATION

Device	Temperature	Package
AM26LS32DC		Ceramic DIP
AM26LS32PC	0 to 70°C	Plastic DIP
MC26LS32D*		SO-16

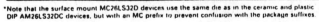

MOTOROLA LINEAR/INTERFACE ICs DEVICE DATA

XR-2206

Monolithic Function Generator

GENERAL DESCRIPTION

The XR-2206 is a monolithic function generator integrated circuit capable of producing high quality sine, square, triangle, ramp, and pulse waveforms of high-stability and accuracy. The output waveforms can be both amplitude and frequency modulated by an external voltage. Frequency of operation can be selected externally over a range of 0.01 Hz to more than 1 MHz.

The circuit is ideally suited for communications, instrumentation, and function generator applications requiring sinusoidal tone, AM, FM, or FSK generation. It has a typical drift specification of 20 ppm/°C. The oscillator frequency can be linearly swept over a 2000:1 frequency range, with an external control voltage, having a very small affect on distortion.

FEATURES

Low-Sine Wave Distortion	0.5%, Typical
Excellent Temperature Stability	20 ppm/°C, Typical
Wide Sweep Range	2000:1, Typical
Low-Supply Sensitivity	0.01%V, Typical
Linear Amplitude Modulation	
TTL Compatible FSK Controls	
Wide Supply Range	10V to 26V
Adjustable Duty Cycle	1% to 99%

APPLICATIONS

Waveform Generation
Sweep Generation
AM/FM Generation
V/F Conversion
FSK Generation
Phase-Locked Loops (VCO)

ABSOLUTE MAXIMUM RATINGS

Power Supply	26V
Power Dissipation	750 mW
Derate Above 25°C	5 mW/°C
Total Timing Current	6 mA
Storage Temperature	−65°C to +150°C

FUNCTIONAL BLOCK DIAGRAM

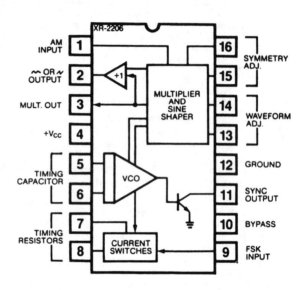

ORDERING INFORMATION

Part Number	Package	Operating Temperature
XR-2206M	Ceramic	−55°C to +125°C
XR-2206N	Ceramic	0°C to +70°C
XR-2206P	Plastic	0°C to +70°C
XR-2206CN	Ceramic	0°C to +70°C
XR-2206CP	Plastic	0°C to +70°C

SYSTEM DESCRIPTION

The XR-2206 is comprised of four functional blocks; a voltage-controlled oscillator (VCO), an analog multiplier and sine-shaper; a unity gain buffer amplifier; and a set of current switches.

The VCO actually produces an output frequency proportional to an input current, which is produced by a resistor from the timing terminals to ground. The current switches route one of the timing pins current to the VCO controlled by an FSK input pin, to produce an output frequency. With two timing pins, two discrete output frequencies can be independently produced for FSK Generation Applications.

XR-2206

ELECTRICAL CHARACTERISTICS

Test Conditions: Test Circuit of Figure 1, V^+ = 12V, T_A = 25°, C = 0.01 μF, R_1 = 100 kΩ, R_2 = 10 kΩ, R_3 = 25 kΩ unless otherwise specified. S_1 open for triangle, closed for sine wave.

PARAMETERS	XR-2206M			XR-2206C			UNITS	CONDITIONS
	MIN	TYP	MAX	MIN	TYP	MAX		
GENERAL CHARACTERISTICS								
Single Supply Voltage	10		26	10		26	V	
Split-Supply Voltage	±5		±13	±5		±13	V	
Supply Current		12	17		14	20	mA	$R_1 \geq 10$ kΩ
OSCILLATOR SECTION								
Max. Operating Frequency	0.5	1		0.5	1		MHz	C = 1000 pF, R_1 = 1 kΩ
Lowest Practical Frequency		0.01			0.01		Hz	C = 50 μF, R_1 = 2 MΩ
Frequency Accuracy		±1	±4		±2		% of f_0	f_0 = 1/R_1C
Temperature Stability		±10	±50		±20		ppm/°C	0°C ≤ T_A ≤ 70°C, R_1 = R_2 = 20 kΩ
Supply Sensitivity		0.01	0.1		0.01		%/V	V_{LOW} = 10V, V_{HIGH} = 20V, R_1 = R_2 = 20 kΩ
Sweep Range	1000:1	2000:1			2000:1		$f_H = f_L$	f_H @ R_1 = 1 kΩ f_L @ R_1 = 2 MΩ
Sweep Linearity								
10:1 Sweep		2			2		%	f_L = 1 kHz, f_H = 10 kHz
1000:1 Sweep		8			8		%	f_L = 100 kHz, f_H = 100 kHz
FM Distortion		0.1			0.1		%	±10% Deviation
Recommended Timing Components								See Note 1, Figure 2.
Timing Capacitor: C	0.001		100	0.001		100	μF	See Figure 4.
Timing Resistors: R_1 & R_2	1		2000	1		2000	kΩ	
Triangle Sine Wave Output								Figure 1, S_1 Open
Triangle Amplitude		160			160		mV/kΩ	Figure 1, S_1 Open
Sine Wave Amplitude	40	60	80		60		mV/kΩ	Figure 1, S_1 Closed
Max. Output Swing		6			6		V p-p	
Output Impedance		600			600		Ω	
Triangle Linearity		1			1		%	
Amplitude Stability		0.5			0.5		dB	For 1000:1 Sweep
Sine Wave Amplitude Stability		4800			4800		ppm/°C	See Note 2.
Sine Wave Distortion								
Without Adjustment		2.5			2.5		%	R_1 = 30 kΩ
With Adjustment		0.4	1.0		0.5	1.5	%	See Figures 6 and 7
Amplitude Modulation								
Input Impedance	50	100		50	100		kΩ	
Modulation Range		100			100		%	
Carrier Suppression		55			55		dB	
Linearity		2			2		%	For 95% modulation
Square-Wave Output								
Amplitude		12			12		V p-p	Measured at Pin 11.
Rise Time		250			250		nsec	C_L = 10 pF
Fall Time		50			50		nsec	C_L = 10 pF
Saturation Voltage		0.2	0.4		0.2	0.6	V	I_L = 2 mA
Leakage Current		0.1	20		0.1	100	μA	V_{11} = 26V
FSK Keying Level (Pin 9)	0.8	1.4	2.4	0.8	1.4	2.4	V	See section on circuit controls
Reference Bypass Voltage	2.9	3.1	3.3	2.5	3	3.5	V	Measured at Pin 10.

Note 1: Output amplitude is directly proportional to the resistance, R_3, on Pin 3. See Figure 2.

Note 2: For maximum amplitude stability, R_3 should be a positive temperature coefficient resistor.

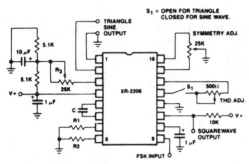

Figure 1. Basic Test Circuit.

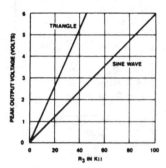

Figure 2. Output Amplitude as a Function of the Resistor, R₃, at Pin 3.

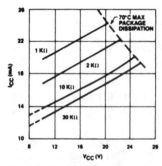

Figure 3. Supply Current versus Supply Voltage, Timing, R.

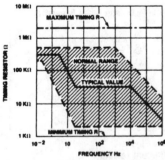

Figure 4. R versus Oscillation Frequency.

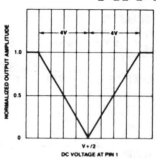

Figure 5. Normalized Output Amplitude versus DC Bias at AM Input (Pin 1).

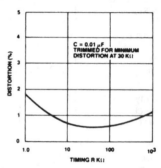

Figure 6. Trimmed Distortion versus Timing Resistor.

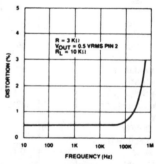

Figure 7. Sine Wave Distortion versus Operating Frequency with Timing Capacitors Varied.

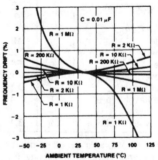

Figure 8. Frequency Drift versus Temperature.

XR-2206

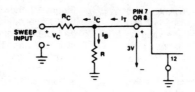

Figure 9. Circuit Connection for Frequency Sweep.

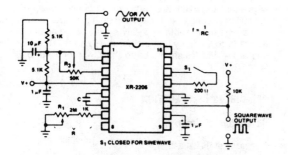

Figure 10. Circuit for Sine Wave Generation without External Adjustment. (See Figure 2 for Choice of R_3).

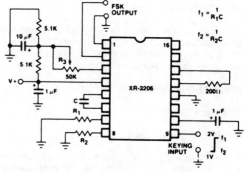

Figure 12. Sinusoidal FSK Generator.

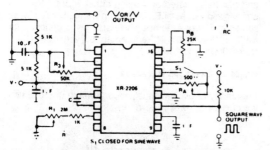

Figure 11. Circuit for Sine Wave Generation with Minimum Harmonic Distortion. (R_3 Determines Output Swing—See Figure 2.)

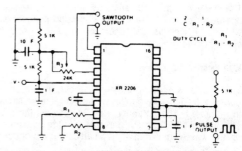

Figure 13. Circuit for Pulse and Ramp Generation.

XR-2206

Frequency-Shift Keying:

The XR-2206 can be operated with two separate timing resistors, R_1 and R_2, connected to the timing Pin 7 and 8, respectively, as shown in Figure 12. Depending on the polarity of the logic signal at Pin 9, either one or the other of these timing resistors is activated. If Pin 9 is open-circuited or connected to a bias voltage $\geq 2V$, only R_1 is activated. Similarly, if the voltage level at Pin 9 is $\leq 1V$, only R_2 is activated. Thus, the output frequency can be keyed between two levels, f_1 and f_2, as:

$$f_1 = 1/R_1 C \text{ and } f_2 = 1/R_2 C$$

For split-supply operation, the keying voltage at Pin 9 is referenced to V^-.

Output DC Level Control:

The dc level at the output (Pin 2) is approximately the same as the dc bias at Pin 3. In Figures 10, 11 and 12, Pin 3 is biased midway between V^+ and ground, to give an output dc level of $\approx V^+/2$.

APPLICATIONS INFORMATION

Sine Wave Generation

Without External Adjustment:

Figure 10 shows the circuit connection for generating a sinusoidal output from the XR-2206. The potentiometer, R_1 at Pin 7, provides the desired frequency tuning. The maximum output swing is greater than $V^+/2$, and the typical distortion (THD) is <2.5%. If lower sine wave distortion is desired, additional adjustments can be provided as described in the following section.

The circuit of Figure 10 can be converted to split-supply operation, simply by replacing all ground connections with V^-. For split-supply operation, R_3 can be directly connected to ground.

With External Adjustment:

The harmonic content of sinusoidal output can be reduced to $\approx 0.5\%$ by additional adjustments as shown in Figure 11. The potentiometer, R_A, adjusts the sine-shaping resistor, and R_B provides the fine adjustment for the waveform symmetry. The adjustment procedure is as follows:

1. Set R_B at midpoint, and adjust R_A for minimum distortion.

2. With R_A set as above, adjust R_B to further reduce distortion.

Triangle Wave Generation

The circuits of Figures 10 and 11 can be converted to triangle wave generation, by simply open-circuiting Pin 13 and 14 (i.e., S_1 open). Amplitude of the triangle is approximately twice the sine wave output.

FSK Generation

Figure 12 shows the circuit connection for sinusoidal FSK signal operation. Mark and space frequencies can be independently adjusted, by the choice of timing resistors, R_1 and R_2; the output is phase-continuous during transitions. The keying signal is applied to Pin 9. The circuit can be converted to split-supply operation by simply replacing ground with V^-.

Pulse and Ramp Generation

Figure 13 shows the circuit for pulse and ramp waveform generation. In this mode of operation, the FSK keying terminal (Pin 9) is shorted to the square-wave output (Pin 11), and the circuit automatically frequency-shift keys itself between two separate frequencies during the positive-going and negative-going output waveforms. The pulse width and duty cycle can be adjusted from 1% to 99%, by the choice of R_1 and R_2. The values of R_1 and R_2 should be in the range of 1 kΩ to 2 MΩ.

PRINCIPLES OF OPERATION

Description of Controls

Frequency of Operation:

The frequency of oscillation, f_0, is determined by the external timing capacitor, C, across Pin 5 and 6, and by the timing resistor, R, connected to either Pin 7 or 8. The frequency is given as:

$$f_0 = \frac{1}{RC} \text{ Hz}$$

and can be adjusted by varying either R or C. The recommended values of R, for a given frequency range, as shown in Figure 4. Temperature stability is optimum for 4 kΩ < R < 200 kΩ. Recommended values of C are from 1000 pF to 100 μF.

Frequency Sweep and Modulation:

Frequency of oscillation is proportional to the total timing current, I_T, drawn from Pin 7 or 8:

$$f = \frac{320 \, I_T \, (mA)}{C \, (\mu F)} \text{ Hz}$$

Timing terminals (Pin 7 or 8) are low-impedance points, and are internally biased at +3V, with respect to Pin 12. Frequency varies linearly with I_T, over a wide range of current values, from 1 μA to 3 mA. The frequency can be controlled by applying a control voltage, V_C, to the activated timing pin as shown in Figure 9. The frequency of oscillation is related to V_C as:

$$f = \frac{1}{RC} 1 + \frac{R}{R_C} (1 - \frac{V_C}{3}) \text{ Hz}$$

XR-2206

where V_C is in volts. The voltage-to-frequency conversion gain, K, is given as:

$$K = \partial f/\partial V_C = - \frac{0.32}{R_C C} \text{ Hz/V}$$

CAUTION: For safety operation of the circuit, I_T should be limited to ≤ 3 mA.

Output Amplitude:

Maximum output amplitude is inversely proportional to the external resistor, R_3, connected to Pin 3 (see Figure 2). For sine wave output, amplitude is approximately 60 mV peak per $k\Omega$ of R_3; for triangle, the peak amplitude is approximately 160 mV peak per $k\Omega$ of R_3. Thus, for example, $R_3 = 50$ $k\Omega$ would produce approximately ±3V sinusoidal output amplitude.

Amplitude Modulation:

Output amplitude can be modulated by applying a dc bias and a modulating signal to Pin 1. The internal impedance at Pin 1 is approximately 100 kΩ. Output amplitude varies linearly with the applied voltage at Pin 1, for values of dc bias at this pin, within ±4 volts of $V^+/2$ as shown in Figure 5. As this bias level approaches $V^+/2$, the phase of the output signal is reversed, and the amplitude goes through zero. This property is suitable for phase-shift keying and suppressed-carrier AM generation. Total dynamic range of amplitude modulation is approximately 55 dB.

CAUTION: AM control must be used in conjunction with a well-regulated supply, since the output amplitude now becomes a function of V^+.

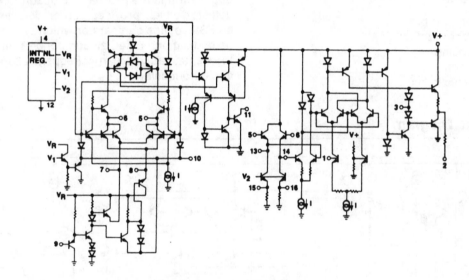

EQUIVALENT SCHEMATIC DIAGRAM

Raytheon

FSK Demodulator/ Tone Decoder

XR-2211

Features

- Wide frequency range — 0.01Hz to 300kHz
- Wide supply voltage range — 4.5V to 20V
- DTL/TTL/ECL logic compatibility
- FSK demodulation with carrier-detector
- Wide dynamic range — 2mV to 3V$_{RMS}$
- Adjustable tracking range — ±1% to ±80%
- Excellent temperature stability — 20ppm/°C typical

Applications

- FSK demodulation
- Data synchronization
- Tone decoding
- FM detection
- Carrier detection

Description

The XR-2211 is a monolithic phase-locked loop (PLL) system especially designed for data communications. It is particularly well suited for FSK modem applications, and operates over a wide frequency range of 0.01Hz to 300kHz. It can accommodate analog signals between 2mV and 3V, and can interface with conventional DTL, TTL and ECL logic families. The circuit consists of a basic PLL for tracking an input signal frequency within the passband, a quadrature phase detector which provides carrier detection, and an FSK voltage comparator which provides FSK demodulation. External components are used to independently set carrier frequency, bandwidth, and output delay.

Schematic Diagram

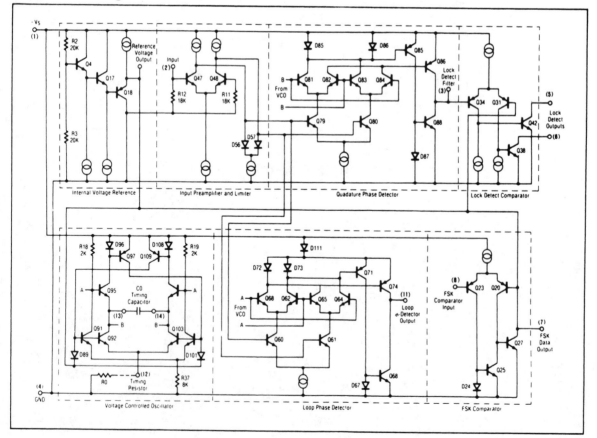

Mask Pattern

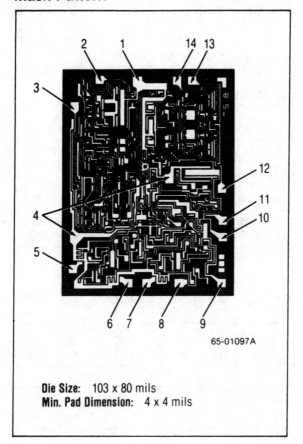

65-01097A

Die Size: 103 x 80 mils
Min. Pad Dimension: 4 x 4 mils

Connection Information

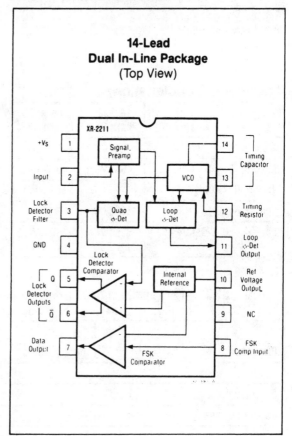

**14-Lead
Dual In-Line Package**
(Top View)

Functional Block Diagram

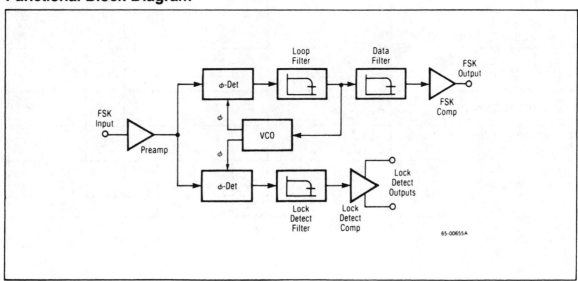

Courtesy of Ratheon Company.

FSK Demodulator/Tone Decoder

XR-2211

Absolute Maximum Ratings

Supply Voltage .... +20V
Input Signal Level .... 3V$_{RMS}$
Storage Temperature
   Range .... −65°C to +150°C

Operating Temperature Range
   XR-2211CN/CP .... 0°C to +75°C
   XR-2211N/P .... −40°C to +85°C
   XR-2211M .... −55°C to +125°C
Lead Soldering Temperature
   (60 Sec) .... +300°C

Thermal Characteristics

	14-Lead Plastic DIP	14-Lead Ceramic DIP
Max. Junction Temp.	125°C	175°C
Max. P$_D$ T$_A$ < 50°C	468mW	1042mW
Therm. Res. θ_{JC}	—	50°C/W
Therm. Res. θ_{JA}	160°C/W	120°C/W
For T$_A$ > 50°C Derate at	6.25mW per °C	8.33mW per °C

Ordering Information

Part Number	Package	Operating Temperature Range
XR-2211CN	Ceramic	0°C to +75°C
XR-2211CP	Plastic	0°C to +75°C
XR-2211N	Ceramic	−40°C to +85°C
XR-2211P	Plastic	−40°C to +85°C
XR-2211M	Ceramic	−55°C to +125°C
XR-2211M/883B*	Ceramic	−55°C to +125°C

*MIL-STD-883, Level B Processing

Electrical Characteristics (Test Conditions +V$_S$ = +12V, T$_A$ = +25°C, R0 = 30kΩ, C0 = 0.033μF. See Figure 1 for component designations.)

Parameters	Test Conditions	XR-2211/M			XR-2211C			Units
		Min	Typ	Max	Min	Typ	Max	
General								
Supply Voltage		4.5		20	4.5		20	V
Supply Current	R0 ≥ 10kΩ		4.0	9.0		5.0	11	mA
Oscillator								
Frequency Accuracy	Deviation from f$_0$ = 1/R0C0		±1.0	±3.0		±1.0		%
Frequency Stability Temperature Coefficient	R1 = ∞		±20	±50		±20		ppm/°C
Power Supply Rejection	+V$_S$ = 12 ±1V		0.05	0.5		0.05		%/V
	+V$_S$ = 5 ±0.5V		0.2			0.2		%/V
Upper Frequency Limit	R0 = 8.2kΩ, C0 = 400pF	100	300			300		kHz
Lowest Practical Operating Frequency	R0 = 2MΩ C0 = 50μF		0.01			0.01		Hz
Timing Resistor, R0 Operating Range		5.0		2000	5.0		2000	kΩ
Recommended Range		15		100	15		100	kΩ

Courtesy of Raytheon Company.

Electrical Characteristics (Continued)

(V_S = +12V, T_A = +25°C, R0 = 30kΩ, C0 = 0.033μF. See Figure 1 for component designations.)

Parameters	Test Conditions	XR-2211/M Min	XR-2211/M Typ	XR-2211/M Max	XR-2211C Min	XR-2211C Typ	XR-2211C Max	Units
Loop Phase Detector								
Peak Output Current	Meas. at Pin 11	±150	±200	±300	±100	±200	±300	μA
Output Offset Current			±1.0			±2.0		μA
Output Impedance			1.0			1.0		MΩ
Maximum Swing	Ref. to Pin 10	±4.0	±5.0		±4.0	±5.0		V
Quadrature Phase Detector								
Peak Output Current	Meas. at Pin 3	100	150			150		μA
Output Impedance			1.0			1.0		MΩ
Maximum Swing			11			11		V_{p-p}
Input Preamp								
Input Impedance	Meas. at Pin 2		20			20		kΩ
Input Signal Voltage Required to Cause Limiting			2.0	10		2.0		mV_{RMS}
Voltage Comparator								
Input Impedance	Meas. at Pins 3 & 8		2.0			2.0		MΩ
Input Bias Current			100			100		nA
Voltage Gain	R_L = 5.1kΩ	55	70		55	70		dB
Output Voltage Low	I_C = 3mA		300			300		mV
Output Leakage Current	V_O = 12V		0.01			0.01		μA
Internal Reference								
Voltage Level	Meas. at Pin 10	4.9	5.3	5.7	4.75	5.3	5.85	V
Output Impedance			100			100		Ω

Courtesy of Ratheon Company.

Description of Circuit Controls

Signal Input (Pin 2)
The input signal is AC coupled to this terminal. The internal impedance at pin 2 is $20k\Omega$. Recommended input signal level is in the range of $10mV_{RMS}$ to $3V_{RMS}$.

Quadrature Phase Detector Output (Pin 3)
This is the high-impedance output of the quadrature phase detector, and is internally connected to the input of lock-detect voltage comparator. In tone detection applications, pin 3 is connected to ground through a parallel combination of R_D and C_D (see Figure 1) to eliminate chatter at the lock-detect outputs. If this tone-detect section is not used, pin 3 can be left open circuited.

Lock-Detect Output, Q (Pin 5)
The output at pin 5 is at a "high" state when the PLL is out of lock and goes to a "low" or conducting state when the PLL is locked. It is an open collector type output and requires a pull-up resistor, R_L, to $+V_S$ for proper operation. In the "low" state it can sink up to 5mA of load current.

Lock-Detect Complement, Q̄ (Pin 6)
The output at pin 6 is the logic complement of the lock-detect output at pin 5. This output is

also an open collector type stage which can sink 5mA of load current in the low or "on" state.

FSK Data Output (Pin 7)
This output is an open collector logic stage which requires a pull-up resistor, R_L, to $+V_S$ for proper operation. It can sink 5mA of load current. When decoding FSK signals the FSK data output will switch to a "high" or off state for low input frequency, and will switch to a "low" or on state for high input frequency. If no input signal is present, the logic state at pin 7 is indeterminate.

FSK Comparator Input (Pin 8)
This is the high-impedance input to the FSK voltage comparator. Normally, an FSK post-detection or data filter is connected between this terminal and the PLL phase-detector output (pin 11). This data filter is formed by R_F and C_F of Figure 1. The threshold voltage of the comparator is set by the internal reference voltage, V_R, available at pin 10.

Reference Voltage, V_R (Pin 10)
This pin is internally biased at the reference voltage level, V_R; $V_R = V+/2 - 650mV$. The DC voltage level at this pin forms an internal reference

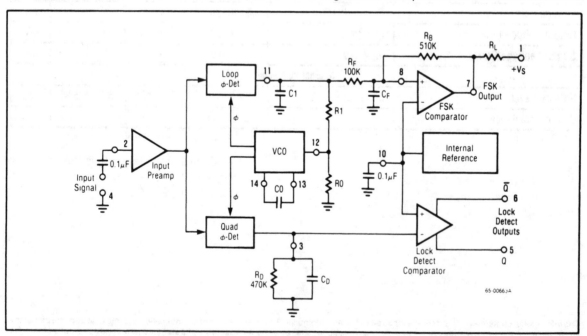

Figure 1. Generalized Circuit Connection for FSK and Tone Detection

for the voltage levels at pin 3, 8, 11, and 12. Pin 10 must be bypassed to ground with a $0.1\mu F$ capacitor.

Loop Phase Detector Output (Pin 11)
This terminal provides a high impedance output for the loop phase-detector. The PLL loop filter is formed by R1 and C1 connected to pin 11 (see Figure 1). With no input signal, or with no phase error within the PLL, the DC level at pin 11 is very nearly equal to V_R. The peak voltage swing available at the phase detector output is equal to $\pm V_R$.

VCO Control Input (Pin 12)
VCO free-running frequency is determined by external timing resistor, R0, connected from this terminal to ground. The VCO free-running frequency, f_0, is given by:

$$f_0(Hz) = \frac{1}{R0C0}$$

where C0 is the timing capacitor across pins 13 and 14. For optimum temperature stability R0 must be in the range of $10k\Omega$ to $100k\Omega$ (see Typical Electrical Characteristics).

This terminal is a low impedance point, and is internally biased at a DC level equal to V_R. The maximum timing current drawn from pin 12 must be limited to $\leq 3mA$ for proper operation of the circuit.

VCO Timing Capacitor (Pins 13 and 14)
VCO frequency is inversely proportional to the external timing capacitor, C0, connected across these terminals. C0 must be non-polarized, and in the range of 200pF to $10\mu F$.

VCO Frequency Adjustment
VCO can be fine tuned by connecting a potentiometer, R_X, in series with R0 at pin 12 (see Figure 2).

VCO Free-Running Frequency, f_0
The XR-2211 does not have a separate VCO output terminal. Instead, the VCO outputs are internally connected to the phase-detector sections of the circuit. However, for setup or adjustment purposes, the VCO free-running frequency can be measured at pin 3 (with C_D disconnected) with no input and with pin 2 shorted to pin 10.

Design Equations
See Figure 1 for Definitions of Components.

1. VCO Center Frequency, f_0:

$$f_0(Hz) = \frac{1}{R0C0}$$

2. Internal Reference Voltage, V_R (measured at pin 10):

$$V_R = \left(\frac{+V_S}{2}\right) - 650mV$$

3. Loop Lowpass Filter Time Constant, τ:

$$\tau = R1C1$$

4. Loop Damping, ζ:

$$\zeta = \left(\sqrt{\frac{C0}{C1}}\right)\left(\frac{1}{4}\right)$$

5. Loop Tracking Bandwidth, $\pm\Delta f/f_0$:

$$\Delta f/f_0 = R0/R1$$

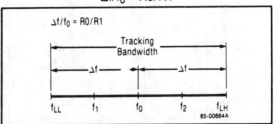

6. FSK Data Filter Time Constant, τ_F:

$$\tau_F = R_F C_F$$

7. Loop Phase Detector Conversion Gain, K_ϕ: (K_ϕ is the differential DC voltage across pins 10 and 11, per unit of phase error at phase-detector input):

$$K_\phi \text{ (in volts per radian)} = \frac{(-2)(V_R)}{\pi}$$

8. VCO Conversion Gain, K0, is the amount of change in VCO frequency per unit of DC voltage change at pin 11:

$$K0 \text{ (in Hertz per volt)} = \frac{-1}{C0R1V_R}$$

9. Total Loop Gain, K_T:

$$K_T \text{ (in radians per second per volt)} = 2\pi K_\phi K0$$
$$= 4/C0R1$$

10. Peak Phase-Detector Current, I_A:

$$I_A \text{ (mA)} = \frac{V_R}{25}$$

Courtesy of Rathcon Company.

Applications

FSK Decoding

Figure 2 shows the basic circuit connection for FSK decoding. With reference to Figures 1 and 2, the functions of external components are defined as follows: R0 and C0 set the PLL center frequency, R1 sets the system bandwidth, and C1 sets the loop filter time constant and the loop damping factor. C_F and R_F form a one pole post-detection filter for the FSK data output. The resistor R_B (= 510kΩ) from pin 7 to pin 8 introduces positive feedback across FSK comparator to facilitate rapid transition between output logic states.

Recommended component values for some of the most commonly used FSK bauds are given in Table 1.

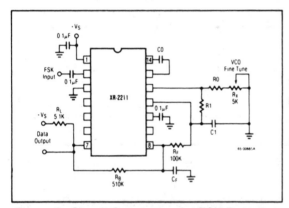

Figure 2. Circuit Connection for FSK Decoding

Table 1. Recommended Component Values for Commonly Used FSK Bands
(See Circuit of Figure 2)

FSK Band	Component Values
300 Baud	C0 = 0.039μF C_F = 0.005μF
f_1 = 1070Hz	C1 = 0.01μF R0 = 18kΩ
f_2 = 1270Hz	R1 = 100kΩ
300 Baud	C0 = 0.022μF C_F = 0.005μF
f_1 = 2025Hz	C1 = 0.0047μF R1 = 18kΩ
f_2 = 2225Hz	R1 = 200kΩ
1200 Baud	C0 = 0.027μF C_F = 0.0022μF
f_1 = 1200Hz	C1 = 0.01μF R0 = 18kΩ
f_2 = 2200Hz	R1 = 30kΩ

Courtesy of Ratheon Company.

Design Instructions

The circuit of Figure 2 can be tailored for any FSK decoding application by the choice of five key circuit components; R0, R1, C0, C1 and C_F. For a given set of FSK mark and space frequencies, f_1 and f_2, these parameters can be calculated as follows:

1. Calculate PLL center frequency, f_0

$$f_0 = \frac{f_1 + f_2}{2}$$

2. Choose a value of timing resistor R0 to be in the range of 10kΩ to 100kΩ. This choice is arbitrary. The recommended value is R0 $\cong$ 20kΩ. The final value of R0 is normally fine-tuned with the series potentiometer, R_X.

3. Calculate value of C0 from Design Equation No. 1 or from Typical Performance Characteristics:

$$C0 = 1/R0f_0$$

4. Calculate R1 to give a Δf equal to the mark-space deviation:
$$R1 = R0\,[f_0/(f_1 - f_2)]$$

5. Calculate C1 to set loop damping. (See Design Equation No. 4.)
   Normally, $\zeta \approx 1/2$ is recommended
   Then: C1 = C0/4 for ζ = 1/2

6. Calculate Data Filter Capacitance, C_F:
   For R_F = 100kΩ, R_B = 510kΩ, the recommended value of C_F is:

$$C_F \text{ (in } \mu F) = \frac{3}{\text{Baud Rate}}$$

Note: All calculated component values except R0 can be rounded off to the nearest standard value, and R0 can be varied to fine-tune center frequency through a series potentiometer, R_X (see Figure 2).

Design Example

75 Baud FSK demodulator with mark/space frequencies of 1110/1170Hz:

Step 1: Calculate f_0:
   f_0 = (1110 + 1170) (1/2) = 1140Hz

Step 2: Choose R0 = 20kΩ (18kΩ fixed resistor in series with 5kΩ potentiometer)

Step 3: Calculate C0 from V_{CO} Frequency vs. Timing Capacitor: C0 = 0.044μF

Step 4: Calculate R1: R1 = R0 (2240/60) = 380kΩ

Step 5: Calculate C1: C1 = C0/4 = 0.011μF

Note: All values except R0 can be rounded off to nearest standard value.

FSK Decoding With Carrier Detect

The lock-detect section of the XR-2211 can be used as a carrier detect option for FSK decoding. The recommended circuit connection for this application is shown in Figure 3. The open-collector lock-detect output, pin 6, is shorted to the data output (pin 7). Thus, the data output will be disabled at "low" state, until there is a carrier within the detection band of the PLL, and the pin 6 output goes "high" to enable the data output.

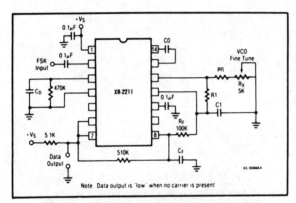

Figure 3. External Connectors for FSK Demodulation With Carrier Detect Capability

The minimum value of the lock-detect filter capacitance C_D is inversely proportional to the capture range, $\pm \Delta f_c$. This is the range of incoming frequencies over which the loop can acquire lock and is always less than the tracking range. It is further limited by C1. For most applications, $\Delta f_c < \Delta f/2$. For $R_D = 470k\Omega$, the approximate minimum value of C_D can be determined by:

$$C_D(\mu F) \geq 16/\text{capture range in Hz}$$

With values of C_D that are too small, chatter can be observed on the lock-detect output as an incoming signal frequency approaches the capture bandwidth. Excessively large values of C_D will slow the response time of the lock-detect output.

Tone Detection

Figure 4 shows the generalized circuit connection for tone detection. The logic outputs, Q and $\bar{Q}$ at pins 5 and 6 are normally at "high" and "low" logic states, respectively. When a tone is present within the detection band of the PLL, the logic state at these outputs becomes reversed for the duration of the input tone. Each logic output can sink 5mA of load current.

Both logic outputs at pins 5 and 6 are open-collector type stages, and require external pull-up resistors R_{L1} and R_{L2} as shown in Figure 4.

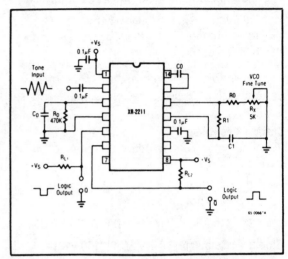

Figure 4. Circuit Connection for Tone Detection

With reference to Figures 1 and 4, the function of the external circuit components can be explained as follows: R0 and C0 set VCO center frequency, R1 sets the detection bandwidth, C1 sets the lowpass-loop filter time constant and the loop damping factor, and R_{L1} and R_{L2} are the respective pull-up resistors for the Q and $\bar{Q}$ logic outputs.

Design Instructions

The circuit of Figure 4 can be optimized for any tone-detection application by the choice of five key circuit components: R0, R1, C0, C1, and C_D. For a given input tone frequency, f_S, these parameters are calculated as follows:

1. Choose R0 to be in the range of 15kΩ to 100kΩ. This choice is arbitrary.
2. Calculate C0 to set center frequency, f_0 equal to f_S: $C0 = 1/R0f_S$.
3. Calculate R1 to set bandwidth $\pm \Delta f$ (see Design Equation No. 5): $R1 = R0(f_0/\Delta f)$

Note: The total detection bandwidth covers the frequency range of $f_0 \pm \Delta f$.

Courtesy of Ratheon Company.

4. Calculate value of C1 for a given loop damping factor:

$$C1 = C0/16\zeta^2$$

Normally $\zeta \approx 1/2$ is optimum for most tone-detector applications, giving C1 = 0.25 C0.

Increasing C1 improves the out-of-band signal rejection, but increases the PLL capture time.

5. Calculate value of filter capacitor C_D. To avoid chatter at the logic output, with R_D = 470kΩ, C_D must be:

$$C_D(\mu F) \geq (16/\text{capture range in Hz})$$

Increasing C_D slows the logic output response time.

Design Examples

Tone detector with a detection band of 1kHz ±20Hz:

Step 1: Choose R0 = 20kΩ (18kΩ in series with 5kΩ potentiometer).

Step 2: Choose C0 for f_0 = 1kHz: C0 = 0.05μF.

Step 3: Calculate R1: R1 = (R0) (1000/20) = 1MΩ.

Step 4: Calculate C1: for ζ = 1/2, C1 = 0.25μF, C0 = 0.013μF.

Step 5: Calculate C_D: C_D = 16/38 = 0.42μF.

Step 6: Fine tune the center frequency with the 5kΩ potentiometer, R_X.

Linear FM Detection

The XR-2211 can be used as a linear FM detector for a wide range of analog communications and telemetry applications. The recommended circuit connection for the application is shown

in Figure 5. The demodulated output is taken from the loop phase detector output (pin 11), through a post detection filter made up of R_F and C_F, and an external buffer amplifier. This buffer amplifier is necessary because of the high impedance output at pin 11. Normally, a non-inverting unity gain op amp can be used as a buffer amplifier, as shown in Figure 5.

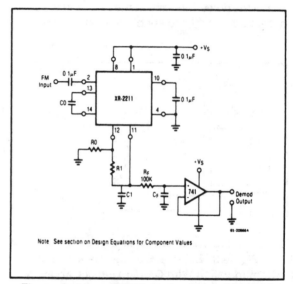

Note See section on Design Equations for Component Values

Figure 5. Linear FM Detector Using XR-2211 and an External Op Amp

The FM detector gain, i.e., the output voltage change per unit of FM deviation, can be given as:

$$V_{OUT} = R1 \, V_R/1.00 \, R0 \text{ Volts/\% deviation}$$

where V_R is the internal reference voltage. For the choice of external components R1, R0, C_D, C1 and C_F, see the section on Design Equations.

Typical Performance Characteristics

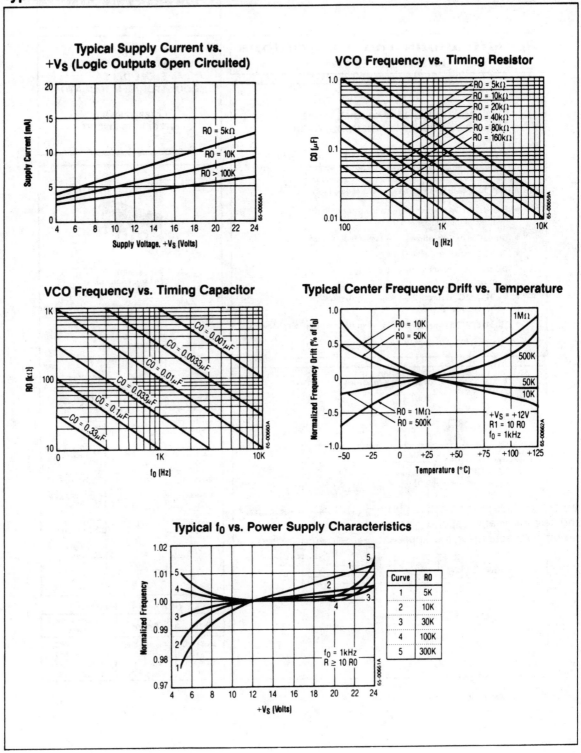

Courtesy of Ratheon Company.

 **MOTOROLA**

MC3417, MC3517
MC3418, MC3518

Specifications and Applications Information

CONTINUOUSLY VARIABLE SLOPE MODULATOR/DEMODULATOR

LASER-TRIMMED INTEGRATED CIRCUIT

CONTINUOUSLY VARIABLE SLOPE DELTA MODULATOR/DEMODULATOR

Providing a simplified approach to digital speech encoding/decoding, the MC3517/18 series of CVSDs is designed for military secure communication and commercial telephone applications. A single IC provides both encoding and decoding functions.

- Encode and Decode Functions on the Same Chip with a Digital Input for Selection
- Utilization of Compatible I^{2}L – Linear Bipolar Technology
- CMOS Compatible Digital Output
- Digital Input Threshold Selectable (V$_{CC}$/2 reference provided on chip)
- MC3417/MC3517 has a 3-Bit Algorithm (General Communications)
- MC3418/MC3518 has a 4-Bit Algorithm (Commercial Telephone)

L SUFFIX
CERAMIC PACKAGE
CASE 620

CVSD BLOCK DIAGRAM

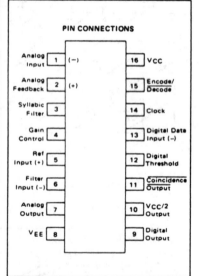

PIN CONNECTIONS

Pin	Name		Pin	Name
1	Analog Input	(−)	16	VCC
2	Analog Feedback	(+)	15	Encode/Decode
3	Syllabic Filter		14	Clock
4	Gain Control		13	Digital Data Input (−)
5	Ref Input (+)		12	Digital Threshold
6	Filter Input (−)		11	Coincidence Output
7	Analog Output		10	VCC/2 Output
8	VEE		9	Digital Output

ORDERING INFORMATION

Device	Package	Temperature Range
MC3417L	Ceramic DIP	0°C to +70°C
MC3418L	Ceramic DIP	0°C to +70°C
MC3517L	Ceramic DIP	−55°C to +125°C
MC3518L	Ceramic DIP	−55°C to +125°C

MC3417, MC3517, MC3418, MC3518

DEFINITIONS AND FUNCTION OF PINS

Pin 1 — Analog Input

This is the analog comparator inverting input where the voice signal is applied. It may be ac or dc coupled depending on the application. If the voice signal is to be level shifted to the internal reference voltage, then a bias resistor between pins 1 and 10 is used. The resistor is used to establish the reference as the new dc average of the ac coupled signal. The analog comparator was designed for low hysteresis (typically less than 0.1 mV) and high gain (typically 70 dB).

Pin 2 — Analog Feedback

This is the non-inverting input to the analog signal comparator within the IC. In an encoder application it should be connected to the analog output of the encoder circuit. This may be pin 7 or a low pass filter output connected to pin 7. In a decode circuit pin 2 is not used and may be tied to $V_{CC}/2$ on pin 10, ground or left open.

The analog input comparator has bias currents of 1.5 μA max, thus the driving impedances of pins 1 and 2 should be equal to avoid disturbing the idle channel characteristics of the encoder.

Pin 3 — Syllabic Filter

This is the point at which the syllabic filter voltage is returned to the IC in order to control the integrator step size. It is an NPN input to an op amp. The syllabic filter consists of an RC network between pins 11 and 3. Typical time constant values of 6 ms to 50 ms are used in voice codecs.

Pin 4 — Gain Control Input

The syllabic filter voltage appears across C_S of the syllabic filter and is the voltage between V_{CC} and pin 3. The active voltage to current (V–I) converter drives pin 4 to the same voltage at a slew rate of typically 0.5 V/μs. Thus the current injected into pin 4 (I_{GC}) is the syllabic filter voltage divided by the R_x resistance. Figure 6 shows the relationship between I_{GC} (x-axis) and the integrating current, I_{Int} (y-axis). The discrepancy, which is most significant at very low currents, is due to circuitry within the slope polarity switch which enables trimming to a low total loop offset. The R_x resistor is then varied to adjust the loop gain of the codec, but should be no larger than 5.0 kΩ to maintain stability.

Pin 5 — Reference Input

This pin is the non-inverting input of the integrator amplifier. It is used to reference the dc level of the output signal. In an encoder circuit it must reference the same voltage as pin 1 and is tied to pin 10.

Pin 6 — Filter Input

This inverting op amp input is used to connect the integrator external components. The integrating current

(I_{Int}) flows into pin 6 when the analog input (pin 1) is high with respect to the analog feedback (pin 2) in the encode mode or when the digital data input (pin 13) is high in the decode mode. For the opposite states, I_{Int} flows out of Pin 6. Single integration systems require a capacitor and resistor between pins 6 and 7. Multipole configurations will have different circuitry. The resistance between pins 6 and 7 should always be between 8 kΩ and 13 kΩ to maintain good idle channel characteristics.

Pin 7 — Analog Output

This is the integrator op amp output. It is capable of driving a 600-ohm load referenced to $V_{CC}/2$ to +6 dBm and can otherwise be treated as an op amp output. Pins 5, 6, and 7 provide full access to the integrator op amp for designing integration filter networks. The slew rate of the internally compensated integrator op amp is typically 0.5 V/μs. Pin 7 output is current limited for both polarities of current flow at typically 30 mA.

Pin 8 — V_{EE}

The circuit is designed to work in either single or dual power supply applications. Pin 8 is always connected to the most negative supply.

Pin 9 — Digital Output

The digital output provides the results of the delta modulator's conversion. It swings between V_{CC} and V_{EE} and is CMOS or TTL compatible. Pin 9 is inverting with respect to pin 1 and non-inverting with respect to pin 2. It is clocked on the falling edge of pin 14. The typical 10% to 90% rise and fall times are 250 ns and 50 ns respectively for V_{CC} = 12 V and C_L = 25 pF to ground.

Pin 10 — $V_{CC}/2$ Output

An internal low impedance mid-supply reference is provided for use of the MC3417/18 in single supply applications. The internal regulator is a current source and must be loaded with a resistor to insure its sinking capability. If a +6 dBmo signal is expected across a 600 ohm input bias resistor, then pin 10 must sink 2.2 V/600 Ω = 3.66 mA. This is only possible if pin 10 sources 3.66 mA into a resistor normally and will source only the difference under peak load. The reference load resistor is chosen accordingly. A 0.1 μF bypass capacitor from pin 10 to V_{EE} is also recommended. The $V_{CC}/2$ reference is capable of sourcing 10 mA and can be used as a reference elsewhere in the system circuitry.

Pin 11 — Coincidence Output

The duty cycle of this pin is proportional to the voltage across C_S. The coincidence output will be low whenever the content of the internal shift register is all 1s or all 0s. In the MC3417 the register is 3 bits long

MC3417, MC3517, MC3418, MC3518

DEFINITIONS AND FUNCTIONS OF PINS (continued)

while the MC3418 contains a 4 bit register. Pin 11 is an open collector of an NPN device and requires a pull-up resistor. If the syllabic filter is to have equal charge and discharge time constants, the value of Rp should be much less than R_S. In systems requiring different charge and discharge constants, the charging constant is $R_S C_S$ while the decaying constant is $(R_S + Rp)C_S$. Thus longer decays are easily achievable. The NPN device should not be required to sink more than 3 mA in any configuration. The typical 10% to 90% rise and fall times are 200 ns and 100 ns respectively for R_L = 4 kΩ to +12 V and C_L = 25 pF to ground.

Pin 12 — Digital Threshold

This input sets the switching threshold for pins 13, 14, and 15. It is intended to aid in interfacing different logic families without external parts. Often it is connected to the $V_{CC}/2$ reference for CMOS interface or can be biased two diode drops above V_{EE} for TTL interface.

Pin 13 — Digital Data Input

In a decode application, the digital data stream is applied to pin 13. In an encoder it may be unused or may be used to transmit signaling message under the control of pin 15. It is an inverting input with respect to pin 9. When pins 9 and 13 are connected, a toggle flip-flop is formed and a forced idle channel pattern

can be transmitted. The digital data input level should be maintained for 0.5 μs before and after the clock trigger for proper clocking.

Pin 14 — Clock Input

The clock input determines the data rate of the codec circuit. A 32K bit rate requires a 32 kHz clock. The switching threshold of the clock input is set by pin 12. The shift register circuit toggles on the falling edge of the clock input. The minimum width for a positive-going pulse on the clock input is 300 ns, whereas for a negative-going pulse, it is 900 ns.

Pin 15 — Encode/$\overline{\text{Decode}}$

This pin controls the connection of the analog input comparator and the digital input comparator to the internal shift register. If high, the result of the analog comparison will be clocked into the register on the falling edge at pin 14. If low, the digital input state will be entered. This allows use of the IC as an encoder/decoder or simplex codec without external parts. Furthermore, it allows non-voice patterns to be forced onto the transmission line through pin 13 in an encoder.

Pin 16 — V_{CC}

The power supply range is from 4.75 to 16.5 volts between pin V_{CC} and V_{EE}.

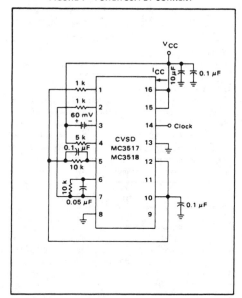

FIGURE 1 — POWER SUPPLY CURRENT

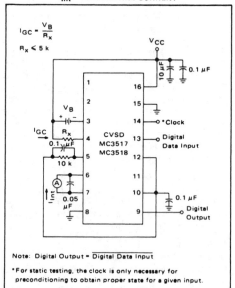

FIGURE 2 — I_{GCR}, GAIN CONTROL RANGE and I_{Int} — INTEGRATING CURRENT

Note: Digital Output = $\overline{\text{Digital Data Input}}$

*For static testing, the clock is only necessary for preconditioning to obtain proper state for a given input.

TP5089 DTMF (TOUCH-TONE) Generator

General Description

The TP5089 is a low threshold voltage, field-implanted, metal gate CMOS integrated circuit. It interfaces directly to a standard telephone keypad and generates all dual tone multi-frequency pairs required in tone-dialing systems. The tone synthesizers are locked to an on-chip reference oscillator using an inexpensive 3.579545 MHz crystal for high tone accuracy. The crystal and an output load resistor are the only external components required for tone generation. A MUTE OUT logic signal, which changes state when any key is depressed, is also provided.

Features

- 3.5V–10V operation when generating tones
- 2V operation of keyscan and MUTE logic
- Static sensing of key closures or logic inputs
- On-chip 3.579545 MHz crystal-controlled oscillator
- Output amplitudes proportional to supply voltage
- High group pre-emphasis
- Low harmonic distortion
- Open emitter-follower low-impedance output
- SINGLE TONE INHIBIT pin

Block Diagram

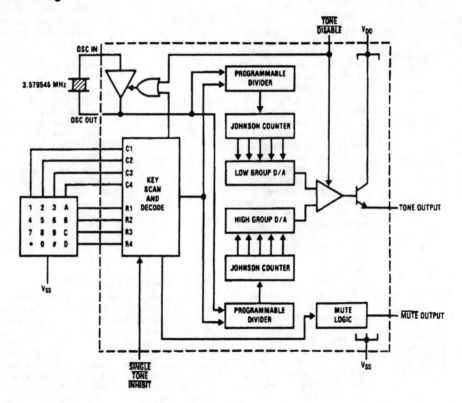

FIGURE 1

Reprinted with permission of National Semiconductor Corporation.

Connection Diagram

Dual-In-Line Package

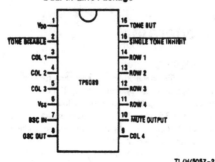

V_{DD}	1	16	TONE OUT
TONE DISABLE	2	15	SINGLE TONE INHIBIT
COL 1	3	14	ROW 1
COL 2	4	13	ROW 2
COL 3	5	12	ROW 3
V_{SS}	6	11	ROW 4
OSC IN	7	10	MUTE OUTPUT
OSC OUT	8	9	COL 4

TP5089

TL/H/5057–2

Top View

Order Number TP5089N
See NS Package N16A

Pin Descriptions

Symbol	Description
V_{DD}	This is the positive voltage supply to the device, referenced to V_{SS}. The collector of the TONE OUT transistor is connected to this pin.
V_{SS}	This is the negative voltage supply. All voltages are referenced to this pin.
OSC IN, OSC OUT	All tone generation timing is derived from the on-chip oscillator circuit. A low cost 3.579545 MHz A-cut crystal (NTSC TV color-burst) is needed between pins 7 and 8. Load capacitors and a feedback resistor are included on-chip for good start-up and stability. The oscillator stops when column inputs are sensed with no valid input having been detected. The oscillator is also stopped when the TONE DISABLE input is pulled to logic low.
Row and Column Inputs	When no key is pushed, pull-up resistors are active on row and column inputs. A key closure is recognized when a single row and a single column are connected to V_{SS}, which starts the oscillator and initiates tone generation. Negative-true logic signals simulating key closures can also be used.
TONE DISABLE Input	The TONE DISABLE input has an internal pull-up resistor. When this input is open or at logic high, the normal tone output mode will occur. When TONE DISABLE input is at logic low, the device will be in the inactive mode, TONE OUT will be at an open circuit state.

Symbol	Description
MUTE Output	The MUTE output is an open-drain N-channel device that sinks current to V_{SS} with any key input and is open when no key input is sensed. The MUTE output will switch regardless of the state of the SINGLE TONE INHIBIT input.
SINGLE TONE INHIBIT Input	The SINGLE TONE INHIBIT input is used to inhibit the generation of other than valid tone pairs due to multiple row-column closures. It has a pull-down resistor to V_{SS}, and when left open or tied to V_{SS} any input condition that would normally result in a single tone will now result in no tone, with all other functions operating normally. When tied to V_{DD}, single or dual tones may be generated, see Table II.
TONE OUT	This output is the open emitter of an NPN transistor, the collector of which is connected to V_{DD}. When an external load resistor is connected from TONE OUT to V_{SS}, the output voltage on this pin is the sum of the high and low group sinewaves superimposed on a DC offset. When not generating tones, this output transistor is turned OFF to minimize the device idle current. Adjustment of the emitter load resistor results in variation of the mean DC current during tone generation, the sinewave signal current through the output transistor, and the output distortion. Increasing values of load resistance decrease both the signal current and distortion.

Functional Description

With no key inputs to the device the oscillator is inhibited, the output transistor is pulled OFF and device current consumption is reduced to a minimum. Key closures are sensed statically. Any key closure activates the MUTE output, starts the oscillator and sets the high group and low group programmable counters to the appropriate divide ratio. These counters sequence two ratioed-capacitor D/A converters through a series of 28 equal duration steps per sine-wave cycle. The two tones are summed by a mixer amplifier, with pre-emphasis applied to the high group tone. The output is an NPN emitter-follower requiring the addition of an external load resistor to V_{SS}. This resistor facilitates adjustment of the signal current flowing from V_{DD} through the output transistor.

The amplitude of the output tones is directly proportional to the device supply voltage.

Reprinted with permission of National Semiconductor Corporation.

TABLE I. Output Frequency Accuracy

Tone Group	Valid Input	Standard DTMF (Hz)	Tone Output Frequency	% Deviation from Standard
Low Group f_L	R1	697	694.8	−0.32
	R2	770	770.1	+0.02
	R3	852	852.4	+0.03
	R4	941	940.0	−0.11
High Group f_H	C1	1209	1206.0	−0.24
	C2	1336	1331.7	−0.32
	C3	1477	1486.5	+0.64
	C4	1633	1639.0	+0.37

TABLE II. Functional Truth Table

SINGLE TONE INHIBIT	TONE DISABLE	ROW	COLUMN	TONE OUT Low	TONE OUT High	MUTE
X	O	O/C	O/C	0V	0V	O/C
X	X	O/C	O/C	0V	0V	O/C
X	0	One	One	V_{OS}	V_{OS}	O
X	1	One	One	f_L	f_H	O
1	1	2 or More	One	—	f_H	O
1	1	One	2 or More	f_L	—	O
1	1	2 or More	2 or More	V_{OS}	V_{OS}	O
0	1	2 or More	One	V_{OS}	V_{OS}	O
0	1	One	2 or More	V_{OS}	V_{OS}	O
0	1	2 or More	2 or More	V_{OS}	V_{OS}	O

Note 1: X is don't care state.

Note 2: V_{OS} is the output offset voltage.

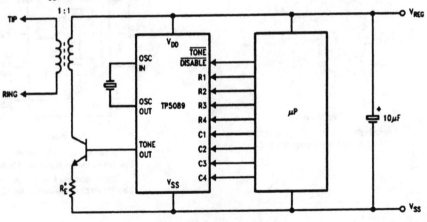

*Adjust R_E for desired tone amplitude.

FIGURE 2. Typical Application

LIFE SUPPORT POLICY

NATIONAL'S PRODUCTS ARE NOT AUTHORIZED FOR USE AS CRITICAL COMPONENTS IN LIFE SUPPORT DEVICES OR SYSTEMS WITHOUT THE EXPRESS WRITTEN APPROVAL OF THE PRESIDENT OF NATIONAL SEMICONDUCTOR CORPORATION. As used herein:

1. Life support devices or systems are devices or systems which, (a) are intended for surgical implant into the body, or (b) support or sustain life, and whose failure to perform, when properly used in accordance with instructions for use provided in the labeling, can be reasonably expected to result in a significant injury to the user.

2. A critical component is any component of a life support device or system whose failure to perform can be reasonably expected to cause the failure of the life support device or system, or to affect its safety or effectiveness.

Reprinted with permission of National Semiconductor Corporation.

**National
Semiconductor
Corporation**

LM567/LM567C Tone Decoder

General Description

The LM567 and LM567C are general purpose tone decoders designed to provide a saturated transistor switch to ground when an input signal is present within the passband. The circuit consists of an I and Q detector driven by a voltage controlled oscillator which determines the center frequency of the decoder. External components are used to independently set center frequency, bandwidth and output delay.

Features

■ 20 to 1 frequency range with an external resistor
■ Logic compatible output with 100 mA current sinking capability

■ Bandwidth adjustable from 0 to 14%
■ High rejection of out of band signals and noise
■ Immunity to false signals
■ Highly stable center frequency
■ Center frequency adjustable from 0.01 Hz to 500 kHz

Applications

■ Touch tone decoding
■ Precision oscillator
■ Frequency monitoring and control
■ Wide band FSK demodulation
■ Ultrasonic controls
■ Carrier current remote controls
■ Communications paging decoders

Connection Diagrams

Metal Can Package

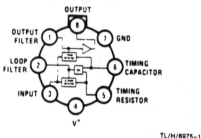

TL/H/6975–1

Top View

**Order Number LM567H or LM567CH
See NS Package Number H08C**

Dual-In-Line and Small Outline Packages

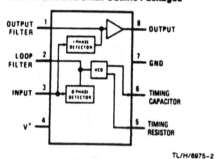

TL/H/6975–2

Top View

**Order Number LM567CM
See NS Package Number M08A
Order Number LM567CN
See NS Package Number N08E**

LM567

Absolute Maximum Ratings

If Military/Aerospace specified devices are required, contact the National Semiconductor Sales Office/Distributors for availability and specifications.

Supply Voltage Pin	9V
Power Dissipation (Note 1)	1100 mW
V_8	15V
V_3	−10V
V_3	$V_4 + 0.5$V
Storage Temperature Range	−65°C to +150°C
Operating Temperature Range	
LM567H	−55°C to +125°C
LM567CH, LM567CM, LM567CN	0°C to +70°C

Soldering Information
Dual-In-Line Package
Soldering (10 sec.)   260°C
Small Outline Package
Vapor Phase (60 sec.)   215°C
Infrared (15 sec.)   220°C
See AN-450 "Surface Mounting Methods and Their Effect on Product Reliability" for other methods of soldering surface mount devices.

Electrical Characteristics AC Test Circuit, T_A = 25°C, V^+ = 5V

Parameters	Conditions	LM567			LM567C/LM567CM			Units
		Min	Typ	Max	Min	Typ	Max	
Power Supply Voltage Range		4.75	5.0	9.0	4.75	5.0	9.0	V
Power Supply Current Quiescent	R_L = 20k		6	8		7	10	mA
Power Supply Current Activated	R_L = 20k		11	13		12	15	mA
Input Resistance		18	20		15	20		kΩ
Smallest Detectable Input Voltage	I_L = 100 mA, f_i = f_o		20	25		20	25	mVrms
Largest No Output Input Voltage	I_C = 100 mA, f_i = f_o	10	15		10	15		mVrms
Largest Simultaneous Outband Signal to Inband Signal Ratio			6			6		dB
Minimum Input Signal to Wideband Noise Ratio	B_n = 140 kHz		−6			−6		dB
Largest Detection Bandwidth		12	14	16	10	14	18	% of f_o
Largest Detection Bandwidth Skew			1	2		2	3	% of f_o
Largest Detection Bandwidth Variation with Temperature			±0.1			±0.1		%/°C
Largest Detection Bandwidth Variation with Supply Voltage	4.75 − 6.75V		±1	±2		±1	±5	%V
Highest Center Frequency		100	500		100	500		kHz
Center Frequency Stability (4.75–5.75V)	0 < T_A < 70 −55 < T_A < +125		35 ± 60 35 ± 140			35 ± 60 35 ± 140		ppm/°C ppm/°C
Center Frequency Shift with Supply Voltage	4.75V − 6.75V 4.75V − 9V		0.5	1.0 2.0		0.4	2.0 2.0	%/V %/V
Fastest ON-OFF Cycling Rate			f_o/20			f_o/20		
Output Leakage Current	V_8 = 15V		0.01	25		0.01	25	μA
Output Saturation Voltage	e_i = 25 mV, I_8 = 30 mA e_i = 25 mV, I_8 = 100 mA		0.2 0.6	0.4 1.0		0.2 0.6	0.4 1.0	V
Output Fall Time			30			30		ns
Output Rise Time			150			150		ns

Note 1: The maximum junction temperature of the LM567 and LM567C is 150°C. For operating at elevated temperatures, devices in the TO-5 package must be derated based on a thermal resistance of 150°C/W, junction to ambient or 45°C/W, junction to case. For the DIP the device must be derated based on a thermal resistance of 110°C/W, junction to ambient. For the Small Outline package, the device must be derated based on a thermal resistance of 160°C/W, junction to ambient.

Note 2: Refer to RETS567X drawing for specifications of military LM567H version.

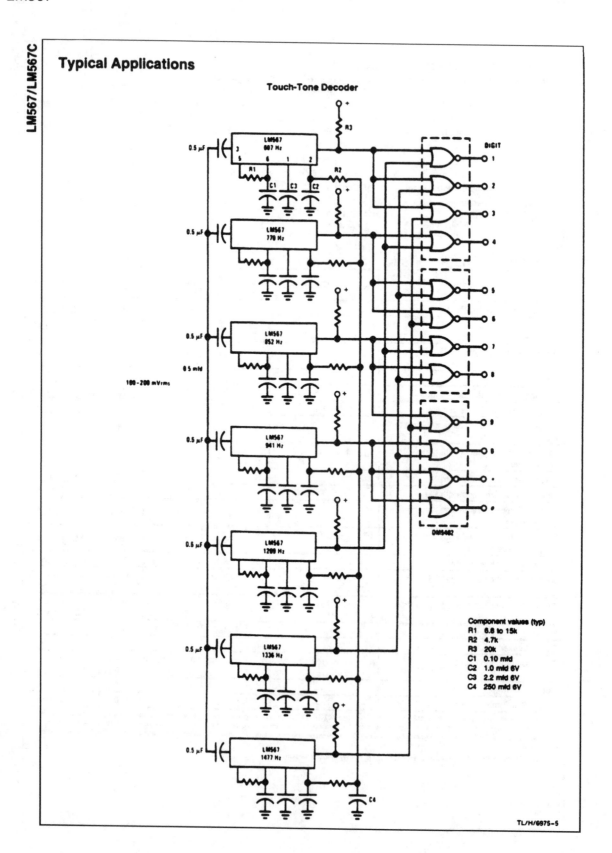

LM567/LM567C

Typical Applications

Touch-Tone Decoder

Typical Applications (Continued)

Oscillator with Quadrature Output

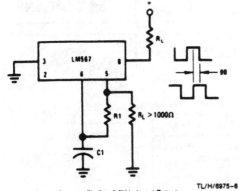

Connect Pin 3 to 2.8V to Invert Output

TL/H/6975-6

Oscillator with Double Frequency Output

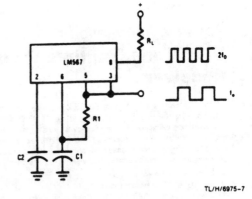

TL/H/6975-7

Precision Oscillator Drive 100 mA Loads

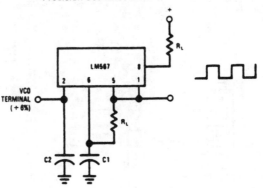

TL/H/6975-8

AC Test Circuit

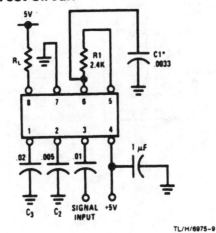

$f_i = 100$ kHz + 5V
*Note: Adjust for $f_o = 100$ kHz.

TL/H/6975-9

Applications Information

The center frequency of the tone decoder is equal to the free running frequency of the VCO. This is given by

$$f_o \approx \frac{1}{1.1\, R_1 C_1}$$

The bandwidth of the filter may be found from the approximation

$$BW = 1070 \sqrt{\frac{V_i}{f_o C_2}} \text{ in \% of } f_o$$

Where:

V_i = Input voltage (volts rms), $V_i \leq 200$ mV

C_2 = Capacitance at Pin 2 (μF)

MC145436

Advance Information

Dual Tone Multiple Frequency Receiver

The MC145436 is a silicon-gate CMOS LSI device containing the filter and decoder for detection of a pair of tones conforming to the DTMF standard with outputs in hexadecimal. Switched capacitor filter technology is used together with digital circuitry for the timing control and output circuits. The MC145436 provides excellent power-line noise and dial tone rejection, and is suitable for applications in central office equipment, PABX, keyphone systems, remote control equipment, and consumer telephony products.

- Single +5 V Power Supply
- Detects All 16 Standard Digits
- Uses Inexpensive 3.579545 MHz Colorburst Crystal
- Provides Guard Time Controls to Improve Speech Immunity
- Output in 4-Bit Hexadecimal Code
- Built-In 60 Hz and Dial Tone Rejection
- Pin Compatible with SSI-204

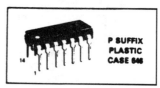

MC145436

P SUFFIX
PLASTIC
CASE 646

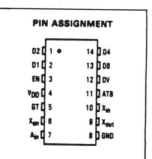

PIN ASSIGNMENT

D2	1	14	D4
D1	2	13	D8
EN	3	12	DV
V_{DD}	4	11	ATB
GT	5	10	X_{in}
X_{en}	6	9	X_{out}
A_{in}	7	8	GND

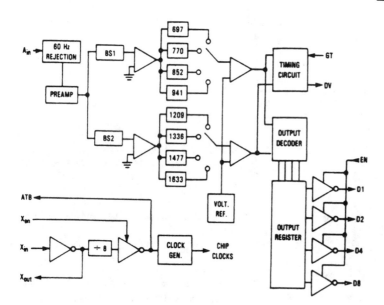

This document contains information on a new product. Specifications and information herein are subject to change without notice.

ABSOLUTE MAXIMUM RATINGS
(Voltages Referenced to GND Unless Otherwise Noted)

Rating	Symbol	Value	Unit
DC Supply Voltage	V_{DD}	-0.5 to $+6.0$	V
Input Voltage, Any Pin Except A_{in}	V_{in}	-0.5 to $V_{DD}+0.5$	V
Input Voltage, A_{in}	V_{in}	$V_{DD}-10$ to $V_{DD}+0.5$	V
DC Current Drain per Pin	I	± 10	mA
Operating Temperature Range	T_A	-40 to $+85$	°C
Storage Temperature Range	T_{stg}	-65 to $+150$	°C

This device contains circuitry to protect the inputs against damage due to high static voltages or electric fields; however, it is advised that normal precautions be taken to avoid applications of any voltage higher than the maximum rated voltages to this high impedance circuit.

For proper operation it is recommended that V_{in} and V_{out} be constrained to the range $V_{SS} \leq (V_{in}$ or $V_{out}) \leq V_{DD}$. Reliability of operation is enhanced if unused inputs are tied to an appropriate logic voltage level (e.g., either V_{SS} or V_{DD}).

ELECTRICAL CHARACTERISTICS
(All Polarities Referenced to $V_{DD}=5.0$ V $\pm 10\%$, $T_A = -40$ to $+85$°C Unless Otherwise Noted)

Parameter		Symbol	Min	Typ	Max	Unit
DC Supply Voltage		V_{DD}	4.5	5	5.5	V
Supply Current ($f_{CLK}=3.58$ MHz)		I_{DD}	—	7	15	mA
Input Current	GT	I_{in}	—	—	200	µA
	EN, X_{in}, X_{en}		—	—	± 1	
Input Voltage Low	EN, GT, X_{en}	V_{IL}	—	—	1.5	V
Input Voltage High	EN, GT, X_{en}	V_{IH}	3.5	—	—	V
High Level Output Current ($V_{OH}=V_{DD}-0.5$ V; Source)	Data, DV	I_{OH}	800	—	—	µA
Low Level Output Current ($V_{OL}=0.4$ V; Sink)	Data, DV	I_{OL}	1.0	—	—	mA
Input Impedance	A_{in}	R_{in}	90	100	—	kΩ
Fanout	ATB	FO	—	—	10	
Input Capacitance	X_{en}, EN	C_{in}	—	6	—	pF

ANALOG CHARACTERISTICS ($V_{DD}=5.0$ V $\pm 10\%$, $T_A = -40$ to $+85$°C)

Parameter	Min	Typ	Max	Unit
Signal Level for Detection (A_{in})	-32	—	-2	dBm
Twist = High Tone/Low Tone	-10	—	10	dB
Frequency Detect Bandwidth (Notes 1 and 2)	$\pm (1.5+2$ Hz)	± 2.5	± 3.5	% f_C
60 Hz Tolerance	—	—	0.8	Vrms
Dial Tone Tolerance (Note 3) (Dial Tone 330 + 440)	—	—	0	dB
Noise Tolerance (Notes 3 and 4)	—	—	-12	dB
Power Supply Noise (Wide Band)	—	—	10	mV p-p
Talk Off (Mitel Tape #CM7290)	—	2	—	Hits

NOTES:
1. f_C is center frequency of bandpass filters.
2. Maximum frequency detect bandwidth of the 1477 Hz filter is $+3.5\%$ to -4%.
3. Referenced to lower amplitude tone.
4. Bandwidth limited (0 to 3.4 kHz) Gaussian noise.

MOTOROLA TELECOMMUNICATIONS DEVICE DATA

MC145436

AC CHARACTERISTICS (V$_{DD}$ = 5.0 V ± 10%, T$_A$ = −40 to +85°C)

Characteristic		Symbol	Min	Typ	Max	Unit
Tone On Time	For Detection	Tone$_{on}$	40	—	—	ms
	For Rejection		—	—	20	
Pause Time	For Detection	Tone$_{off}$	40	—	—	ms
	For Rejection		—	—	20	
Detect Time	GT = 0	t$_{det}$	22	—	40	ms
	GT = 1		32	—	50	
Release Time	GT = 0	t$_{rel}$	28	—	40	ms
	GT = 1		18	—	30	
Data Setup Time		t$_{su}$	7	—	—	µs
Data Hold Time		t$_h$	4.2	4.6	5	ms
Pulse Width	GT	t$_{w(GT)}$	18	—	—	µs
DV Reset Lag Time		t$_{lag(DV)}$	—	—	5	ms
Enable High to Output Data Valid		t$_{EHDV}$	—	200	—	ns
Enable Low to Output High-Z		t$_{ELDZ}$	—	150	—	ns

TIMING

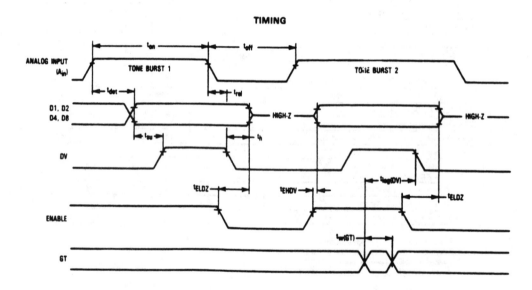

MC145436

PIN DESCRIPTION

D1, D2, D4, D8—DATA OUTPUT

These digital outputs provide the hexadecimal codes corresponding to the detected digit (see Table 1). The digital outputs become valid after a tone pair has been detected, and are cleared when a valid pause is timed. These output pins are high impedance when Enable is at a logic 0.

EN—ENABLE

Outputs D1, D2, D4, D8 are enabled when EN is at a logic 1, and high impedance (disabled) when EN is at a logic 0.

GT—GUARD TIME

The Guard Time control input provides two sets of detected time and release time, both within the allowed ranges of tone on and tone off. A longer tone detect time rejects signals too short to be considered valid. With GT = 1, talk off performance is improved, since it reduces the probability that tones simulated by speech will maintain signal conditions long enough to be accepted. In addition, a shorter release time reduces the probability that a pause simulated by an interruption in speech will be detected as a valid pause. On the other hand, a shorter tone detect time with a long release time would be appropriate for an extremely noisy environment where fast acquisition time and immunity to drop-outs would be required. In general, the tone signal time generated by a telephone is 100 ms, nominal, followed by a pause of about 100 ms. A high-to-low, or low-to-high transition on the GT pin resets the internal logic, and the MC145436 is immediately ready to accept a new tone input.

X_en—OSCILLATOR ENABLE

A logic 1 on X_{en} enables the on-chip crystal oscillator. When using alternate time base from the ATB pin, X_{en} should be tied to GND.

A_in—ANALOG INPUT

This pin accepts the analog input, and is internally biased so that the input signal may be ac coupled. The input may be dc coupled so long as it does not exceed the positive supply. (See Figure 1.)

X_in/X_out—OSCILLATOR IN AND OSCILLATOR OUT

These pins connect to an internal crystal oscillator. In operation, a parallel resonant crystal is connected from X_{in} to X_{out}, as well as a 1 MΩ resistor in parallel with the crystal. When using the alternate clock source from ATB, X_{in} should be tied to V_{DD}.

ATB—ALTERNATE TIME BASE

This pin serves as a frequency reference when more than one MC145436 is used, so that only one crystal is required for multiple MC145436s. In this case, all ATB pins should be tied together as shown in Figure 2. When only one MC145436 is used, this pin should be left unconnected. The output frequency of ATB is 447.4 kHz.

DV—DATA VALID

DV signals a detection by going high after a valid tone pair is sensed and decoded at output pins D1, D2, D4, D8. DV remains high until a loss of the current DTMF signal occurs, or until a transition in GT occurs.

V_DD—POSITIVE POWER SUPPLY

The digital supply pin, which is connected to the positive side of the power supply.

GND—GROUND

Ground return pin is typically connected to the system ground.

Table 1. Hexadecimal Codes

Digit	Output Code			
	D8	D4	D2	D1
1	0	0	0	1
2	0	0	1	0
3	0	0	1	1
4	0	1	0	0
5	0	1	0	1
6	0	1	1	0
7	0	1	1	1
8	1	0	0	0
9	1	0	0	1
0	1	0	1	0
*	1	0	1	1
#	1	1	0	0
A	1	1	0	1
B	1	1	1	0
C	1	1	1	1
D	0	0	0	0

OPERATIONAL INFORMATION

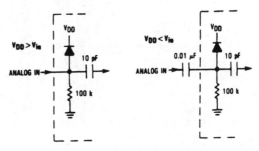

Figure 1. Analog Input

MC145436

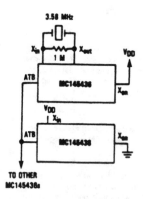

Figure 2. Multiple MC145436s

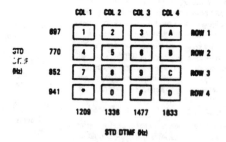

Figure 3. 4 x 4 Keyboard Matrix

Signetics

Linear Products

NE5533/5533A
NE/SA/SE5534/5534A
Dual and Single Low Noise Op Amp

Product Specification

DESCRIPTION

The 5533/5534 are dual and single high-performance low noise operational amplifiers. Compared to other operational amplifiers, such as TL083, they show better noise performance, improved output drive capability and considerably higher small-signal and power bandwidths.

This makes the devices especially suitable for application in high quality and professional audio equipment, in instrumentation and control circuits and telephone channel amplifiers. The op amps are internally compensated for gain equal to, or higher than, three. The frequency response can be optimized with an external compensation capacitor for various applications (unity gain amplifier, capacitive load, slew rate, low overshoot, etc.) If very low noise is of prime importance, it is recommended that the 5533A/5534A version be used which has guaranteed noise specifications.

FEATURES

- Small-signal bandwidth: 10MHz
- Output drive capability: 600Ω, $10V_{RMS}$ at $V_S = \pm 18V$
- Input noise voltage: $4nV/\sqrt{Hz}$
- DC voltage gain: 100000
- AC voltage gain: 6000 at 10kHz
- Power bandwidth: 200kHz
- Slew rate: $13V/\mu s$
- Large supply voltage range: ± 3 to $\pm 20V$

PIN CONFIGURATIONS

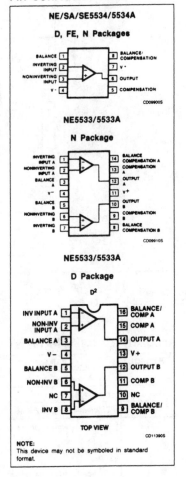

NE/SA/SE5534/5534A

D, FE, N Packages

BALANCE	1	8 BALANCE/COMPENSATION
INVERTING INPUT	2	7 V-
NONINVERTING INPUT	3	6 OUTPUT
V-	4	5 COMPENSATION

CD09900S

NE5533/5533A

N Package

INVERTING INPUT A	1	14 BALANCE/COMPENSATION A
NONINVERTING INPUT A	2	13 COMPENSATION A
BALANCE A	3	12 OUTPUT A
V-	4	11 V+
BALANCE B	5	10 OUTPUT B
NONINVERTING B	6	9 COMPENSATION B
INVERTING B	7	8 BALANCE/COMPENSATION B

CD09910S

NE5533/5533A

D Package

D²

INV INPUT A	1	16 BALANCE/COMP A
NON-INV INPUT A	2	15 COMP A
BALANCE A	3	14 OUTPUT A
V-	4	13 V+
BALANCE B	5	12 OUTPUT B
NON-INV B	6	11 COMP B
NC	7	10 NC
INV B	8	9 BALANCE/COMP B

TOP VIEW

CD113905

NOTE:
This device may not be symboled in standard format.

ORDERING INFORMATION

DESCRIPTION	TEMPERATURE RANGE	ORDER CODE
14-Pin Plastic DIP	0 to +70°C	NE5533N
16-Pin Plastic SO package	0 to +70°C	NE5533AD
14-Pin Plastic DIP	0 to +70°C	NE5533AN
16-Pin Plastic SO package	0 to +70°C	NE5533D
8-Pin Plastic SO package	0 to +70°C	NE5534D
8-Pin Hermetic Cerdip	0 to +70°C	NE5534FE
8-Pin Plastic DIP	0 to +70°C	NE5534N
8-Pin Plastic SO package	0 to +70°C	NE5534AD
8-Pin Hermetic Cerdip	0 to +70°C	NE5534AFE
8-Pin Plastic DIP	0 to +70°C	NE5534AN
8-Pin Plastic DIP	−40°C to +85°C	SA5534N
8-Pin Plastic DIP	−40°C to +85°C	SA5534AN
8-Pin Hermetic Cerdip	−55°C to +125°C	SE5534AFE
8-Pin Plastic DIP	−55°C to +125°C	SE5534N
8-Pin Hermetic Cerdip	−55°C to +125°C	SE5534AFE
8-Pin Plastic DIP	−55°C to +125°C	SE5534AN

Dual and Single Low
Noise Op Amp

EQUIVALENT SCHEMATIC

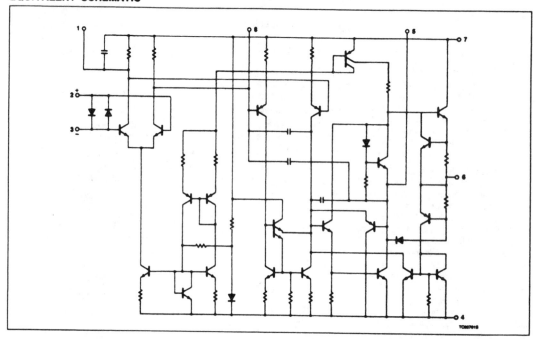

ABSOLUTE MAXIMUM RATINGS

SYMBOL	PARAMETER	RATING	UNIT
V_S	Supply voltage	± 22	V
V_{IN}	Input voltage	± V supply	V
V_{DIFF}	Differential input voltage[1]	± 0.5	V
T_A	Operating temperature range SE SA NE	 −55 to +125 −40 to +85 0 to +70	 °C °C °C
T_{STG}	Storage temperature range	−65 to +150	°C
T_J	Junction temperature	150	°C
P_D	Power dissipation at 25°C[2] 5533N, 5534N, 5534FE	 800	 mW
	Output short-circuit duration[3]	Indefinite	
T_{SOLD}	Lead soldering temperature (10sec max)	300	°C

NOTES:
1. Diodes protect the inputs against over voltage. Therefore, unless current-limiting resistors are used, large currents will flow if the differential input voltage exceeds 0.6V. Maximum current should be limited to ± 10mA.
2. For operation at elevated temperature, derate packages based on the following junction-to-ambient thermal resistance:
   8-pin ceramic DIP 150°C/W
   8-pin plastic DIP 105°C/W
   8-pin plastic SO 160°C/W
   14-pin ceramic DIP 100°C/W
   14-pin plastic DIP 80°C/W
   16-pin plastic SO 90°C/W
3. Output may be shorted to ground at V_S = ± 15V, T_A = 25°C. Temperature and/or supply voltages must be limited to ensure dissipation rating is not exceeded.

Dual and Single Low Noise Op Amp

NE5533/5533A
NE/SA/SE5534/5534A

DC ELECTRICAL CHARACTERISTICS $T_A = 25°C$, $V_S = \pm 15V$, unless otherwise specified. [1, 2]

SYMBOL	PARAMETER	TEST CONDITIONS	SE5534/5534A			NE5533/5533A/5534/5534A			UNIT
			Min	Typ	Max	Min	Typ	Max	
V_{OS}	Offset voltage	Over temperature		0.5	2		0.5	4	mV
					3			5	mV
$\Delta V_{OS}/\Delta T$				5			5		$\mu V/°C$
I_{OS}	Offset current	Over temperature		10	200		20	300	nA
					500			400	nA
$\Delta I_{OS}/\Delta T$				200			200		$pA/°C$
I_B	Input current	Over temperature		400	800		500	1500	nA
					1500			2000	nA
$\Delta I_B/\Delta T$				5			5		$nA/°C$
I_{CC}	Supply current per op amp	Over temperature		4	6.5		4	8	mA
					9			10	mA
V_{CM}	Common mode input range		± 12	± 13		± 12	± 13		V
CMRR	Common mode rejection ratio		80	100		70	100		dB
PSRR	Power supply rejection ratio			10	50		10	100	$\mu V/V$
A_{VOL}	Large-signal voltage gain	$R_L \geqslant 600\Omega$, $V_O = \pm 10V$	50	100		25	100		V/mV
		Over temperature	25			15			V/mV
V_{OUT}	Output swing 5534 only	$R_L \geqslant 600\Omega$	± 12	± 13		± 12	± 13		V
		Over temperature	± 10	± 12		± 10	± 12		V
		$R_L \geqslant 600\Omega$, $V_S = \pm 18V$	± 15	± 16		± 15	± 16		V
		$R_L \geqslant 2k\Omega$	± 13	± 13.5		± 13	± 13.5		V
		Over temperature	± 12	± 12.5		± 12	± 12.5		V
R_{IN}	Input resistance		50	100		30	100		$k\Omega$
I_{SC}	Output short circuit current			38			38		mA

NOTES:
1. For NE5533/5533A/5534/5534A, $T_{MIN} = 0°C$, $T_{MAX} = 70°C$.
2. For SE5534/5534A, $T_{MIN} = -55°C$, $T_{MAX} = +125°C$.

AC ELECTRICAL CHARACTERISTICS $T_A = 25°C$, $V_S = 15V$, unless otherwise specified.

SYMBOL	PARAMETER	TEST CONDITIONS	SE5534/5534A			NE5533/5533A/5534/5534A			UNIT
			Min	Typ	Max	Min	Typ	Max	
R_{OUT}	Output resistance	$A_V = 30dB$ closed-loop $f = 10kHz$, $R_L = 600\Omega$, $C_C = 22pF$		0.3			0.3		Ω
	Transient response	Voltage-follower, $V_{IN} = 50mV$ $R_L = 600\Omega$, $C_C = 22pF$, $C_L = 100pF$							
t_R	Rise time			20			20		ns
	Overshoot			20			20		%
	Transient response	$V_{IN} = 50mV$, $R_L = 600\Omega$ $C_C = 47pF$, $C_L = 500pF$							
t_R	Rise time			50			50		ns
	Overshoot			35			35		%
A_V	Gain	$f = 10kHz$, $C_C = 0$		6			6		V/mV
		$f = 10kHz$, $C_C = 22pF$		2.2			2.2		V/mV
BW	Gain bandwidth product	$C_C = 22pF$, $C_L = 100pF$		10			10		mHz
SR	Slew rate	$C_C = 0$		13			13		V/μs
		$C_C = 22pF$		6			6		V/μs
	Power bandwidth	$V_{OUT} = \pm 10V$, $C_C = 0$		200			200		kHz
		$V_{OUT} = \pm 10V$, $C_C = 22pF$		95			95		kHz
		$V_{OUT} = \pm 14V$, $R_L = 600\Omega$ $C_C = 22pF$, $V_{CC} = \pm 18V$		70			70		kHz

ELECTRICAL CHARACTERISTICS $T_A = 25°C$, $V_S = 15V$, unless otherwise specified.

SYMBOL	PARAMETER	TEST CONDITIONS	5533/5534			5533A/5534A			UNIT
			Min	Typ	Max	Min	Typ	Max	
V_{NOISE}	Input noise voltage	$f_O = 30Hz$		7			5.5	7	nV/$\sqrt{Hz}$
		$f_O = 1kHz$		4			3.5	4.5	nV/$\sqrt{Hz}$
I_{NOISE}	Input noise current	$f_O = 30Hz$		2.5			1.5		pA/$\sqrt{Hz}$
		$f_O = 1kHz$		0.6			0.4		pA/$\sqrt{Hz}$
	Broadband noise figure	$f = 10Hz - 20kHz$, $R_S = 5k\Omega$					0.9		dB
	Channel separation	$f = 1kHz$, $R_S = 5k\Omega$		110			110		dB

Dual and Single Low
Noise Op Amp

TYPICAL PERFORMANCE CHARACTERISTICS

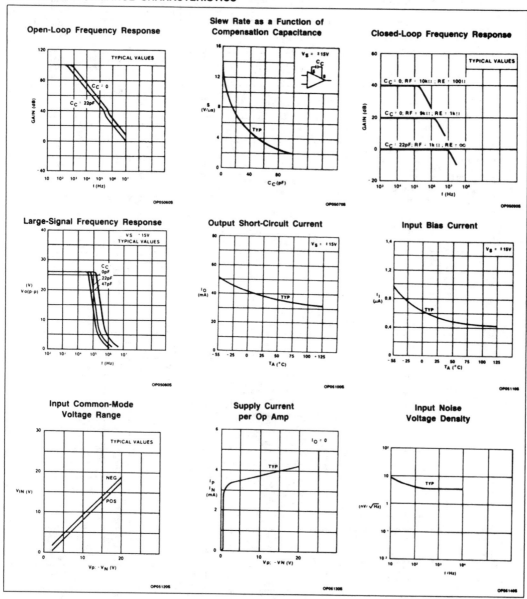

Dual and Single Low
Noise Op Amp

TYPICAL PERFORMANCE CHARACTERISTICS (Continued)

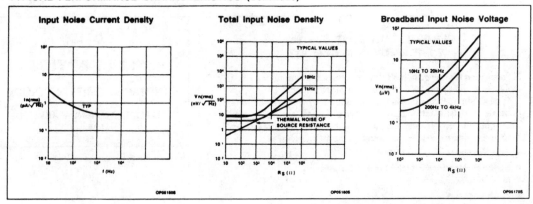

TEST LOAD CIRCUITS

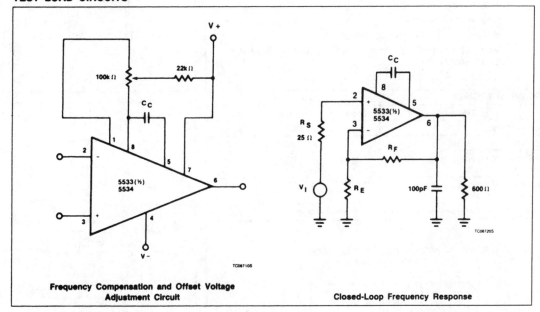

Frequency Compensation and Offset Voltage Adjustment Circuit

Closed-Loop Frequency Response

 **MOTOROLA**

FIBER OPTIC LOW COST SYSTEM
FLCS INFRARED-EMITTING DIODE

... designed for low cost, medium frequency, short distance Fiber Optic Systems using 1000 micron core plastic fiber.

Typical applications include: high isolation interconnects, disposable medical electronics, consumer products, and microprocessor controlled systems such as coin operated machines, copy machines, electronic games, industrial clothes dryers, etc.

- Fast Response — > 10 MHz
- Spectral Response Matched to FLCS Detectors: MFOD71, 72, 73
- FLCS Package
  - Low Cost
  - Includes Connector
  - Simple Fiber Termination and Connection
  - Easy Board Mounting
  - Molded Lens for Efficient Coupling
  - Mates with 1000 Micron Core Plastic Fiber (DuPont OE1040, Eska SH4001)

FLCS LINE
FIBER OPTICS
INFRARED-EMITTING DIODE
GALLIUM ALUMINUM ARSENIDE
820 nm

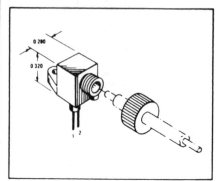

MAXIMUM RATINGS

Rating	Symbol	Value	Unit
Reverse Voltage	V_R	6.0	Volts
Forward Current	I_F	150	mA
Total Power Dissipation @ T_A = 25°C Derate above 25°C	$P_D(1)$	150 2.5	mW mW/°C
Operating and Storage Junction Temperature Range	T_J, T_{stg}	– 40 to + 85	°C

(1) Measured with the device soldered into a typical printed circuit board.

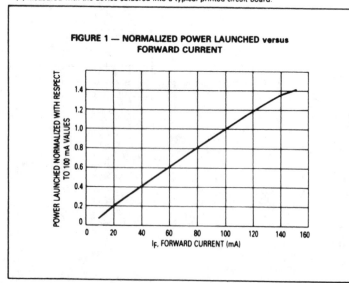

FIGURE 1 — NORMALIZED POWER LAUNCHED versus FORWARD CURRENT

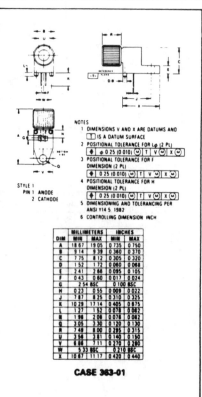

NOTES
1. DIMENSIONS V AND X ARE DATUMS AND ⊥ IS A DATUM SURFACE
2. POSITIONAL TOLERANCE FOR Lφ (2 PL)
3. POSITIONAL TOLERANCE FOR F DIMENSION (2 PL)
4. POSITIONAL TOLERANCE FOR H DIMENSION (2 PL)
5. DIMENSIONING AND TOLERANCING PER ANSI Y14.5, 1982
6. CONTROLLING DIMENSION INCH

STYLE 1
PIN 1 ANODE
    2 CATHODE

DIM	MILLIMETERS MIN	MILLIMETERS MAX	INCHES MIN	INCHES MAX
A	18.67	19.05	0.735	0.750
B	9.14	9.39	0.360	0.370
C	7.75	8.12	0.305	0.320
D	1.52	1.72	0.060	0.068
E	2.41	2.66	0.095	0.105
F	0.43	0.60	0.017	0.024
G	2.54 BSC		0.100 BSC	
H	0.23	0.55	0.009	0.022
J	7.87	8.25	0.310	0.325
K	10.29	17.14	0.405	0.675
L	1.27	1.52	0.078	0.082
N	1.96	2.08	0.078	0.082
Q	3.05	3.30	0.120	0.130
R	7.49	8.00	0.295	0.315
U	3.56	3.81	0.140	0.150
V	6.86	7.11	0.270	0.280
W	5.33 BSC		0.210 BSC	
X	10.67	11.17	0.420	0.440

CASE 363-01

MFOE71

ELECTRICAL CHARACTERISTICS (T$_A$ = 25°C unless otherwise noted)

Characteristic	Fig. No.	Symbol	Min	Typ	Max	Unit
Reverse Breakdown Voltage (I$_R$ = 100 μA)	—	V$_{(BR)R}$	2.0	4.0	—	Volts
Forward Voltage (I$_F$ = 100 mA)	—	V$_F$	—	1.5	2.0	Volts

OPTICAL CHARACTERISTICS (T$_A$ = 25°C unless otherwise noted)

Characteristic	Fig. No.	Symbol	Min	Typ	Max	Unit
Power Launched	2, 4	P$_L$	110	165	—	μW
Optical Rise and Fall Time	3	t$_r$, t$_f$	—	25	35	ns
Peak Wavelength (I$_F$ = 100 mA)	1	λ$_P$	—	820	—	nm

For simple fiber termination instructions, see the MFOD71, 72 and 73 data sheet.

FIGURE 2 — POWER LAUNCHED TEST SET

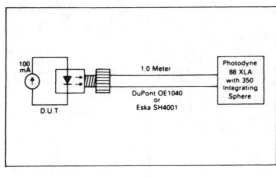

FIGURE 3 — POWER LAUNCHED (P$_L$) versus FIBER LENGTH

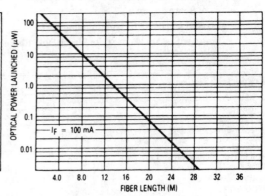

FIGURE 4 — OPTICAL RISE AND FALL TIME TEST SET (10%–90%)

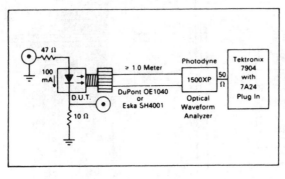

FIGURE 5 — TYPICAL SPECTRAL OUTPUT versus WAVELENGTH

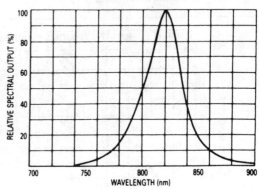

**MOTOROLA**

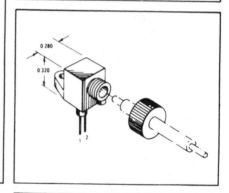

MFOD71
MFOD72
MFOD73

FIBER OPTIC LOW COST SYSTEM
FLCS DETECTORS

... designed for low cost, short distance Fiber Optic Systems using 1000 micron core plastic fiber.

Typical applications include: high isolation interconnects, disposable medical electronics, consumer products, and microprocessor controlled systems such as coin operated machines, copy machines, electronic games, industrial clothes dryers, etc.

- Fast PIN Photodiode: Response Time <5.0 ns
- Standard Phototransistor
- High Sensitivity Photodarlington
- Spectral Response Matched to MFOE71 LED
- Annular Passivated Structure for Stability and Reliability
- FLCS Package
  - Includes Connector
  - Simple Fiber Termination and Connection (Figure 4)
  - Easy Board Mounting
  - Molded Lens for Efficient Coupling
  - Mates with 1000 Micron Core Plastic Fiber (DuPont OE1040, Eska SH4001)

FLCS LINE

FIBER OPTICS
DETECTORS

MAXIMUM RATINGS (T$_A$ = 25°C unless otherwise noted)

Rating		Symbol	Value	Unit
Reverse Voltage	MFOD71	V$_R$	100	Volts
Collector-Emitter Voltage	MFOD72 MFOD73	V$_{CEO}$	30 60	Volts
Total Power Dissipation @ T$_A$ = 25°C MFOD71 Derate above 25°C MFOD72/73 Derate above 25°C		P$_D$	100 1.67 150 2.5	mW mW/°C mW mW/°C
Operating and Storage Junction Temperature Range		T$_J$, T$_{stg}$	−40 to +85	°C

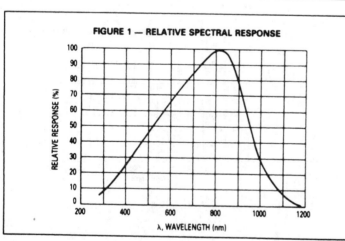

FIGURE 1 — RELATIVE SPECTRAL RESPONSE

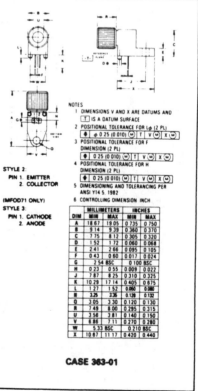

STYLE 2:
PIN 1. EMITTER
2. COLLECTOR

(MFOD71 ONLY)
STYLE 3:
PIN 1. CATHODE
2. ANODE

NOTES
1. DIMENSIONS V AND X ARE DATUMS AND [T] IS A DATUM SURFACE
2. POSITIONAL TOLERANCE FOR L⌀ (2 PL)
3. POSITIONAL TOLERANCE FOR F DIMENSION (2 PL)
4. POSITIONAL TOLERANCE FOR H DIMENSION (2 PL)
5. DIMENSIONING AND TOLERANCING PER ANSI Y14.5, 1982
6. CONTROLLING DIMENSION INCH

DIM	MILLIMETERS		INCHES	
	MIN	MAX	MIN	MAX
A	18.67	19.05	0.735	0.750
B	9.14	9.39	0.360	0.370
C	7.75	8.12	0.305	0.320
D	1.52	1.72	0.060	0.068
E	2.41	2.66	0.095	0.105
F	0.43	0.60	0.017	0.024
G	2.54 BSC		0.100 BSC	
H	0.23	0.55	0.009	0.022
J	7.87	8.25	0.310	0.325
K	10.29	17.14	0.405	0.675
L	1.27	1.52	0.050	0.060
N	3.25	3.35	0.120	0.132
Q	3.05	3.30	0.120	0.130
R	7.49	8.00	0.295	0.315
U	3.56	3.81	0.140	0.150
V	6.86	7.11	0.270	0.280
W	5.33 BSC		0.210 BSC	
X	10.67	11.17	0.420	0.440

CASE 363-01

MFOD71, MFOD72, MFOD73

STATIC ELECTRICAL CHARACTERISTICS (T_A = 25°C unless otherwise noted)

Characteristic	Symbol	Min	Typ	Max	Unit
Dark Current (V_R = 20 V, R_L = 1.0 MΩ) T_A = 25°C T_A = 85°C	I_D	— 	0.06 10	10 —	nA
Reverse Breakdown Voltage (I_R = 10 μA)	$V_{(BR)R}$	50	100	—	Volts
Forward Voltage (I_F = 50 mA)	V_F	—	—	1.1	Volts
Series Resistance (I_F = 50 mA)	R_s	—	8.0	—	ohms
Total Capacitance (V_R = 20 V; f = 1.0 MHz)	C_T	—	3.0	—	pF

OPTICAL CHARACTERISTICS (T_A = 25°C)

Responsivity (V_R = 5.0 V, Figure 2)	R	0.15	0.2	—	μA/μW
Response Time (V_R = 5.0 V, R_L = 50 Ω)	$t_{(resp)}$	—	5.0	—	ns

MFOD72/MFOD73

STATIC ELECTRICAL CHARACTERISTICS

Collector Dark Current (V_{CE} = 10 V)		I_D	—	—	100	nA
Collector-Emitter Breakdown Voltage (I_C = 10 mA)	MFOD72 MFOD73	$V_{(BR)CEO}$	30 60	— —	— —	Volts

OPTICAL CHARACTERISTICS (T_A = 25°C unless otherwise noted)

Responsivity (V_{CC} = 5.0 V, Figure 2)	MFOD72 MFOD73	R	80 1,000	125 1,500	— —	μA/μW
Saturation Voltage (λ = 820 nm, V_{CC} = 5.0 V) (P_{in} = 10 μW, I_C = 1.0 mA) (P_{in} = 1.0 μW, I_C = 2.0 mA)	 MFOD72 MFOD73	$V_{CE(sat)}$		 0.25 0.75	 0.4 1.0	Volts
Turn-On Time	R_L = 2.4 kΩ, P_{in} = 10 μW,	MFOD72	t_{on}	—	10	—	μs
Turn-Off Time	λ = 820 nm, V_{CC} = 5.0 V		t_{off}	—	60	—	μs
Turn-On Time	R_L = 100 Ω, P_{in} = 1.0 μW,	MFOD73	t_{on}	—	125	—	μs
Turn-Off Time	λ = 820 nm, V_{CC} = 5.0 V		t_{off}	—	150	—	μs

TYPICAL COUPLED CHARACTERISTICS

FIGURE 2 — RESPONSIVITY TEST CONFIGURATION

FIGURE 3 — DETECTOR CURRENT versus FIBER LENGTH

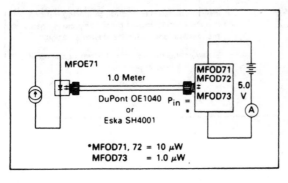

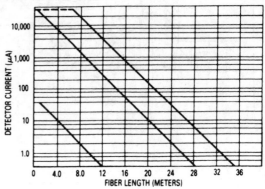

MFOD71, MFOD72, MFOD73

FLCS WORKING DISTANCES

The system length achieved with a FLCS emitter and detector using the 1000 micron core fiber optic cable depends upon the forward current through the LED and the Responsivity of the detector chosen. Each emitter/detector combination will work at any cable length up to the maximum length shown.

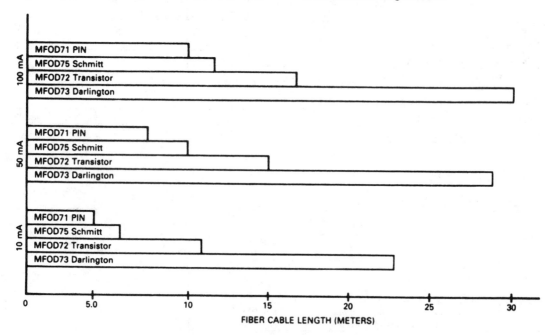

FIBER CABLE LENGTH (METERS)

FIGURE 4 — FO CABLE TERMINATION AND ASSEMBLY

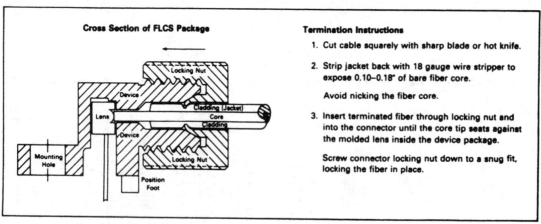

Cross Section of FLCS Package

Termination Instructions

1. Cut cable squarely with sharp blade or hot knife.

2. Strip jacket back with 18 gauge wire stripper to expose 0.10–0.18" of bare fiber core.

   Avoid nicking the fiber core.

3. Insert terminated fiber through locking nut and into the connector until the core tip seats against the molded lens inside the device package.

   Screw connector locking nut down to a snug fit, locking the fiber in place.

MFOD71, MFOD72, MFOD73

Input Signal Conditioning

The following circuits are suggested to provide the desired forward current through the emitter.

TTL TRANSMITTERS

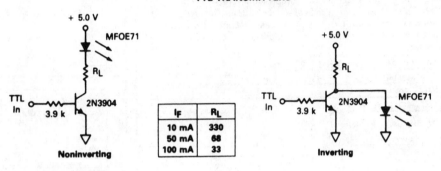

I_F	R_L
10 mA	330
50 mA	68
100 mA	33

Noninverting

Inverting

Output Signal Conditioning

The following circuits are suggested to take the FLCS detector output and condition it to drive TTL with an acceptable bit error rate.

TTL RECEIVERS

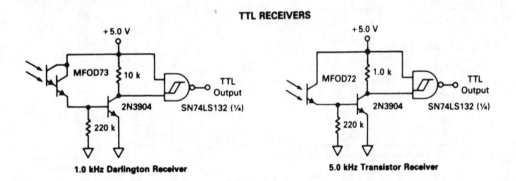

1.0 kHz Darlington Receiver

5.0 kHz Transistor Receiver

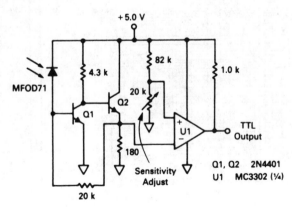

1.0 MHz PIN Receiver

 **MOTOROLA**

INFRARED-EMITTING DIODE

... designed for infrared remote control applications for use with the MRD701 phototransistor in optical slotted coupler / interrupter module applications, and for industrial processing and control applications such as light modulators, shaft or position encoders, end of tape detectors.

- Continuous P_O = 2.5 mW (Typ) @ I_F = 50 mA
- Low Cost, Miniature, Clear Plastic Package
- Package Designed for Accurate Positioning
- Lens Molded into Package
- Narrow Spatial Radiation Pattern

INFRARED-EMITTING DIODE

PN GALLIUM ARSENIDE

940 nm

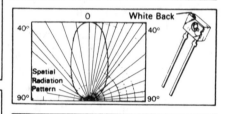

MAXIMUM RATINGS

Rating	Symbol	Value	Unit
Reverse Voltage	V_R	6.0	Volts
Forward Current — Continuous	I_F	50	mA
Total Power Dissipation @ T_A = 25°C Derate above 25°C	$P_D(1)$	150 2.0	mW mW/°C
Operating and Storage Junction Temperature Range	T_J, T_{stg}	-40 to +100	°C

THERMAL CHARACTERISTICS

Characteristic	Symbol	Max	Unit
Thermal Resistance Junction to Ambient	$R_{\theta JA}(1)$	350	°C/W

(1)Measured with the device soldered into a typical printed circuit board.

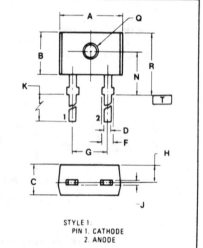

STYLE 1:
PIN 1. CATHODE
2. ANODE

NOTES:
1. DIMENSIONS A, B AND C ARE DATUMS.
2. POSITIONAL TOLERANCE FOR D DIMENSION:
⌖ .25 (0.010) Ⓜ T A Ⓜ C Ⓜ
3. POSITIONAL TOLERANCE FOR Q DIAMETER:
⌖ ⌀.25 (0.010) Ⓜ A Ⓜ B Ⓜ
4. ⊥ IS SEATING PLANE.
5. DIMENSIONING AND TOLERANCING PER ANSI Y14.5, 1973.

DIM	MILLIMETERS		INCHES	
	MIN	MAX	MIN	MAX
A	3.43	4.60	0.135	0.185
B	2.79	3.30	0.110	0.130
C	2.03	3.18	0.080	0.125
D	0.43	0.60	0.017	0.024
F	1.14	1.40	0.045	0.055
G	2.54 BSC		0.100 BSC	
H	1.52 BSC		0.060 BSC	
J	0.23	0.56	0.009	0.022
K	12.83	19.05	0.505	0.750
N	3.05	3.30	0.120	0.130
Q	0.76	1.52	0.030	0.060
R	3.81	4.60	0.150	0.185

CASE 349-01

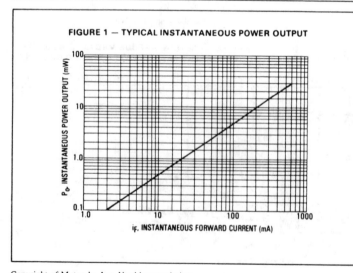

FIGURE 1 — TYPICAL INSTANTANEOUS POWER OUTPUT

MLED71

ELECTRICAL CHARACTERISTICS (T_A = 25°C unless otherwise noted)

Characteristic	Fig. No.	Symbol	Min	Typ	Max	Unit
Reverse Leakage Current (V_R = 6.0 V, R_L = 1.0 Megohm)	—	I_R	—	50	—	nA
Reverse Breakdown Voltage (I_R = 100 μA)	—	$V_{(BR)R}$	6.0	—	—	Volts
Instantaneous Forward Voltage (I_F = 50 mA)	2	v_F	—	1.3	1.8	Volts
Total Capacitance (V_R = 0 V, f = 1.0 MHz)	—	C_T	—	25	—	pF

OPTICAL CHARACTERISTICS (T_A = 25°C unless otherwise noted)

Characteristic	Fig. No.	Symbol	Min	Typ	Max	Unit
Continuous Power Out, Note 1 (I_F = 50 mA)	—	P_O	2.0	2.5	—	mW
Instantaneous Power Out, Note 1 (I_F = 100 mA, 100 pps –100 μs pw)	1	P_O	—	5.0	—	mW
Radiant Intensity (I_F = 100 mA)		I_O	—	3.5	—	mW/st
Optical Turn-On Turn-Off Time	—	t_{on}, t_{off}	—	1.0	—	μs

Note 1. Power out measurements were made using a SPECTRA-1000 photometer with an integrated sphere.

TYPICAL ELECTRICAL CHARACTERISTICS

FIGURE 2 — FORWARD CHARACTERISTICS

FIGURE 3 — POWER OUTPUT versus JUNCTION TEMPERATURE

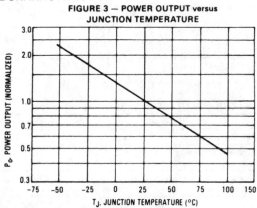

TYPICAL COUPLED CHARACTERISTICS USING MLED71 EMITTER AND MRD701 PHOTOTRANSISTOR DETECTOR

FIGURE 4 — CONTINUOUS MRD701 COLLECTOR LIGHT CURRENT versus DISTANCE FROM MLED71

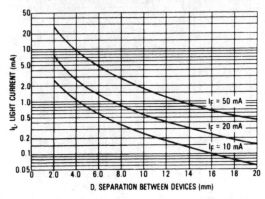

FIGURE 5 — INSTANTANEOUS MRD701 COLLECTOR LIGHT CURRENT versus MLED71 FORWARD CURRENT

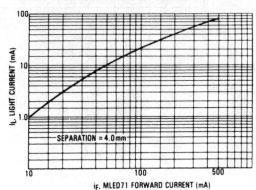

 **MOTOROLA**

MRD721

PIN SILICON PHOTO DIODE

PIN PHOTO DIODE

100 VOLT

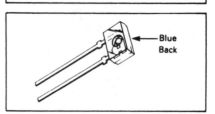
Blue Back

. . . designed for application in laser detection, light demodulation, detection of visible and near infrared light-emitting diodes, shaft or position encoders, switching and logic circuits, or any design requiring radiation sensitivity, ultra high-speed, and stable characteristics.

- Ultra Fast Response — (<1.0 ns Typ)
- Sensitive Throughout Visible and Near Infrared Spectral Range for Wide Application
- Annular Passivated Structure for Stability and Reliability
- Economical, Low Profile, Miniature Plastic Package
- Lense Molded Into Package
- Designed for Automatic Handling and Accurate Positioning

MAXIMUM RATINGS (T_A = 25°C unless otherwise noted)

Rating	Symbol	Value	Unit
Reverse Voltage	V_R	100	Volts
Total Power Dissipation @ T_A = 25°C Derate above 25°C	P_D	100 1.33	mW mW/°C
Operating and Storage Junction Temperature Range	T_J, T_{stg}	−40 to +100	°C

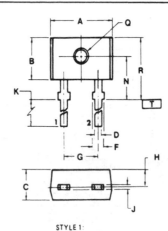

STYLE 1:
PIN 1. CATHODE
2. ANODE

NOTES:
1. DIMENSIONS A, B AND C ARE DATUMS.
2. POSITIONAL TOLERANCE FOR D DIMENSION:
3. POSITIONAL TOLERANCE FOR Q DIAMETER:
4. IS SEATING PLANE.
5. DIMENSIONING AND TOLERANCING PER ANSI Y14.5, 1973.

DIM	MILLIMETERS		INCHES	
	MIN	MAX	MIN	MAX
A	3.43	4.60	0.135	0.185
B	2.79	3.30	0.110	0.130
C	2.03	3.18	0.080	0.125
D	0.43	0.60	0.017	0.024
F	1.14	1.40	0.045	0.055
G	2.54 BSC		0.100 BSC	
H	1.52 BSC		0.060 BSC	
J	0.23	0.56	0.009	0.022
K	12.83	19.05	0.505	0.750
N	3.05	3.30	0.120	0.130
Q	0.76	1.52	0.030	0.060
R	3.81	4.60	0.150	0.185

CASE 349-01

FIGURE 1 — TYPICAL OPERATING CIRCUIT

+V

H

V_{signal}

50 Ω

MRD721

STATIC ELECTRICAL CHARACTERISTICS (T_A = 25°C unless otherwise noted)

Characteristic	Fig. No.	Symbol	Min	Typ	Max	Unit
Dark Current (V_R = 20 V, R_L = 1.0 MΩ; Note 2) T_A = 25°C T_A = 100°C	3 and 4	I_D		— —	0.06 14	10 —	nA
Reverse Breakdown Voltage (I_R = 10 μA)	—	$V_{(BR)R}$	100	200	—	Volts
Forward Voltage (I_F = 50 mA)	—	V_F	—	—	1.1	Volts
Series Resistance (I_F = 50 mA)	—	R_S	—	8.0	—	ohms
Total Capacitance (V_R = 20 V; f = 1.0 MHz)	5	C_T	—	3.0	—	pF

OPTICAL CHARACTERISTICS (T_A = 25°C)

Characteristic	Fig. No.	Symbol	Min	Typ	Max	Unit
Light Current (V_R = 20 V, Note 1)	2	I_L	1.5	4.0	—	μA
Sensitivity (V_R = 20 V, Note 3)	— —	$S(\lambda = 0.8\ \mu m)$ $S(\lambda = 0.94\ \mu m)$	— —	5.0 1.2	— —	$\mu A/mW/cm^2$
Response Time (V_R = 20 V, R_L = 50 Ω)	—	$t_{(resp)}$	—	1.0	—	ns
Wavelength of Peak Spectral Response	6	λ_S	—	0.8	—	μm

NOTES: 1. Radiation Flux Density (H) equal to 5.0 mW/cm^2 emitted from a tungsten source at a color temperature of 2870 K.
     2. Measured under dark conditions. (H ≈ 0)
     3. Radiation Flux Density (H) equal to 0.5 mW/cm^2

MRD721

FIGURE 2 — IRRADIATED VOLTAGE — CURRENT CHARACTERISTIC

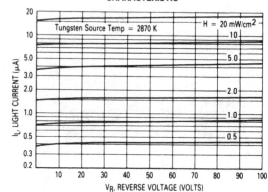

FIGURE 3 — DARK CURRENT versus TEMPERATURE

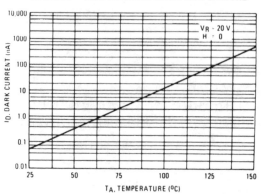

FIGURE 4 — DARK CURRENT versus REVERSE VOLTAGE

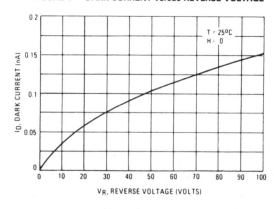

FIGURE 5 — CAPACITANCE versus VOLTAGE

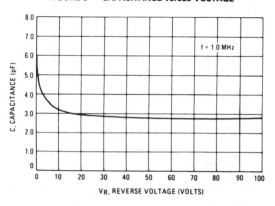

FIGURE 6 — RELATIVE SPECTRAL RESPONSE

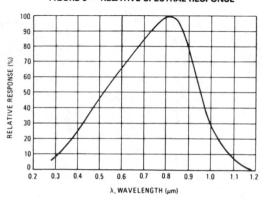

Copyright of Motorola, Inc. Used by permission.

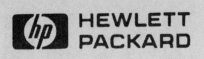

HEWLETT PACKARD

T-1 3/4 (5mm) LOW PROFILE SOLID STATE LAMPS

RED ● HLMP-3200 SERIES
HIGH EFFICIENCY RED ● HLMP-3350 SERIES
YELLOW ● HLMP-3450 SERIES
HIGH PERFORMANCE GREEN ● HLMP-3550 SERIES

TECHNICAL DATA   JANUARY 1984

Features

- **HIGH INTENSITY**
- **LOW PROFILE: 5.8mm (0.23 in) NOMINAL**
- **T-1¾ DIAMETER PACKAGE**
- **LIGHT OUTPUT CATEGORIES**
- **DIFFUSED AND NON-DIFFUSED TYPES**
- **GENERAL PURPOSE LEADS**
- **IC COMPATIBLE/LOW CURRENT REQUIREMENTS**
- **RELIABLE AND RUGGED**

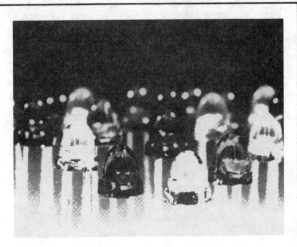

Description

The HLMP-3200 Series are Gallium Arsenide Phosphide Red Light Emitting Diodes with a red diffused lens.

The HLMP-3350 Series are Gallium Arsenide Phosphide on Gallium Phosphide High Efficiency Red Light Emitting Diodes.

The HLMP-3450 Series are Gallium Arsenide Phosphide on Gallium Phosphide Yellow Light Emitting Diodes.

The HLMP-3550 Series are Gallium Phosphide Green Light Emitting Diodes.

The Low Profile T-1¾ package provides space savings and is excellent for backlighting applications.

Package Dimensions

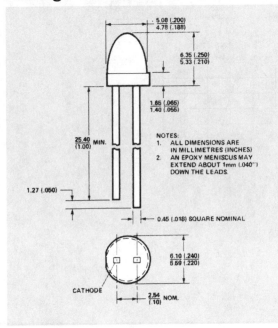

NOTES:
1. ALL DIMENSIONS ARE IN MILLIMETRES (INCHES).
2. AN EPOXY MENISCUS MAY EXTEND ABOUT 1mm (.040'') DOWN THE LEADS.

Part Number HLMP-	Application	Lens	Color
3200	Indicator — General Purpose	Tinted Diffused Wide Angle	Red
3201	Indicator — High Brightness		
3350	Indicator — General Purpose	Tinted Diffused Wide Angle	High Efficiency Red
3351	Indicator — High Brightness		
3365	General Purpose Point Source	Tinted Non-diffused Narrow Angle	
3366	High Brightness Annunciator		
3450	Indicator — General Purpose	Tinted Diffused Wide Angle	Yellow
3451	Indicator — High Brightness		
3465	General Purpose Point Source	Tinted Non-diffused Narrow Angle	
3466	High Brightness Annunciator		
3553	Indicator — General Purpose	Tinted Diffused Wide Angle	Green
3554	Indicator — High Brightness		
3567	General Purpose Point Source	Tinted Non-diffused Narrow Angle	
3568	High Brightness Annunciator		

Absolute Maximum Ratings at $T_A = 25°C$

Parameter	3200 Series	3350 Series	3450 Series	3550 Series	Units
Peak Forward Current	1000	90	60	90	mA
Average Forward Current[1]	50	25	20	25	mA
DC Current[2]	50	30	20	30	mA
Power Dissipation[3]	100	135	85	135	mW
Operating Temperature Range	−55 to +100	−55 to +100	−55 to +100	−40 to +100	C°
Storage Temperature Range				−55 to +100	
Lead Soldering Temperature [1.6 mm (0.063 in.) from body]	260°C for 5 seconds				

NOTES:

1. See Figure 5 (Red), 10 (High Efficiency Red), 15 (Yellow) or 20 (Green) to establish pulsed operating conditions.
2. For High Efficiency Red and Green Series derate linearly from 50°C at 0.5 mA/°C. For Red and Yellow Series derate linearly from 50°C at 0.2 mA/°C.
3. For High Efficiency Red and Green Series derate power linearly from 25°C at 1.8 mW/°C. For Red and Yellow Series derate power linearly from 50°C at 1.6 mW/°C.

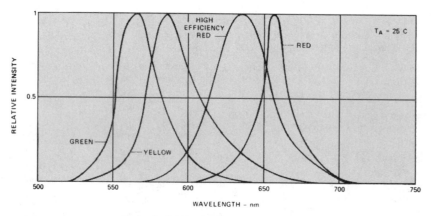

Figure 1. Relative Intensity versus Wavelength.

RED HLMP-3200 SERIES
Electrical Specifications at $T_A = 25°C$

Symbol	Description	Device HLMP–	Min.	Typ.	Max.	Units	Test Conditions
I_v	Axial Luminous Intensity	3200	1.0	2.0		mcd	I_F = 20mA (Fig. 3)
		3201	2.0	4.0			
$2\theta_{1/2}$	Included Angle Between Half Luminous Intensity Points			60		deg.	Note 1 (Fig. 6)
λ_{PEAK}	Peak Wavelength			655		nm	Measurement @ Peak (Fig. 1)
λ_d	Dominant Wavelength			648		nm	Note 2
τ_s	Speed of Response			15		ns	
C	Capacitance			100		pF	V_F = 0; f = 1 MHz
θ_{JC}	Thermal Resistance			125		°C/W	Junction to Cathode Lead 1.6 mm (0.063 in.) from Body
V_F	Forward Voltage		1.4	1.6	2.0	V	I_F = 20mA (Fig. 2)
V_{BR}	Reverse Breakdown Voltage		3	10		V	I_R = 100µA
η_v	Luminous Efficacy			55		lm/W	Note 3

Notes: 1. $\theta_{1/2}$ is the off-axis angle at which the luminous intensity is half the axial luminous intensity. 2. Dominant wavelength, λ_d, is derived from the CIE chromaticity diagram and represents the single wavelength which defines the color of the device. 3. Radiant Intensity I_e, in watts/steradian may be found from the equation $I_e = I_v/\eta_v$, where I_v is the luminous intensity in candelas and η_v is the luminous efficacy in lumens/watt.

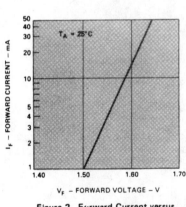

Figure 2. Forward Current versus Forward Voltage.

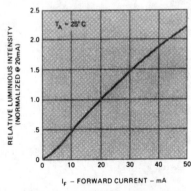

Figure 3. Relative Luminous Intensity versus Forward Current.

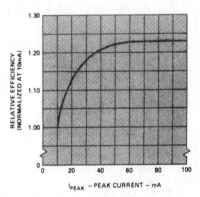

Figure 4. Relative Efficiency (Luminous Intensity per Unit Current) versus Peak Current.

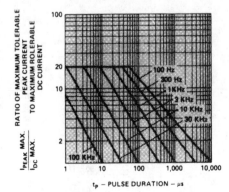

Figure 5. Maximum Tolerable Peak Current versus Pulse Duration. (I_{DC} MAX as per MAX Ratings)

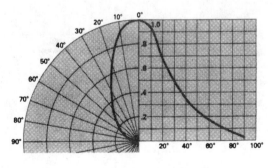

Figure 6. Relative Luminous Intensity versus Angular Displacement.

AMP

OPTIMATE
Fiber Optic Cables
(Continued)

Dimensioning:
Unless otherwise specified, dimensions are in millimeters and inches.

Values in brackets are equivalent U.S. customary units.

Plenum Grade—Dual Channel

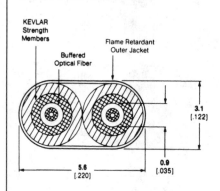

KEVLAR Strength Members

Buffered Optical Fiber

Flame Retardant Outer Jacket

3.1 [.122]

5.6 [.220]

0.9 [.035]

Specifications

No. of Fibers	2
Fiber Type	Glass Multimode
Core Diameter	See Chart, Below
Cladding Diameter	See Chart, Below
Cable Weight	34 kg/km
Max. Installation Load	1000 N [224.8 lb]
Max. Operational Load	200 N [45.0 lb]
Min. Bend Radius (Unloaded)	10.0 cm [3.94 in]
Operating Temperature Range	0°C to +50°C

Plastic Grade—Single Channel

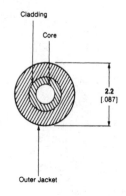

Cladding

Core

2.2 [.087]

Outer Jacket

Specifications

No. of Fibers	1
Fiber Type	Plastic Multimode
Core Diameter	See Chart, Below
Cladding Diameter	See Chart, Below
Cable Weight	4 kg/km
Tensile Strength—Break	10 to 12 kg
Min. Bend Radius (Unloaded)	2.0 cm [.787 in]
Operating Temperature Range	−40°C to +85°C

Note: See page 49 for undercarpet fiber optic cables

Cable Grade	Channels	Core Dia. μm	Cladding Dia. μm	Max Attenuation 850/1300 nm	Min Bandwidth 850 nm	Part Numbers Length - m(ft)			
						25 [82.03]	50 [164.05]	100 [328.1]	1000 [3281.0]
Light Duty	Single	50	125	4.0/2.5 dB/km	400 MHz-km	501110-1	501110-2	501110-3	501110-4
	Single	85	125	5.0/3.5 dB/km	200 MHz-km	501111-1	501111-2	501111-3	501111-4
	Single	100	140	5.0/4.0 dB/km	100 MHz-km	501112-1	501112-2	501112-3	501112-4
	Dual	50	125	4.0/2.5 dB/km	400 MHz-km	501113-1	501113-2	501113-3	501113-4
	Dual	85	125	5.0/3.5 dB/km	200 MHz-km	501114-1	501114-2	501114-3	501114-4
	Dual	100	140	5.0/4.0 dB/km	100 MHz-km	501115-1	501115-2	501115-3	501115-4
Heavy Duty	Dual	50	125	4.0/2.5 dB/km	400 MHz-km	501116-1	501116-2	501116-3	501116-4
	Dual	85	125	5.0/3.5 dB/km	200 MHz-km	501117-1	501117-2	501117-3	501117-4
	Dual	100	140	5.0/4.0 dB/km	100 MHz-km	501118-1	501118-2	501118-3	501118-4
Plenum Duty	Dual	50	125	5.0/3.5 dB/km	400 MHz-km	501119-1	501119-2	501119-3	501119-4
	Dual	85	125	5.0/3.5 dB/km	200 MHz-km	501120-1	501120-2	501120-3	501120-4
	Dual	100	140	7.0/5.5 dB/km	100 MHz-km	501121-1	501121-2	501121-3	501121-4
Plastic	Single	980	1000	—	—	501232-1	501232-2	501232-3	501232-4

OPTIMATE DNP
(Dry Non-Polish)
Connectors

Features

- Performance and reliability at low cost
- No adhesive
- No polishing
- Simple field assembly
- Designed for low cost plastic fibers
- Quick connect/disconnect with audible snap action
- Repeatable coupling efficiency
- Dual position polarized
- Designer Kit available

Technical Documents

AMP Instruction Sheet:
IS 2974—Dry Non-Polish Connectors

AMP Product Specification:
108-45000

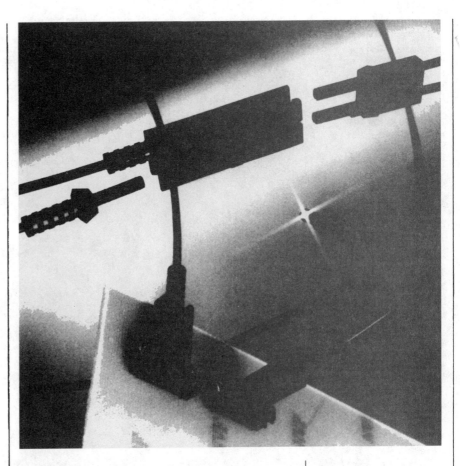

The AMP OPTIMATE DNP Fiber Optic Connectors provide for the assembly of both dual and single channel cables as well as the interface to emitters and detectors. The product family includes the following:

- Splice
- Single Position Plug
- Single Position Device Mount (Honeywell Low-Cost Plastic Sweetspot™)
- Single Position Device Mount (TO-92)
- Single Position Bulkhead Receptacle
- Dual Position Bulkhead Receptacle
- Dual Position Plug

When used in conjunction with low cost plastic fiber of 1000 microns in diameter and low cost active devices, the AMP family of OPTIMATE DNP

Connectors can provide a reliable and cost effective electro-optics system.

Optical Characteristics

Insertion Loss:
2 dB, Ref: FOTP 34 Method C

Environmental Characteristics

Temperature Range:
−40°C to +60°C

Mechanical Characteristics

Insertion Force:
13.34 N [3 lb] (plug to receptacle)

Cable Retention in Plug:
8.89N [12 lb]

Materials

Retention Clip:
Copper Alloy, plated

Splice:
Copper Alloy, plated

Connectors:
Thermoplastic

AMP

OPTIMATE DNP
(Dry Non-Polish)
Connectors
(Continued)

Dimensioning:
Dimensions are in millimetres and inches.
Values in brackets are equivalent U.S.
customary units.

Features

■ Performance and reliability at low cost

■ No adhesive

■ No polishing

■ Simple field assembly

■ Designed for low cost plastic fibers

■ Quick connect/ disconnect with audible snap action

■ Low loss—less than 2 dB

■ Repeatable coupling efficiency

■ Semiautomatic application

Technical Documents

IS 2974—Dry Non-Polish Connectors

AMP Product Specification: 108-45000

Use With:

Single Position
Device Mounts, pp. 28 & 29

Single Position
Bulkhead Receptacle, p. 30

Dual Position
Bulkhead Receptacle, p. 31

AMP Plastic Fiber
Optic Cable, p. 37

Splice and Retention Clip
Splice—Part No. 228051-1
Retention Clip—Part No. 228046-1

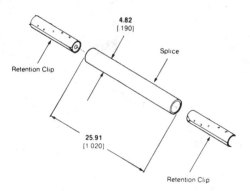

Single Position
Plug Assembly
Part No. 228087-1[1]

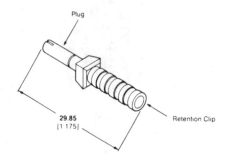

Dual Position
Plug Assembly
Part No. 228088-1[2]

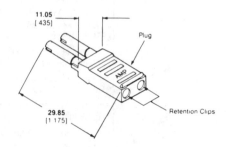

[1] Cutting Tool Fixture Part No. 228837-1 required for proper termination

[2] Cutting Tool Fixture Part No. 228836-1 required for proper termination

![AMP]

OPTIMATE DNP
(Dry Non-Polish)
Connectors
(Continued)

Dimensioning:
Dimensions are in millimetres and inches.
Values in brackets are equivalent U.S.
customary units.

Single Position Device Mounts

For TO-92 Device—
Part No. 228040-1

For .085 x .180 Device—
Part No. 228709-1

Features

■ Performance and reliability at low cost

■ No adhesive

■ No polishing

■ Simple field assembly

■ Designed for low cost plastic fibers

■ Quick connect/disconnect with audible snap action

■ Low loss—less than 2 dB

■ Repeatable coupling efficiency

■ Semiautomatic application

Technical Documents

IS 2974—Dry Non-Polish Connectors

AMP Product Specification: 108-45000

Use With:

Single Position Plug Assembly p. 27

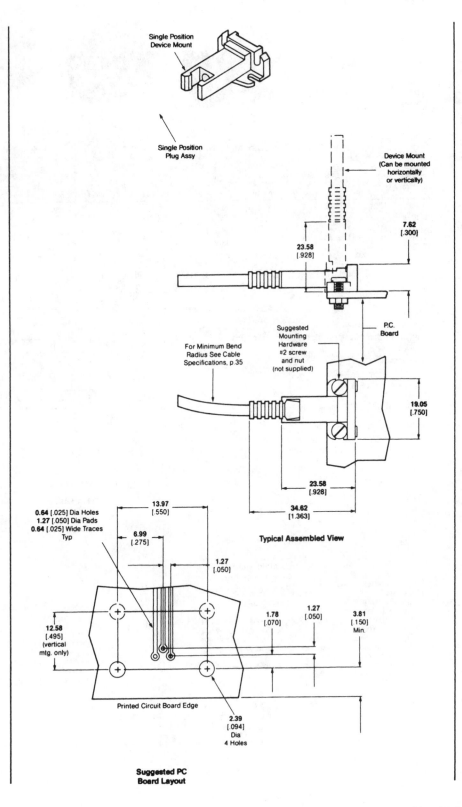

APPENDIX: MANUFACTURER DATA SHEETS   385

ATTENUATORS/SWITCHES *Coaxial*

BI-PHASE *1 MHz to 2 GHz*

ZFAS

ZMAS

ZAS

MODEL NO.	FREQUENCY MHz		INSERTION LOSS dB (±20 mA)				MAX. INPUT PWR dBm (±20 mA)		IN-OUT ISOLATION, dB (0 mA)						BI-PHASE X̄ (±20 mA) Typ.				CASE STYLE	C O N N E C T I O N	PRICE $ ea. Qty. (1-9)
	IN	CON	Mid-Band m		Total Range		1 dB compr.	no damage	L		M		U		Δ AMP (dB)		Phase(deg) deviation from 180°				
	f_L-f_U		Typ.	Max.	Typ.	Max.			Typ.	Min.	Typ.	Min.	Typ.	Min.	m	Total Range	m	Total Range	Note B		
ZMAS-1*	5-450	DC-0.05	3.5	4	3.5	4.7	20	30	65	50	55	40	35	25	0.10	0.1	0.5	1.2	M21	cp	66.95
ZMAS-3*	1-200	DC-0.05	1.4	2	1.6	2.5	15	30	65	50	50	40	50	35	0.10	0.1	0.5	1.0	M21	cp	67.95
ZAS-1*	5-450	DC-0.05	3.5	4	3.5	4.7	20	30	65	50	55	40	35	25	0.10	0.1	0.5	1.2	M22	cp	59.95
ZAS-3*	1-200	DC-0.05	1.4	2	1.6	2.5	15	30	65	50	50	40	50	35	0.10	0.1	0.5	1.0	M22	cp	59.95
▲ ZFAS-2000**	100-2000	DC-0.5	4.2	6.5	5.4	7.5	19✻	25	30	22	—	—	26	20	0.3	0.4	5.0	8.0	K18	cn	64.95

L = low range (f_L to $10f_L$)   M = mid range ($10f_L$ to $f_U/2$)   U = upper range ($f_U/2$ to f_U)
m = mid band ($2f_L$ to $f_U/2$)

suggested control port biasing configuration

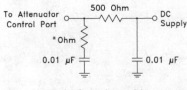

*50 ohm for 50 ohm models
75 ohm for 75 ohm models

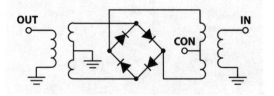

NSN GUIDE
MCL NO.	NSN
ZAS-3B	5985-01-267-2832
ZMAS-1	5985-01-140-4291

NOTES:
* Recommended for electronic attenuator
** Recommended for bi-phase modulator
▲ Available only with SMA connectors.
✻ +15 dBm from 100-800 MHz
A. General Quality Control Procedures, Environmental Specifications, Hi-Rel and MIL description are given in General Information (Section 0).
B. Connector types and case mounted options, case finishes are given in section 0, see "Case styles & Outline Drawings".
C. Prices and specifications subject to change without notice.
1. Absolute maximum power, voltage and current ratings: 1a. Control current, 30mA
2. Performance specifications apply for input power up to 10 dB below stated 1dB compression.

coaxial connections
see case style outline drawings
PORT	cn	cp
INPUT	2	3
OUTPUT	1	1
CONTROL	3	2
GND EXT.	—	—
CASE GND	—	—
NOT USED	—	—

060913

Mini-Circuits®
INTERNET http://www.minicircuits.com
P.O. Box 350166, Brooklyn, New York 11235-0003 (718) 934-4500 Fax (718) 332-4661
Distribution Centers NORTH AMERICA 800-654-7949 • 417-335-5935 • Fax 417-335-5945 • EUROPE 44-1252-832600 • Fax 44-1252-837010
Mini-Circuits ISO 9001 & ISO 14001 Certified
184

DIGITAL STEP ATTENUATORS 50Ω Precision

☐ Mini-Circuits®

TTL CONTROL, PIN DIODE 10 MHz to 1 GHz

TOAT

ZFAT

ZSAT

MODEL NO.	FREQUENCY MHz		PRIMARY ATTENUATION STEPS dB @TTL CONTROL PORT			ATTENUATION dB LOGIC STATE*		VSWR (:1)			CASE STYLE	CONNECTION	PRICE $ ea. Qty. (1-9)
	f_L	f_U	#1	#2	#3	(1,1,1)** Nom.	(0,0,0) Max.	L	M	U	Note B		
TOAT-R512	10	1000	0.5±0.18	1±0.25	2±0.25	3.5	4.0	1.6	1.4	1.5	QQ96	cq	64.45
TOAT-124	10	1000	1±0.25	2±0.25	4±0.3	7.0	4.0	1.6	1.4	1.5	QQ96	cq	64.45
TOAT-3610	10	1000	3±0.3	6±0.4	10±0.4	19.0	4.0	1.6	1.4	1.5	QQ96	cq	64.45
TOAT-4816	10	1000	4±0.4	8±0.4	16±0.5	28.0	4.0	1.6	1.4	1.5	QQ96	cq	64.45
TOAT-51020	10	1000	5±0.4	10±0.4	20±0.5	35.0	4.0	1.6	1.4	1.5	QQ96	cq	64.45
ZFAT-R512	10	1000	0.5±0.18	1±0.25	2±0.25	3.5	4.0	1.6	1.4	1.5	SSS173	-	89.95
ZFAT-124	10	1000	1±0.25	2±0.25	4±0.3	7.0	4.0	1.6	1.4	1.5	SSS173	-	89.95
ZFAT-3610	10	1000	3±0.3	6±0.4	10±0.4	19.0	4.0	1.6	1.4	1.5	SSS173	-	89.95
ZFAT-4816	10	1000	4±0.4	8±0.4	16±0.5	28.0	4.0	1.6	1.4	1.5	SSS173	-	89.95
ZFAT-51020	10	1000	5±0.4	10±0.4	20±0.5	35.0	4.0	1.6	1.4	1.5	SSS173	-	89.95
			SIX CONTROL PORTS										
ZSAT-31R5	10	1000	(1) 0.5±0.18 (4) 4±0.3	(2) 1±0.25 (5) 8±0.4	(3) 2±0.25 (6)16±0.5	31.5	7.0	2.0	1.5	1.6	AR214	-	119.00

L = 10 to 100 MHz   M = 100 to 500 MHz   U = 500 to 1000 MHz

features
- wide frequency band, 10-1000 MHz
- excellent step accuracy, 0.2 dB typ.
- excellent VSWR, 1.3 typ.
- low DC current, 6 mA typ.
- operates over -55° to 100 °C
- small case, 0.6" dia., TO-8

NOTES:
* For ZSAT-31R5: Total attenuation (1,1,1,1,1,1)
  Thru-Loss (0,0,0,0,0,0)
** Total attenuation above thru-loss.
A. General Quality Control Procedures, Environmental Specifications, Hi-Rel and MIL description are given in section 0, see "Mini-Circuits Guarantees Quality" article.
B. Connector types and case mounted options, case finishes are given in section 0, see "Case Styles & Outline Drawings".
C. Prices and Specifications subject to change without notice.
1. Absolute maximum power, voltage and current rating:
   1a. Input power, 15 dBm
   1b. DC voltage, 5.5 Volts
   1c. TTL, 5.5 Volts
   1d. Storage temperature -55°C to +125°C for TOAT models.
2. Step accuracy is specified for basic steps. For combination of steps accuracy is additive.
3. Thru-loss is minimum insertion loss with all attenuation elements bypassed (All TTL controls state are Low).
4. For optimum operation of TOAT models, ensure the device case is properly connected to the ground plane (of PC board).

ADDITIONAL SPECIFICATIONS
DC Voltage +5V
DC current 12mA max.
Switching Time (50% TTL to within specified accuracy of the next-selected attenuation step, and to within 0.1 dB of steady-state Thru-Loss) 10 μs typ., 15μs max.
TTL input High Threshold 2V min.
TTL input Low Threshold 0.8V max.
TTL Toggle Rate: 50 kHz typ.
1dB compression: 0 dBm (10-100MHz)
                  +10dBm (100-1000MHz)
For ZSAT-31R5:
   1dB compression: +10 dBm (10-100 MHz)
                     +15 dBm (100-1000 MHz)

Logic function:
TTL High activates associated in-line attenuation
TTL Low bypasses this attenuation

NSN GUIDE

MCL NO.	NSN
TOAT-124	5985-01-416-9021
TOAT-51020	5985-01-416-9020

pin connections
see case style outline drawing

PORT	cq
RF IN	4
RF OUT	11
TTL CONTROL #1	2
TTL CONTROL #2	3
TTL CONTROL #3	1
+5V DC	12
CASE GND	5,6,7,8,9,10

060713

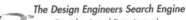

Coaxial
Voltage Controlled Oscillator ZX95-100

Linear Tuning   50 to 100 MHz

Features
- Linear Tuning
- Octave Bandwidth
- Low Phase Noise
- Low Pushing & Pulling
- Excellent Harmonic Suppression
- Protected by US Patent 6,790,049

Applications
- R & D
- Lab
- Instrumentation
- Test Equipment

CASE STYLE: GB956

Connectors	Model	Price	Qty.
SMA	ZX95-100-S	$37.95 ea.	(1-9)

Electrical Specifications

MODEL NO.	FREQ. (MHz)		POWER OUTPUT (dBm)	PHASE NOISE dBc/Hz SSB at offset frequencies, kHz Typ.				TUNING			3 dB MODULATION BANDWIDTH (MHz)	NON-HARMONIC SPURIOUS (dBc)	HARMONICS (dBc)		PULLING pk-pk @ 12 dBr (MHz)	PUSHING (MHz/V)	DC OPERATING POWER	
								VOLTAGE RANGE (V)	SENSI-TIVITY (MHz/V)	PORT CAP (pF)								
	Min.	Max.	Typ.	1	10	100	1000	Min.	Max.	Typ.	Typ.	Typ.	Typ.	Typ.	Max.	Typ.	Typ.	Vcc (volts)	Current (mA) Max.
ZX95-100	50	100	10.0	-86	-110	-131	-151	0.5	17.0	3.5-4.5	550	0.18	-90	-33.0	-24.0	0.4	0.1	12.0	20

Maximum Ratings

Operating Temperature	-55°C to 85°C
Storage Temperature	-55°C to 100°C
Absolute Max. Supply Voltage (Vcc)	13V
Absolute Max. Tuning Voltage (Vtune)	18V
All specifications	50 ohm system

Outline Drawing

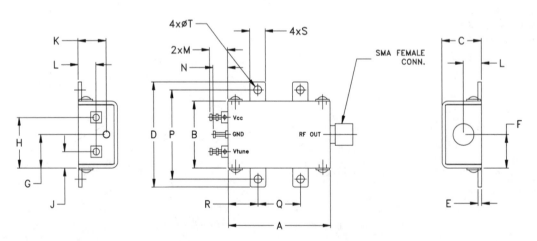

Outline Dimensions ($\frac{Inch}{mm}$)

A	B	C	D	E	F	G	H	J	K	L	M	N	P	Q	R	S	T	wt.
1.20	.75	.46	1.18	.04	.38	.45	.57	.18	.33	.21	.22	.18	1.00	.50	.35	.18	.09	grams
30.50	19.10	11.6	30.0	1.0	9.6	11.4	14.5	4.7	8.3	5.3	5.6	4.6	25.4	12.7	8.9	4.6	2.3	35.0

Mini-Circuits®
P.O. Box 350166, Brooklyn, New York 11235-0003 (718) 934-4500 Fax (718) 332-4661
INTERNET http://www.minicircuits.com
Distribution Centers NORTH AMERICA 800-654-7949 • 417-335-5935 • Fax 417-335-5945 • EUROPE 44-1252-832600 • Fax 44-1252-837010
Mini-Circuits ISO 9001 & ISO 14001 Certified

REV. OR.
M96760
EDR-7306/1
ZX95-100
RAV/URJ
060122
page 1 of 2

Performance Data & Curves* ZX95-100

V TUNE	TUNING SENS (MHz/V)	FREQUENCY (MHz)			POWER OUTPUT (dBm)			Icc (mA)	HARMONICS (dBc)			FREQ. PUSHING (MHz/V)	FREQ. PULLING (MHz)	PHASE NOISE (dBc/Hz) at offsets				FREQ OFFSET (kHz)	PHASE NOISE at 75 MHz (dBc/Hz)
		-55°C	+25°C	+85°C	-55°C	+25°C	+85°C		F2	F3	F4			1 kHz	10 kHz	100 kHz	1 MHz		
0.0	5.0	45.2	44.1	44.0	8.0	7.5	6.6	16	-37.0	-38.2	-52.0	0.04	0.05	-84.1	-105.4	-126.9	-150.0	1.00	-85.4
1.0	4.6	49.3	48.9	49.1	9.7	9.4	8.7	15	-37.2	-39.6	-52.6	0.10	0.11	-87.4	-112.9	-132.1	-154.4	2.00	-93.0
2.0	3.7	53.4	53.2	53.4	10.1	10.0	9.4	15	-37.5	-38.0	-52.1	0.02	0.06	-88.6	-113.2	-132.8	-153.7	3.50	-99.0
3.0	3.5	56.9	56.9	57.0	10.2	10.0	9.5	15	-30.8	-37.3	-50.4	0.03	0.03	-88.8	-112.1	-132.4	-153.2	6.00	-104.4
5.0	3.6	63.7	63.8	64.0	10.2	10.0	9.5	15	-28.9	-37.0	-46.5	0.08	0.07	-86.0	-111.0	-131.1	-151.2	8.50	-108.1
6.0	3.7	67.3	67.4	67.6	10.2	9.9	9.4	15	-29.4	-36.7	-44.9	0.08	0.06	-85.8	-110.1	-130.5	-150.7	10.00	-108.6
7.0	3.8	71.1	71.2	71.3	10.1	9.9	9.3	15	-30.2	-36.4	-43.3	0.06	0.03	-85.7	-108.8	-129.9	-150.3	20.80	-115.6
8.0	3.9	74.9	75.0	75.2	10.1	9.7	9.1	15	-31.0	-36.2	-42.2	0.04	0.01	-86.9	-108.8	-129.7	-149.5	35.50	-120.4
9.0	3.9	78.9	78.8	79.1	10.0	9.6	9.0	15	-31.8	-36.1	-41.5	0.02	0.01	-84.8	-109.4	-128.9	-149.5	60.70	-125.5
10.0	3.9	83.0	82.7	82.9	9.9	9.5	8.8	15	-32.5	-36.0	-40.9	0.01	0.05	-84.5	-108.4	-129.4	-149.5	86.70	-128.3
11.0	3.9	87.1	86.6	86.8	9.8	9.3	8.6	15	-33.3	-35.7	-40.8	0.04	0.14	-86.3	-109.9	-130.1	-149.9	100.00	-129.6
12.0	4.1	91.3	90.6	90.7	9.6	9.1	8.4	15	-34.2	-35.6	-40.6	0.06	0.20	-85.5	-110.7	-130.4	-150.2	211.60	-135.8
13.0	4.2	95.5	94.7	94.7	9.4	8.9	8.1	15	-35.3	-35.5	-39.8	0.07	0.27	-86.9	-110.0	-130.6	-150.7	361.50	-140.4
14.0	4.4	99.9	99.0	98.9	9.2	8.6	7.7	15	-36.5	-35.1	-39.2	0.10	0.41	-85.6	-108.5	-129.3	-150.2	507.50	-143.4
15.0	4.6	104.5	103.4	103.3	9.0	8.3	7.4	15	-37.3	-34.8	-38.2	0.15	0.39	-86.7	-107.9	-128.5	-149.7	600.00	-144.8
16.0	4.7	109.2	108.1	107.9	8.7	8.0	7.0	15	-37.1	-34.2	-37.0	0.23	0.48	-87.8	-106.0	-127.2	-148.9	851.60	-148.1
17.0	4.8	114.0	112.8	112.5	8.5	7.6	6.6	15	-36.3	-33.3	-35.9	0.34	0.59	-81.9	-104.1	-126.0	-147.6	1000.00	-149.7

*at 25°C unless mentioned otherwise

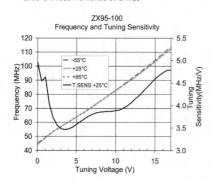

ZX95-100
Frequency and Tuning Sensitivity

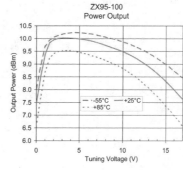

ZX95-100
Power Output

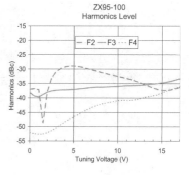

ZX95-100
Harmonics Level

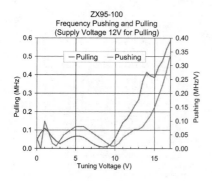

ZX95-100
Frequency Pushing and Pulling
(Supply Voltage 12V for Pulling)

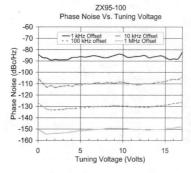

ZX95-100
Phase Noise Vs. Tuning Voltage

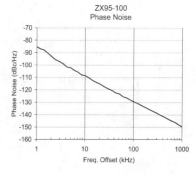

ZX95-100
Phase Noise

page 2 of 2

Mini-Circuits®
INTERNET http://www.minicircuits.com
P.O. Box 350166, Brooklyn, New York 11235-0003 (718) 934-4500 Fax (718) 332-4661
Distribution Centers NORTH AMERICA 800-654-7949 • 417-335-5935 • Fax 417-335-5945 • EUROPE 44-1252-832600 • Fax 44-1252-837010
Mini-Circuits ISO 9001 & ISO 14001 Certified

Coaxial
Adapter, SMA-M to BNC-F

SM-BF50+

50Ω   DC to 2 GHz

CASE STYLE: DJ1024

Maximum Ratings

Operating Temperature	-55°C to 100°C
Storage Temperature	-55°C to 100°C

Features
• flat response
• excellent VSWR
• low cost adapters, available from stock
• brass body, nickel plated

Applications
• interconnection of RF cables and equipment
• testing

Connectors		Model	Price	Qty.
Conn1	**Conn2**			
SMA-M	**BNC-F**	SM-BF50+	$3.95 ea.	(1-49)

*+ RoHS compliant in accordance
with EU Directive (2002/95/EC)*

The +Suffix has been added in order to identify RoHS
Compliance. See our web site for RoHS Compliance
methodologies and qualifications.

Outline Drawing

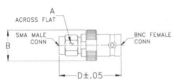

Outline Dimensions ($\frac{inch}{mm}$)

A	B	C	D	E	wt
.312	.53	--	1.10	--	grams
7.92	13.46	--	27.94	--	8.6

Electrical Specifications T_{AMB}=25°C

FREQUENCY (GHz)	INSERTION LOSS (dB)	VSWR (:1) Max.
f_L-f_U	Typ.	DC-2 GHz
DC-2	0.05	1.20

Typical Performance Data

Frequency (MHz)	Insertion Loss (dB)	VSWR (:1) SMA-M	VSWR (:1) BNC-F
1.00	0.00	1.00	1.00
10.00	0.00	1.00	1.00
20.00	0.00	1.00	1.00
40.00	0.01	1.00	1.01
50.00	0.01	1.00	1.01
60.00	0.01	1.00	1.01
70.00	0.01	1.00	1.01
80.00	0.01	1.00	1.01
90.00	0.01	1.00	1.01
100.00	0.02	1.00	1.01
200.00	0.02	1.00	1.01
400.00	0.03	1.01	1.02
600.00	0.04	1.01	1.03
800.00	0.05	1.01	1.04
1000.00	0.04	1.01	1.05
1200.00	0.05	1.01	1.05
1400.00	0.05	1.01	1.06
1600.00	0.06	1.02	1.07
1800.00	0.05	1.02	1.07
2000.00	0.07	1.03	1.08

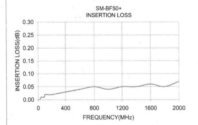

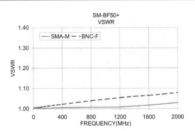

Mini-Circuits®   P.O. Box 350166, Brooklyn, New York 11235-0003 (718) 934-4500 Fax (718) 332-4661
Distribution Centers NORTH AMERICA 800-654-7949 • 417-335-5935 • Fax 417-335-5945 • EUROPE 44-1252-832600 • Fax 44-1252-837010

INTERNET http://www.minicircuits.com

Mini-Circuits ISO 9001 & ISO 14001 Certified

REV. A
M98898
SM-BF50+
ED-11362/4
RS/TD/CP
061030

FREQUENCY MIXERS *Coaxial*

LEVEL 7 *500 Hz to 10 GHz*

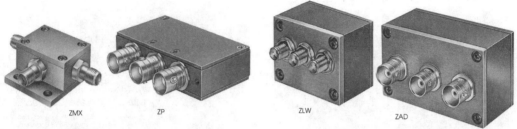

ZMX   ZP   ZLW   ZAD

+7 dBm LO, up to +1 dBm RF

MODEL NO.	FREQUENCY MHz LO/RF f_L-f_U	FREQUENCY MHz IF	CONVERSION LOSS dB Mid-Band m $\bar{x}$	σ	Max.	Total Range Max.	LO-RF ISOLATION, dB L Typ.	Min.	M Typ.	Min.	U Typ.	Min.	LO-IF ISOLATION, dB L Typ.	Min.	M Typ.	Min.	U Typ.	Min.	CASE STYLE Note B	CONNECTION	PRICE $ Qty. (1-9)
ZMX-7GR	3700-7000	DC-1000	5.0	.30	—	8.2	30 (typ.) 20 (min.)						36 (typ.) 20 (min.)						BU413	af	71.95
ZMX-10G	3700-10000	DC-2000	5.0	.10	—	8.5	37 (typ.) 20 (min.)*						17 (typ.) 8 (min.)						BU413	ad	81.95
ZP-1	2-600	DC-600	5.85	.10	7.0	8.0	60	50	42	30	37	25	60	45	47	30	36	22	GG60	ag	39.95
ZP-2	50-1000	DC-1000	5.85	.10	7.5	9.0	58	40	47	30	42	25	50	35	44	20	29	18	GG60	ag	39.95
† ZP-3	0.15-400	DC-400	4.7	.10	7.0	8.0	60	50	46	30	35	25	60	40	47	25	35	20	GG60	ag	39.95
ZP-5	20-1500	DC-1000	5.7	.10	9.0	8.0	54	40	42	30	39	25	40	25	32	18	23	8	GG60	ag	47.95
ZP-5X	1-1500	1-1000	5.9	.10	7.0	9.0	60	40	40	20	28	17	60	45	45	25	38	20	GG60	hg	47.95
ZP-10514	.2-500	DC-500	5.18	.10	7.0	8.5	55	45	50	35	35	30	50	40	36	30	30	20	GG60	ag	62.95
ZLW-1	.5-500	DC-500	5.81	.08	7.0	8.5	50	45	45	30	35	25	45	35	40	25	30	20	M21	ae	51.95
ZLW-1W	1-750	DC-750	5.74	.05	7.5	8.5	50	45	45	30	35	25	45	30	40	25	30	20	M21	ae	56.95
ZLW-1-1	.1-500	DC-500	4.82	.07	7.5	8.5	50	45	45	30	35	25	45	30	40	25	30	20	M21	ae	53.95
ZLW-2	1-1000	DC-1000	5.68	.08	7.5	9.5	55	45	40	25	35	20	50	40	40	25	30	20	M21	ae	56.95
ZLW-3	.025-200	DC-200	4.61	.06	7.5	8.5	60	50	45	35	35	25	45	35	40	30	30	20	M21	ae	53.95
ZLW-5	5-1500	10-600	5.81	.08	7.5	8.5	55	40	35	25	30	20	50	40	35	25	30	20	M21	ae	61.95
ZLW-6	.003-100	DC-100	4.58	.05	7.5	8.5	60	50	45	30	35	25	60	45	40	25	30	25	M21	ae	64.95
ZLW-11	5-2000	10-600	6.85	.10	8.5	9.0	50	45	35	25	30	20	45	40	30	20	25	15	M21	ae	71.95
ZAD-1	.5-500	DC-500	5.24	.10	7.0	8.5	50	45	45	30	35	25	45	35	40	25	30	20	M22	ae	43.95
ZAD-1-1	.1-500	DC-500	4.83	.04	7.5	8.5	50	45	45	30	35	25	45	30	40	25	30	20	M22	ae	44.95
ZAD-2	1-1000	.5-500	5.66	.07	7.5	8.5	45	30	35	20	30	20	40	30	35	20	30	20	M22	ad	49.95
ZAD-3	.025-200	DC-200	4.61	.06	7.5	8.5	60	50	45	35	35	25	45	35	40	30	30	20	M22	ae	45.95
ZAD-6	.003-100	DC-100	4.65	.08	7.5	8.5	60	50	45	30	35	25	60	45	40	25	30	20	M22	ae	51.95
ZAD-8	.0005-10	DC-10	5.79	.05	7.5	8.5	60	50	50	40	45	35	60	50	50	40	45	35	M22	ae	54.95
ZAD-11	5-2000	10-600	7.12	.12	8.5	9.0	50	45	35	25	30	20	45	40	30	20	25	15	M22	ae	61.95

L = low range (f_L to 10 f_L)   M = mid range (10 f_L to f_U/2)   U = upper range (f_U/2 to f_U)

m = mid band (2f_L to f_U/2)

NOTES:

$\bar{x}$ Average of conversion loss at center of mid-band frequency (f_L+f_U/4)
σ   Standard deviation
▲   Available only with SMA connectors
▼   When ordering, specify BNC or SMA connectors (ZFM-2000, ZFM-4212 SMA only.)
†   Phase detection, positive polarity
*   15 dB min. 8.5 to 10 GHz
A.   General Quality Control Procedures, Environmental Specifications, Hi-Rel and MIL description are given in section 0, see "Mini-Circuits Guarantees Quality" article.
B.   Connector types and case mounted options, case finishes are given in section 0, see "Case Styles & Outline Drawings".
C.   Prices and Specifications subject to change without notice.
1.   Absolute maximum power, voltage and current ratings:
   1a.   RF power, 50mW
   1b.   Peak IF current, 40mA

Mini-Circuits®
INTERNET http://www.minicircuits.com
P.O. Box 350166, Brooklyn, New York 11235-0003 (718) 934-4500 Fax (718) 332-4661
Distribution Centers NORTH AMERICA 800-654-7949 • 417-335-5935 • Fax 417-335-5945 • EUROPE 44-1252-832600 • Fax 44-1252-837010
Mini-Circuits ISO 9001 & ISO 14001 Certified

050112

94

ZEM

ZFM ▼

ZAM ▲

+7 dBm LO, up to +1 dBm RF

MODEL NO.	FREQUENCY MHz LO/RF f_L-f_U	FREQUENCY MHz IF	CONVERSION LOSS dB Mid-Band m $\bar{x}$	σ	Max.	CONVERSION LOSS dB Total Range Max.	LO-RF ISOLATION, dB L Typ.	L Min.	M Typ.	M Min.	U Typ.	U Min.	LO-IF ISOLATION, dB L Typ.	L Min.	M Typ.	M Min.	U Typ.	U Min.	CASE STYLE Note B	CONNECTION	PRICE $ Qty. (1-9)
ZEM-2B	10-1000	DC-1000	5.74	.07	7.0	8.5	55	50	30	25	25	20	55	45	30	20	25	20	V37	ad	59.95
ZEM-4300	300-4300	DC-1000	6.65	.06	—	9.5	40	20	—	—	30	17	15	8	—	—	15	8	V37	af	79.95
ZFM-1W	10-750	DC-750	5.42	.14	7.0	8.0	50	45	45	30	35	25	45	40	40	25	27	20	K18	ad	51.95
ZFM-2	1-1000	DC-1000	5.72	.06	7.5	8.5	50	45	40	25	30	25	45	40	35	25	25	20	K18	ad	53.95
ZFM-3	0.04-400	DC-400	4.78	.03	7.0	8.0	60	50	50	35	35	25	55	40	45	30	35	25	K18	ad	61.95
† ZFM-4	5-1250	DC-1250	5.70	.34	7.5	8.5	50	45	40	30	30	25	45	40	35	25	25	20	K18	ad	61.95
ZFM-5X	1-1500	1-1000	5.9	.10	7.0	9.0	60	40	40	20	28	17	60	45	45	25	38	20	K18	ae	59.95
ZFM-11	1-2000	5-600	7.03	.17	8.5	9.0	50	45	35	25	25	20	45	40	27	20	25	20	K18	ad	89.95
ZFM-12	800-1250	50-90	5.67	.12	—	7.5	35	25	35	25	35	25	30	20	30	20	30	20	K18	ad	79.95
ZFM-2000	100-2000	DC-600	7.49	.20	9.5	9.5	—	—	37	20	—	—	—	—	—	—	30	20	K18	ad	71.95
ZFM-4212	2000-4200	DC-1300	5.44	.088	—	8.5	—	—	25	17	—	—	—	—	18	10	—	—	K18	ad	54.95
ZAM-42	1500-4200	DC-500	5.67	.11	—	8.5	25	14	25	14	25	14	18	10	18	10	18	10	F14	af	54.95

L = low range (f_L to 10 f_L) M = mid range (10 f_L to f_U/2) U = upper range (f_U/2 to f_U)
m = mid band (2f_L to f_U/2)

NSN GUIDE

MCL NO.	NSN
ZAD-1	5895-01-455-4088
ZAD-1B(BNC)	5985-00-280-7750
ZAD-4B	5895-01-127-0376
ZAD-6B	5895-01-344-7843
ZEM-2B	5895-01-235-7834
ZFM-1W	5895-01-412-3037
ZFM-2	4935-01-230-3782
ZFM-3	5895-01-257-9523
ZFM-3 (SMA)	5895-01-214-7362
ZFM-3B	5895-01-381-9289
ZFM-11(SMA)	6625-01-415-2182
ZLW-1	5895-01-394-4973
ZLW-1W	5895-00-607-7010
ZLW-2	6920-01-037-1974
ZLW-2B	5840-01-186-8398
ZP-3	5985-00-105-9756
ZP-10514	6625-01-108-6156
ZP-10514(BNC)	5895-01-384-7453

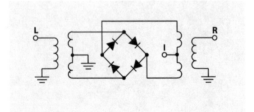

coaxial connections
see case style outline drawings

PORT	ad	ae	af	ag	hg
LO	1	1	2	L	L
RF	2	3	1	R	X
IF	3	2	3	X	R
GND EXT.	—	—	—	—	—
CASE GND	—	—	—	—	—
NOT USED	—	—	—	—	—

The Design Engineers Search Engine
Provides Actual Data Instantly
At: http://www.minicircuits.com

In Stock... Immediate Delivery
For Custom Versions Of Standard Models
Consult Our Applications Dept.

*Available
Tape & Reel*

95

040205

LOW-NOISE AMPLIFIERS 50Ω

BROADBAND, LINEAR 0.1 to 3000 MHz

AMP

MAN

ZFL

ZX60 (GA955)

ZX60 (GC957)

up to +16 dBm output

MODEL NO.	FREQ. (MHz) $f_L - f_U$	NF (dB) Typ.	GAIN (dB) Min.	GAIN Flatness Max. ±m	GAIN Total range	MAXIMUM POWER (dBm) Output (1 dB Comp.)	MAXIMUM POWER (dBm) Input (no damage)	INTERCEPT POINT (dBm) IP3 Typ.	VSWR Typ. In	VSWR Typ. Out	DC POWER Volt (V)	DC POWER Current (mA)	CASE STYLE Note B	CONNECTION	PRICE $ ea. Qty. (1-9)
AMP-15	5-1000	2.8	13	±0.6	±1.2	+8	+13	+22	2:1	2:1	15	29	PP120	cd	52.20
AMP-75	5-500	2.4	19	±0.4	±1.0	+12	+13	+28	2:1	2:1	15	31	PP120	cd	52.20
AMP-76	5-500	3.1	26	±0.7	±1.0	+13.5	+6	+28	2:1	2:1	15	71	PP120	cd	83.45
AMP-77	5-500	3.3	15	±0.4	±1.0	+16	+13	+32	2:1	2:1	15	56	PP120	cd	60.45
MAN-1LN**	0.5-500	3.0	28	±0.5	±1.4	+7	+15	+18	1.8:1	1.8:1	12	60	A05	cc	22.20
MAN-1HLN	10-500	3.7	10	±0.5	±0.8	+15	+15	+30	1.8:1	1.8:1	12	70	A06	cc	22.20
ZFL-500HLN	10-500	3.8	19	—	±0.4	+16	+15	+30	2:1	2:1	15	110	Y460	—	99.95
ZFL-500LN*	0.1-500	2.9	24	—	±0.5	+5	+5	+14	1.5:1	1.6:1	15	60	Y460	—	79.95
ZFL-1000LN	0.1-1000	2.9	20	—	±0.5	+3	+5	+14	1.5:1	2:1	15	60	Y460	—	89.95
NEW ZX60-1215LN	800-1000	0.4	14	—	±1.3	+12.5	+13	+26	1.65:1	1.40:1	12	50	GA955	—	149.95
	1000-1400	0.4	11	—	±2.5	+12.5	+13	+27.5	1.70:1	1.40:1	12	50			
NEW ZX60-1614LN	1217-1620	0.5	11	—	±2.0	+13.5	+13	+30	1.3:1	1.3:1	12	50	GA955	—	149.95
NEW ZX60-3011	400-1000	1.4	12	—	±.70	+19.5	+15	+31	1.7:1	1.6:1	12	120	GC957	—	139.95
	1000-1700	1.5	11	—	±1.0	+19.5	+15	+31	1.7:1	1.6:1	12	120			
	1700-2400	1.7	9	—	±1.0	+18.5	+15	+31	1.7:1	1.6:1	12	120			
	2400-3000	1.8	7.5	—	±.70	+18.0	+15	+31	1.7:1	1.6:1	12	120			

m = mid range (2 f_L to f_U/2)

NOTES:
* VSWR 1.6:1 maximum from 0.1 to 0.2 MHz. Also available with BNC connectors.
** Below 5 MHz, 1 dB compression point decreases to 6.5 dBm.
▲ Available only with SMA connectors
B. Connector types and case mounted options, case finishes are given in section 0, see "Case styles & outline drawings".
C. Prices and specifications subject to change without notice.
D. For Quality Control Procedures see Table of Contents, Section 0, "Mini-Circuits Guarantees Quality" article. For Environmental Specifications see Amplifier Selection Guide.
1. Absolute maximum power, voltage and current rating:
   1a. AMP models, 17V DC.   1b. MAN models, 12.5V DC.
   1c. ZQL models, 17V DC.   1d. ZX60, 15VDC   1e. ZQLSC, 36VDC
2. Open load is not recommended, potentially can cause damage. With no load, derate max input power by 20 dB.
3. ZEL and TO models, NF specified at room temperature, increases to 2 dB typical at +85 deg.C. (TO-0812LN increases to 1.6 dB max.)
4. ZHL models, NF specified at room temperature, increases to 2.3 dB maximum at +65 deg.C.

NSN GUIDE

MCL NO.	NSN
AMP-15	5996-01-350-9550
AMP-75	5996-01-350-9551
AMP-77	5996-01-350-9549
ZEL-1724LN	5996-01-450-0781
ZFL-1000LN	5996-01-412-3031
ZHL-0812HLN	5996-01-453-2464

060713

Mini-Circuits®

INTERNET http://www.minicircuits.com
P.O. Box 350166, Brooklyn, New York 11235-0003 (718) 934-4500 Fax (718) 332-4661
Distribution Centers NORTH AMERICA 800-654-7949 • 417-335-5935 • Fax 417-335-5945 • EUROPE 44-1252-832600 • Fax 44-1252-837010
Mini-Circuits ISO 9001 & ISO 14001 Certified

172

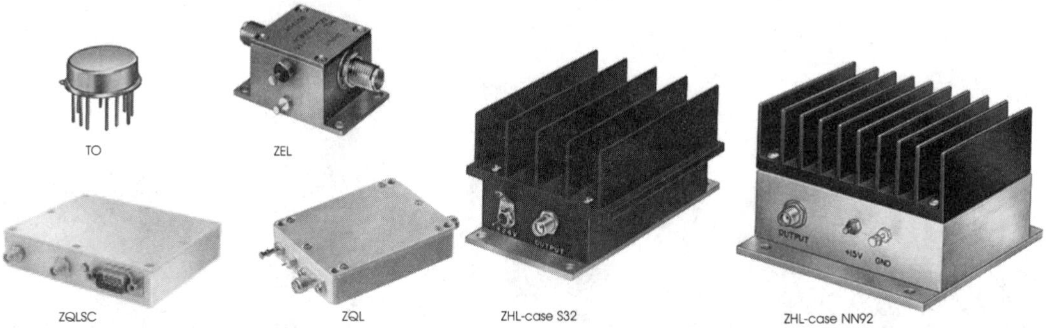
TO
ZEL
ZQLSC
ZQL
ZHL-case S32
ZHL-case NN92

up to +27 dBm output

MODEL NO.	FREQ. (MHz) f_L-f_U	NF (dB) Max.	GAIN (dB) Min.	GAIN Flatness Max.	MAXIMUM POWER (dBm) Output (1 dB Comp.) Typ.	MAXIMUM POWER Input (no damage)	INTERCEPT POINT (dBm) IP3 Typ.	VSWR Max. In	VSWR Out	DC POWER Volt (V)	DC POWER Current (mA)	CASE STYLE Note B	CONNECTION	PRICE $ ea. Qty. (1-9)
TO-0812LN	800-1200	1.2	20	±1.0	+8	+10	+22.5	2.5:1	2.5:1	15	70	QQ96	ce	203.50
TO-1217LN	1200-1700	1.6	20	±1.0	+10	+13	+25	2.5:1	2.5:1	15	70	QQ96	ce	203.50
TO-1724LN	1700-2400	1.6	20	±1.0	+10	+13	+22	2.5:1	2.5:1	15	70	QQ96	ce	203.50
▲ ZEL-0812LN	800-1200	1.5	20	±1.0	+8	+13	+18	2.5:1	2.5:1	15	70	EEE132	—	274.95
▲ ZEL-1217LN	1200-1700	1.5	20	±1.0	+10	+13	+25	2.5:1	2.5:1	15	70	EEE132	—	274.95
▲ ZEL-1724LN	1700-2400	1.5	20	±1.0	+10	+13	+22	2.5:1	2.5:1	15	70	EEE132	—	274.95
▲ ZHL-0812MLN	800-1200	1.6	28	±1.0	+20	0	+33	2.5:1	2.5:1	15	300	S32	—	295.00
▲ ZHL-1217MLN	1200-1700	1.5	30	±1.0	+20	0	+34	2.5:1	2.5:1	15	300	S32	—	295.00
▲ ZHL-1724MLN	1700-2400	1.5	28	±1.0	+20	0	+32	2.5:1	2.5:1	15	300	S32	—	295.00
▲ ZHL-0812HLN	800-1200	1.5	30	±1.0	+26	+10	+36	2.4:1	2.4:1	15	725	NN92	—	399.50
▲ ZHL-1217HLN	1200-1700	1.5	30	±1.0	+26	+10	+36	2.4:1	2.4:1	15	725	NN92	—	399.50
▲ ZHL-1724HLN	1700-2400	1.5	30	±1.0	+26	+10	+36	2.4:1	2.4:1	15	725	NN92	—	399.50
NEW ZQLSC-1100	600-824	1.1	17.5	±1.0	+18.5	+10	+32.5	2.0:1	2.0:1	24	185	GZ1067	pt	295.00
	824-849	0.7	17.5	±0.3	+19	+10	+34	1.7:1	1.7:1	24	185			
	850-915	0.8	17.0	±0.4	+19	+10	+35	1.8:1	1.8:1	24	185			
	915-1100	1.0	15.0	±1.2	+19	+10	+35.5	1.8:1	1.8:1	24	185			
NEW ZQLSC-2400	1400-1850	1.15	13.5	±2.5	+19	+10	+37	1.7:1	1.6:1	24	185	GZ1067	pt	295.00
	1850-2000	1.25	12.5	±1.0	+20.5	+10	+35	1.5:1	1.6:1	24	185			
	2000-2400	1.65	10.5	±2.0	+20	+10	+34	1.6:1	1.9:1	24	185			
								Typ.						
ZQL-900LNW	800-900	1.6	13	±1.6	+21	+10	+35	1.2:1	1.1:1	15	160	CW686	—	229.00
ZQL-900LN	824-849	1.3	15	±0.5	+21	+10	+35	1.2:1	1.1:1	15	160	CW686	—	229.00
ZQL-1900LNW	1700-2000	1.6	14	±1.8	+18.5	+10	+37	1.15:1	1.25:1	15	160	CW686	—	249.00
ZQL-1900LN	1850-1910	1.5	15	±0.5	+19	+10	+37	1.15:1	1.25:1	15	160	CW686	—	249.00
ZQL-900MLNW	800-900	1.7	22	±2.2	+23	+3	+41	1.3:1	1.4:1	15	230	CW686	—	249.00
ZQL-900MLN	824-849	1.5	25.5	±0.5	+24.5	+3	+41	1.3:1	1.4:1	15	230	CW686	—	249.00
ZQL-1900MLNW	1800-2000	1.6	23	±2.0	+25	+3	+41	*1.4:1	1.25:1	15	310	CW686	—	265.00
ZQL-1900MLN	1850-1910	1.5	25	±0.7	+26	+3	+41	*1.25:1	1.2:1	15	310	CW686	—	265.00
ZQL-2700MLNW	2200-2400	1.3	25	±1.0	+25	+3	+38	1.25:1	1.15:1	15	325	CW686	—	281.95
	2200-2700	1.5	25	±2.3	+25	+3	+38	1.25:1	1.15:1	15	325			

*measured at 1900 MHz.

pin connections

PORT	cc	cd	ce	pt
RF IN	1	2	5	J1
RF OUT	8	4	11	J2
DC	5	1	2	5
TTL ALARM OUTPUT	—	—	—	1
GND TO TEST ALARM, normally open	—	—	—	7*,9*
CASE GND	2,3,4,6	3	1,3,4,6,7,8,9,10,12	2,4
NOT USED	7	—	—	3,6,8

*Grounding Pin 7 will sink 75mA of current through Pin 7 creating a high-current alarm condition inside the amplifiers. A red LED and TTL high output will occur. Pin 7 floats at +4.3V typ. when open.

*Grounding Pin 9 will sink 2mA of current through pin 9 and creating a low-current alarm condition inside the amplifier. A red LED and TTL high output will occur. Pin 9 floats at about +0.6V typ. when open.

Alarm Functions for ZQLSC Series

Normal:	TTL low output (0 to 0.8V), green LED
Alarm:	TTL high output (4 to 5V), red LED
DC & alarm connector:	9-pin male D-sub

060713

The Design Engineers Search Engine
Provides Actual Data Instantly
At: http://www.minicircuits.com

In Stock... Immediate Delivery
For Custom Versions Of Standard Models
Consult Our Applications Dept.
Available Tape & Reel

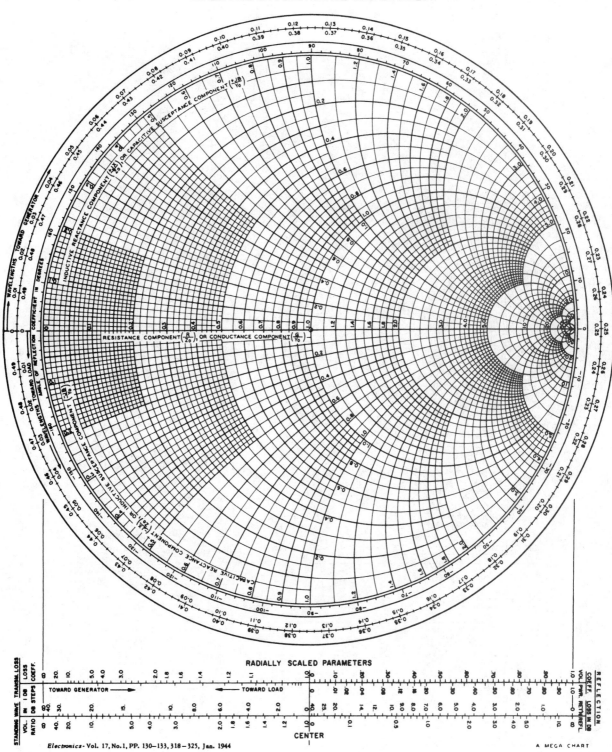

IMPEDANCE OR ADMITTANCE COORDINATES

RADIALLY SCALED PARAMETERS

TOWARD GENERATOR →   ← TOWARD LOAD

CENTER

Electronics- Vol. 17, No. 1, PP. 130–133, 318–325, Jan. 1944

A MEGA CHART

SMITH CHART FORM 82BSPR (2-49) | KAY ELECTRIC COMPANY, PINE BROOK, N.J. ©1949 PRINTED IN U.S.A. | DATE

Sunersedes G.R. Form 5301-7560 N

IMPEDANCE OR ADMITTANCE COORDINATES

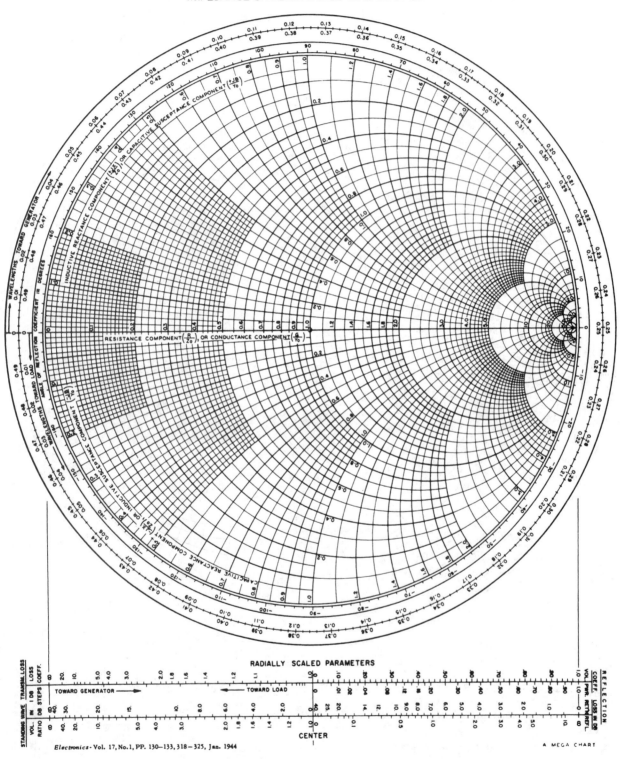

RADIALLY SCALED PARAMETERS

TOWARD GENERATOR ⟶ ⟵ TOWARD LOAD

CENTER

Electronics - Vol. 17, No.1, PP. 130—133, 318—325, Jan. 1944

A MEGA CHART

NAME	TITLE	DWG. NO.
		DATE
SMITH CHART FORM 82BSPR (2-49)	KAY ELECTRIC COMPANY, PINE BROOK, N.J. ©1949 PRINTED IN U.S.A.	

Sunersedes G.R. Form 5301 7560 N

IMPEDANCE OR ADMITTANCE COORDINATES

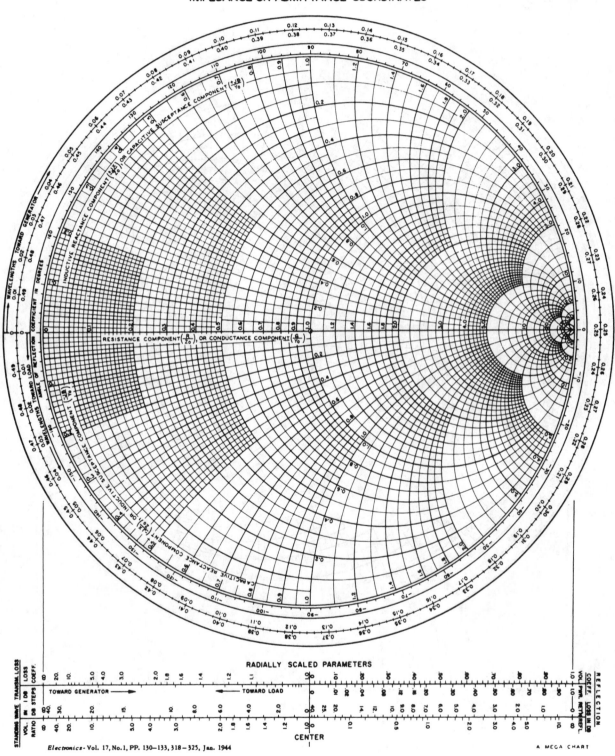

RADIALLY SCALED PARAMETERS

Electronics- Vol. 17, No.1, PP. 130–133, 318 – 325, Jan. 1944

A MEGA CHART

IMPEDANCE OR ADMITTANCE COORDINATES

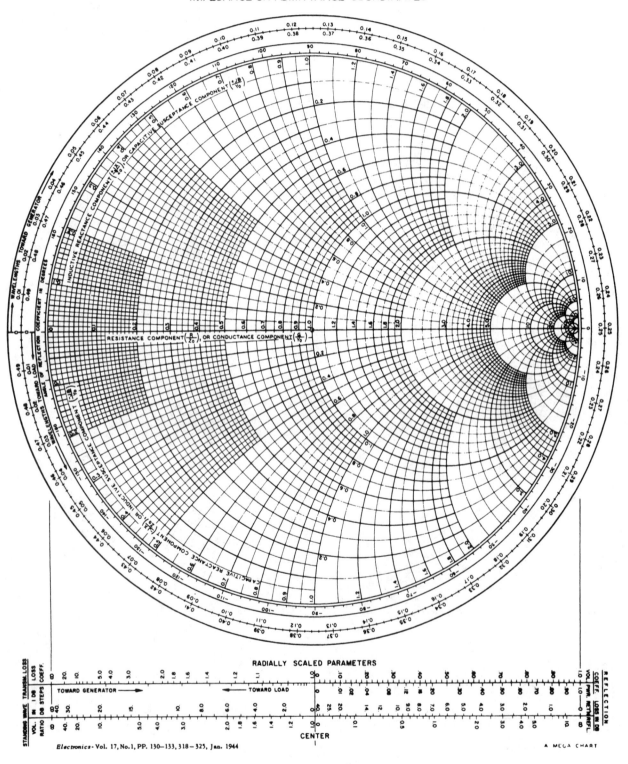

RADIALLY SCALED PARAMETERS

Electronics- Vol. 17, No.1, PP. 130—133, 318—325, Jan. 1944

A MEGA CHART

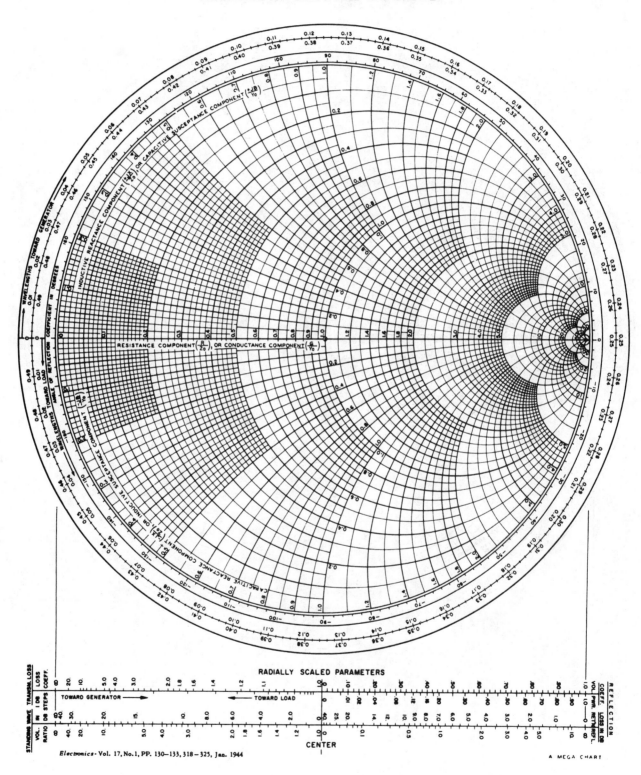

IMPEDANCE OR ADMITTANCE COORDINATES

NAME		TITLE		DWG. NO.	
				DATE	

SMITH CHART FORM 82B5PR (2-49)   KAY ELECTRIC COMPANY   PINE BROOK, N.J. ©1949   PRINTED IN U.S.A.

Sunersedes G.R. Form 5301 7560 N

RESISTANCE COMPONENT $\left(\frac{R}{Z_0}\right)$, OR CONDUCTANCE COMPONENT $\left(\frac{G}{Y_0}\right)$

RADIALLY SCALED PARAMETERS

TOWARD GENERATOR ⟶   ⟵ TOWARD LOAD

CENTER

Electronics- Vol. 17, No.1, PP. 130—133, 318—325, Jan. 1944

A MEGA CHART